Studies in Economic Reform and Social Justice

Natural Resources, Taxation, and Regulation: Unusual Perspectives on a Classic Topic

Laurence S. Moss, ed.

The Series
Studies in Economic Reform and Social Justice

Laurence S. Moss, Series Editor

Robert V. Andelson, ed.
Land-Value Taxation Around the World

Critics of Henry George, 2nd edition, volume 1 (2003)

Critics of Henry George, 2nd edition, volume 2 (2004)

J. A. Giacalone and C. W. Cobb, eds.
The Path to Justice: Following in the Footsteps of Henry George

Christopher K. Ryan
Harry Gunnison Brown
An Orthodox Economist and His Contributions

Philip Day
Land: The Elusive Quest for Social Justice, Taxation Reform and a Sustainable Planetary Environment

Laurence S. Moss, ed.
Natural Resources, Taxation, and Regulation: Unusual Perspectives on a Classic Topic

Studies in Economic Reform and Social Justice

Natural Resources, Taxation, and Regulation: Unusual Perspectives on a Classic Topic

Laurence S. Moss, ed.

350 Main Street, Malden, MA 02148-5020, USA
9600 Garsington Road, Oxford OX4 2DQ, UK
550 Swanston Street, Carlton, Victoria 3053, Australia

First published 2006 by Blackwell Publishing Ltd.

Library of Congress Cataloging-in-Publication Data

Natural resources, taxation and regulation : unusual perspectives on a classic problem / edited by Laurence S. Moss.
p. cm. — (Studies in economic reform and social justice)
Includes bibliographical references and index.
ISBN-13: 978-1-4051-5995-1 (hardback)
ISBN-13: 978-1-4051-5996-8 (pbk.)
1. Natural resources—Management. 2. Natural resources—Government policy. 3. Ecology. 4. Environmental economics. I. Moss, Laurence S., 1944–
HC85.N39 2005
333.7—dc22

2006018481

A catalogue record for this title is available from the Library of Congress.

Set in 10 on 13pt Garamond Light
by SNP Best-set Typesetter Ltd., Hong Kong
Printed and bound in Singapore

For further information on
Blackwell Publishing, visit our website:
http://www.blackwellpublishing.com

Portrait of Grover Pease Osborne (courtesy of Gerald F. Vaughn)

Contents

RETHINKING THE CONCEPTUAL FOUNDATIONS OF NATURAL RESOURCE ECONOMICS

Editor's Introduction

William L. Anderson and Jacquelynne W. McLellan persuade us that the leading newspapers can be and often are quite stubborn and biased about their reporting of scientific information: Stubborn in the sense that once the leading newspapers fix their sights on a certain line of reasoning and embrace a certain type of policy solution, they get stuck and cannot argue in reverse; and biased in that they favor always a statist or bureaucratic solution to the problems that they report. They champion more government jobs and more paperwork. In its cheerleading for more government intervention, the print media remains blind to the issue as to whether the perceived benefits of the recommended policy will exceed the costs. Without this type of calculation, the media contributes to the wastefulness of government without in any way "curing" the underlying problem that led to the call for more government.

This media bias in the environmental/resource management debate was brought out quite clearly after 1987. The National Acid Precipitation Assessment Program report reversed the conventional wisdom about the causes of acid rain and the subsequent poisoning of fish in lakes and streams. A blue-ribbon committee of about 3,000 scientists were asked to assess the damage from acid rain. They concluded in their report to the U.S. Congress that smokestack emissions from power plants could not be the main culprit in causing acid rain and its drastic consequences. Incredibly, the worst cases of acidic showering of the lakes were in geographic places where no smokestack emissions source could be found! The media did not (or would not) recant. It was as if many newspapers simply ignored the report completely or else mentioned the report only in passing. Rather than report the new scientific *consensus* among the experts, six leading newspapers, including the *New York Times* and the *Los Angeles Times*, ignored the new evidence completely.

The newspaper editorial writers put on their mental blinders and continued their campaign to sponsor more government interventions and regulation of the power industry regardless of the prospective benefits and with great indifference to the costs. It seems that once the media fixes its mind on government interventions the die has been cast forever more. The authors call this phenomena the "statist quo," and question whether the media is ever able to report certain types of news fairly and objectively. Anderson and McLellan pioneered an original empirical approach in which they track down and analyze the relative frequency with which several leading U.S. newspapers beat the acid rain drum for more government intervention as the scientific community locked into a more nuanced understanding of the problem and backed away from the populist rants against the power industry. Most disturbingly, the news coverage failed to report the scientific literature that challenged the original theories of lake acidification and in large part exonerated the power companies from blame for acidification.

Of course, legislation often gets passed into law when it has the strong support of certain lobby groups. The Sean Alley and John Marangos chapter offers a comparison between lobbying efforts in the United States and in Australia in support of policies that benefit farmers. In the United States, the farm lobby—the American Farm Bureau Federation (AFBF)—got its subsidies for its farmers and protection from foreign imports. In Australia, the National Farmers Federation (NFF) has responded to two warring constituencies. Some years the graziers (sheep farmers and animal herders) have gained the upper hand, and at other times the crop farmers have gotten control and power of that lobbying organization. After a long and checkered history that Alley and Marangos document in some detail, the NFF has come out in favor of "protecting biodiversity" and at the same time favoring free international trade in farm products and the removal of trade barriers. The NFF urges the elimination of Australian farm subsidies even when that policy results in certain farm bankruptcies within Australia. This relative laissez faire attitude among the farm lobby in Australia contrasts dramatically with the aspirations of the AFBF in the United States, which fights for small farmer life-support at the taxpayers' expense. Why the difference between the political programs of the AFBF and the NFF?

Alley and Marangos offer a modest explanation. The ultimate positions of each organization seem to have arisen spontaneously over time. The current policy positions are the result of what happened previously in the history of each region. This evolutionary process is "path dependent." The policies and politics of previous decades have given shape to the present line-up of political ideas. It may be reading a bit too much into their chapter to say that these authors view historical outcomes as often defying any a priori principles or logic. They are what they are, and all that can be said about the logic of lobby group politics is that history and institutions seems to have mattered more in these case studies than any abstract principles about farmers and the government interventions that they support.

Our next essay highlights the historical evolution of land and forestry management policies among the developed nations and especially in the United States. Robert H. Nelson reflects on his experiences as an economist in government, committed to implementing policy. Sometime during the 1960s, the United States rediscovered the importance of scientific land management. The early Progressive Period ideology of "scientific management" gave way in the 1970s to environmental impact statements and rational management to optimize the benefit-cost ratio(s) of successive policy initiatives. Nelson is a veteran operative in these deliberations, who conscientiously and honestly tried to come up with the then-fashionable benefit/cost ratios. He joined the Office of Police Analysis in the Office of the Secretary of Interior's Bureau of Land Management (BLM) in 1971 after earning his Ph.D. in economics from Princeton University.

In 10 years' time, the initial romance with benefit-cost analysis was over, and the BLM had grown indifferent to benefit-cost studies. Nelson uses the animal grazing program on federally owned lands as an illustrative example of how far that agency had come from caring about the impact of its decision on the gross domestic product of the nation. The benefit-cost analysis required by the regulations had become mere "window dressing" for policy decisions that were made for other reasons, and not unrelated to the lobby groups pressing the agency on behalf of certain special interests.

The "benefit-cost" requirements of the laws morphed into the offensive tools wielded by the environmental movement lobbyists to hammer the government agencies to get their way. The debate shifted away from the calculation of net benefits to the sustainability of the ecosystem. Today, the U.S. Forest Service (to use a specific example) has moved far away from its earlier mission of promoting multiple uses of the nation's forests to "ecosystem protection." Ecosystem protection can be used to simply root out humans completely as the bureaucracies try to preserve forests in some pristine state that allegedly existed before the Europeans even thought about stepping into the ecosystem of the New World. By pointing toward a public good that no particular person really thinks about or values, the modern government agencies are in a position to command enormous power over resource management. In this contemporary world, government economists should abandon any pretense of problem solving. Success in government is all about preaching and setting aspirations that are ill-defined and romantic in spirit. Nelson ends his chapter on this somber note.

The research team of John Loomis, Lindsey Ellingson, Armando Gonzalez-Caban, and Andy Seidl remind us that economists in government can still conduct statistical surveys and infer valuation amounts from these results. This team tried to determine the maximum amount that different individuals would be willing and able to pay for the service of what is called "mechanical fuel reduction." Mechanical fuel reduction is when the Forest Service manages the burning and elimination of tender wood and brush that might, if left unattended, might spark up into a massive and frighteningly uncontrolled forest fire. This contingent valuation method was employed to place a dollar value on the nonmarket benefits of mechanical fire fuel reduction as perceived by different ethnic groups in the population. Somewhat surprisingly, Hispanic-Spanish respondents placed the highest "willingness to pay" amount on the proposed Forest Service program.

Buoyed by their results, the researchers plan to explore in subsequent writings how ethnic groups divide on the valuation of other sorts of nonmarket public goods provisioning. The idea is to assign a dollar value of aggregate total benefit that a targeted group of beneficiaries can expect to enjoy on the assumption (often challenged by other economists) that the subjective meaning of a unit of currency is the same for all members of the group. The optimism displayed by this research group about the positive role economists play in government contrasts with the apparent pessimism of Robert H. Nelson and his experiences at the BLM.

Our next essay is by Gerald F. Vaughn. Vaughn has provided the portrait of Grover Pease Osborne that appears as the frontispiece of this *AJES Supplement* volume. Vaughn greatly admires Osborne, who he claims was the author of the first American textbook on resource economics. This field started during the Progressive Period in the United States (from 1890 to the 1920s). Vaughn uses his historical descriptions of the pioneering Osborne book to launch his own policy program for taxing nonrenewable resources and presumably discouraging their consumption while substitute technologies are making their way into the market place.

Erling Røed Larsen of Norway offers a case analysis of Norway's experiences with oil exploration and recovery and the international marketing of this resource for cash. Unlike many resource-dependent regions of the world, Norway did not end up with blossoming oil export revenues filling corrupt government officials' bank accounts while the average Norwegians experienced deprivation and poverty. There was no "Dutch disease" in Norway, at least not in the usual sense of "rent seeking" and political corruption gone wild. Still, Larsen contends, Norway is not out of the woods! As the politicians promise special interest groups that their favorite extravagant programs will get financed out of global sales of oil, the oil bonanza will give out. In the end, it may be necessary to inflate the currency and run the budget deficits that result in a lost of international competitiveness, de-industrialization, and an eventual currency crisis. Norway's future prospects remain a matter of concern.

It has become a staple of college teaching to explain how the material resource content of a unit of GDP has been falling during the last half-century or so as the developed regions become more service economy–based and less industrialized. The new economy has resulting in a sharp movement toward what media pundits call a "weightless world" of GDP production. Intellectual property rights are more important today than steel output or pig iron production. Technological innovation has often contributed to this important substitution of synthetic substances for scarcer natural resources. A case in point is the switch from copper wiring in communications to fiber optics made from strands of glass or "beach sand." There has never been a movement to prevent the utilization of beach sand and, to the best of my knowledge, no one of any reputation has started a conservationist movement to "stop using Nature's sands." These optimistic insights have informed a "resource optimist school of policy experts," of which the late Julian Simon was a leading and most erudite representative (Simon 1996 [1977]).

In his article, Jonathan Perraton offers to document these alleged trends. Part of his research does back up the broad patterns described in the above paragraph. The natural resource content of a unit of GDP has been falling for the developed regions of the world! Still, Perraton claims that this development is not enough to completely eliminate the West's dependence on certain resources and their primary producers. Developed nations may soon expect a revival of economic demand for their products, with rising resource prices in this 21st century offsetting the apparent secular decline in resource prices that characterized the last part of the 20th century. Unlike Simon, who expected these trends to continue ad infinitum, Perraton forecasts a reverse movement in the near future.

Many economists prefer to lump land along with capital goods into one aggregate measure of "capital." This may be a mistake. It is a mistake in the important sense that the approach camouflages selected policy problems along with other important market phenomena. That is why we have included a section titled "Georgist Perspectives on Resource Utilization and Financing."

Professor C. Lowell Harriss provides the keynote article in this section with his eloquent reminder that both prudence and our ethical senses about fairness require that we do indeed make a greater use of the severance tax when self-interested organizations remove resources from the public commons (privatized for administrative convenience only) and then sell them on the open market so others can consume them in production.

How the taxation of property is actually carried out also matters. It is critically important that the tax czars impose a different rate of taxation on the unimproved value of, say, land and a different and much lower rate of taxation (if any taxation at all) on the improvements and investments of entrepreneurially minded men and women. What is called "two-tiered taxation" is not always politically expedient in certain tax jurisdictions. Much depends on whether a particular state's constitution authorizes the legislation to install such a two-tier system. Even when the state constitution does allow two-tiered taxation, the interest groups line up around the reform in peculiar ways, as Robert Andrew Peters explains in his important essay.

Most curiously, American farmers line up against "two-tier taxation" for reasons that include their interest in preserving open spaces. In Pennsylvania, the farm lobby has been effective in trumping the desires of local governments to erase the burden on improvements. This chapter lines up nicely with the findings of several others. It teaches us what seems so logically compelling to a technically trained economist may not at all be politically acceptable to a democratic regime in which interest groups align around issues in surprising ways. We have seen this also in the Alley and Marangos essay, claiming that policy positions of the lobby groups are "path dependent" and often do not fit into stereotypical categories at all.

In the next essay, the most accomplished modern "Georgist" theoretician, Mason Gaffney, offers a seminal analysis of "tax bias" and some rudimentary observations about how to detect tax bias. As Gaffney points out in his opening sentence, the strategy is to infer the direction of the tax bias by tracking the differential effects on the present valuations of the rival uses for land. This essay is something of a capstone summary of Gaffney's many earlier essays on this and related subjects in which he broke new ground modernizing Georgist insights. Finally, we conclude this collection of Georgist materials with a useful but insightfully annotated bibliography by Jeffrey J. Smith and Thomas A. Gihring on a topic called "value capture."

A word or two is needed to motivate interest in this article since it might at first seem to deal with a topic far removed from the central concerns of the this volume and thus be skipped over. As Fred Foldvary makes clear in his chapter in this volume, the spatial location of economic activity is often critically important to the financial success of that activity. In market settings, people will be able and willing to pay premiums to obtain the rights to locations nearest to certain attractions such as a fire station, a modern shopping mall, or a meritorious public museum and monuments. One such "attractor" is access to a well-functioning modern transit system.

Modern transit systems often are financed by contributions from the general tax fund and certain guaranteed bond issues that are earmarked to be paid off over many years by the revenues provided from the pricing of the transit service. In their annotated bibliography, Smith and Gihring document that many researchers from distant regions of the world now concur that it is perfectly feasible to finance the production of transit systems in a different and (they argue) more equitable way. Why not use a "property tax" that varies depending on the location of the property to the attractor that makes that location more valuable? It might be useful to think of this tax as a "betterment tax" imposed by the municipality on the property owners who find their property values surging vis a viz their spatial location to the transit systems. The ethics of this alternative method of transit system pricing my be persuasive among those who care about the injustice of taxing the general population for such attractive amenities enjoyed by the few. Such a perspective is "Georgist" and that is why this useful bibliography has been included in this section.

The Georgist perspective is notoriously overlooked among modern economists. Despite its sensible theoretical underpinnings, a chorus of hoots, hollers, and jeers often greets economists who have retained an appreciation for the older ideas and distinctions originally found in classical-school economic theory. The prejudice began with the apparent unanimous decision among leading neoclassical economists such as Irving Fisher and the later Chicago school theorists to lump together land with capital goods in a singular aggregate measure of some sort. The growth and development literature is famous for the utilization of Robert Solow's model, which does away with the land/capital goods distinction completely. The clock can not be rewound without a thorough understanding of why land or space needs to be kept separate from capital goods. Professor Fred Foldvary's article opens our last thematic section but it could serve just as well as a conclusion for the Georgist section as well. Foldvary pleads for the taxonomy of the factors and why a more fruitful understanding of economic phenomena really depends on returning to the older classical school approach.

Two additional conceptual chapters help bring this volume to a close. We offer Professors Malte Faber and Ralph Winkler's interesting extension of "neo-Austrian" capital theory to the problem of sustainability of an economy contained with a broader ecologically sensitive framework. Capital theory is "Austrian" when it emphasizes both the time-consuming nature of setting up and maintaining routine production processes and also emphasizes the heterogeneity of capital goods themselves. Such an approach has the important feature of allowing investigators to extend the important work of Bernholz and Faber to ecological economics. Ecological economics is presented by many as an "alternative" to neoclassical economics because it takes into consideration the physical or natural world's constraints (such as the laws of thermodynamics) on economic activity and its willingness to embrace radical uncertainty and complexity issues having to do with exceedingly long time frames. Along the way, ecological economics includes discussions about intergenerational equity and responsibility as well.

And last but certainly not least, Professor Richard J. Brazee and L. Martin Cloutier offer an elegant reconciliation of Lewis Cecil Gray's numerical examples with Harold Hotelling's mathematical modeling about how nonrenewable resources should be utilized for maximum economic effect. It turns out that by reworking the broad conceptual foundations of the approaches the authors were able to reconcile Hotelling's famous strictures about nonrenewable resource utlization with Gray's ideas about resource management. Hotelling's elegant rules about the optimal utilization of resources can be shown to be only a special case of Gray's more nuanced description of the background economic conditions.

The field of natural resource economics is a broad one, and the 14 essays included in this volume scope out major landmarks that exist in this vast territory. Hopefully, this will not be the last or final statement about what modern analysis can accomplish when it illuminates many of these important problems. This book is a robust collection and wide ranging in its inclusion of topics and conceptual approaches.

References

Simon, Julian. (1996) [1977]. *The Ultimate Resource 2* Princeton, New Jersey.

The Ideology of Environment and Resource Utilization Debates

Newspaper Ideological Bias or "Statist Quo"?

The Acid (Rain) Test

By William L. Anderson and Jacquelynne W. McLellan*

Absract. Throughout the 1980s and early 1990s, acid rain was an important topic of public debate. Newspapers ramped up coverage in the early 1980s, which then peaked in the mid-1980s and died off slowly in the late 1980s and early 1990s, to be rarely seen again. The question asked is whether the tone of the acid rain coverage was an example of alleged "liberal bias" of journalists, or if it was due to other factors. This paper examines various explanations of newspaper behavior, including one given by the late Warren Brookes concluding that the real bias of reporters tends toward the expansion of government, or the *statist quo*. In our paper, we examine the coverage of acid rain before and after the election of George Bush in 1988, an event that led directly to the passage of the the Clean Air Act Amendments of 1990, which had strong acid rain provisions. Our statistical tests, while mixed, give credence to the "statist quo" hypothesis, especially where newspapers of "national stature" are concerned.

*The authors teach Economics for Managers and Decision Making Analysis in the M.B.A. program at Frostburg State University. They would like to thank Randy Beard, John Jackson, Henry Thompson, William Shughart II, Lawrence Moore, and an anonymous referee for their helpful insights, comments, and suggestions. All errors are their own.

American Journal of Economics and Sociology, Vol. 65, No. 3 (July, 2006).

I

Introduction

For 30 years, debate has swirled around the alleged "liberal bias" of U.S. journalists, and especially journalists who write for publications such as *Time, Newsweek*, the *New York Times*, and the *Washington Post.* Sutter (2001) has written that a veritable "cottage industry" has sprung up as a number of authors, including Efron (1971), Lichter, Rothman, and Lichter (1986), and Goldberg (2003), have written books alleging and documenting what they say are incidents of "liberal bias" in news organizations.

Lichter, Rothman, and Lichter (1986) interviewed and surveyed a large number of journalists in print and broadcast media, and their study concludes that the journalists tended to have economic, political, and social views much more "liberal" or left of center than the beliefs of average Americans. The authors conclude that those views determine the slant of their news accounts.

The problem with such conclusions, says Sutter, is that a personal outlook of a journalist does not necessarily translate into biased news coverage. He writes, "Unfortunately, we cannot simply 'test' the news and determine once and for all if a liberal bias exists" (2001: 432).

Sutter (2001), Weaver (1994), and Brookes (1991) look to an alternative hypothesis: that journalists may have a bias toward expansion of government powers. Brookes says journalists do not favor the "status quo" as much as the "*statist* quo" (emphasis added). Brookes reasons that reporters mostly inform readers on issues of government, and the more the state expands, the better it is for journalists.

Crain and Tollison (1997) write that the need for the monitoring services of the press increases as government expands. It is in the self-interest of those in journalism, they write, to promote news that encourages expansion in the powers of the state.

Government has expanded its reach greatly in the past three decades in environmental regulation. Regardless of the individual biases of journalists on environmental matters, it would seem that if the "statist quo" viewpoint were true, the press would champion growth of government environmental regulation. We believe that the issue of acid rain allows for a test case for the "growth of government" view of journalism.

In this paper, we analyze how newspapers covered the acid rain debate during the 1980s and early 1990s. First, we describe how six major U.S. newspapers reported the issue relative to the scientific developments that occurred during the 1980s. Second, we test the patterns of acid rain news coverage and interpret those results according to an economic point of view, especially testing Brookes's "statist quo" description of journalism. While the results are mixed, it is clear, controlling for other factors, that once it became apparent that the U.S. government would attempt to "stop" acid rain through passage of new clean air legislation, the press lost much of its interest in the story, moving on to other issues.

II

Journalism and Economic Analysis

IF THE BROOKES THESIS HOLDS, benefits would exist for the press when state power expands. Stigler (1971) writes that the growth of government regulation allows outsiders to gain access to the affairs of private enterprises. This additional access also would provide more opportunities for journalists.

Brookes (1991) writes that journalists favor the "statist quo" because it provides opportunities not afforded to them when business decisions are made within a market setting. Using the example of reporters on the "energy beat" losing their jobs after the deregulation of oil prices in early 1981, he notes that once oil markets returned to a pattern consistent with other free markets, the "need" for energy alternatives like solar power suddenly disappeared.

He says that journalists mostly cover government agencies, local, state, and national, concluding that journalists have a vested interest in the maintenance and growth of government programs. Using the example of oil and the "energy beat," Brookes finds that government price and allocation controls provided opportunities for reporters to advance their careers by writing on the "energy crisis" and "energy futures," even though the controls actually were economically harmful. Therefore, the press tended to report favorably upon the controls themselves, mostly blaming oil companies for the gasoline lines and other problems that accompanied fuel shortages.

"Like it or not, that [the loss of 'energy beat' jobs] makes journalistic incentives very clear: the more government, the more power [and jobs] the news media will have," writes Brookes. He further says the press "became one of [the] principal beneficiaries" of "this massive explosion of government" (1991: 15).

In rejecting the Pigovian or "public interest" interpretation of government, McCormick and Tollison (1981: 3–4) write: "Economists in the public-choice tradition have reacted to the Pigovian approach at a very general level essentially by stressing that it is not a very *believable* theory of government action and, moreover, that it is flawed by the unwarranted assumption that government can be called upon to correct imperfect markets in a perfect and costless manner." As noted earlier, although economists interpret and model actions of politicians and members of interest groups, few economists have attempted to examine journalism in the same way. While journalism helps facilitate the actions of interest groups, legislators, and public officials, most individuals model the profession as Pigovian.

Those who claim that a "liberal" bias exists in reporting believe that such "bias" violates the high standards of journalism. According to the *Canons of Journalism* (1922), journalists are expected to shed all bias, preconceived ideas, and personal interests before pursuing and writing a news story.

However, economic theory permits us to look at journalism from an alternative perspective, one that examines the perceived marginal costs and benefits to action taken by journalists. In this view, journalists are not guided only by the *Canons of Journalism*, but also by their own career interests.

Acid rain news coverage did not accurately reflect the debates within the scientific community, instead repeating what some scientists were saying in the early 1980s. Among the nation's major newspapers, only the *Washington Post* published a story on the dissenting scientists, and that only after the Clean Air Act Amendments of 1990 were law. Therefore, the inaccuracy of acid rain stories created only a minor stir, and most news organizations simply refused to acknowledge that credible alternative explanations even existed in the first place.

III

Acid Rain: The Scientific Issues

THE APPLIED SUBJECT OF THIS PAPER measures news coverage of acid rain from six major U.S. newspapers during the 1980s. The coverage period begins in 1980 and continues until 1993, three years after the passage of the 1990 Amendments to the Clean Air Act, which included provisions to reduce acid rain in the United States.

Robert Angus Smith, an English chemist, discovered that rainfall in London and other industrial cities had become very acidic by 1852, and linked what he called "acid rain" to the burning of raw coal to power the factories and electric generators of industrial England (LaBastille 1981). However, the rainfall, which had a pH factor considerably lower than the 5.5 pH of "normal" rain, was not seen to be an environmental crisis until the 1970s, when scientists in the United States, Canada, and Scandinavia found that acidic surface waters of northern lakes could no longer support aquatic life.

Paulos (1995) writes that popular press coverage often precedes detailed scientific analysis, and acid rain was no exception. The National Academy of Sciences predicted a 100-fold increase in acid lakes by 1990 if SO_2 emissions were not severely curtailed (National Research Council 1981). However, only a small number of scientific studies had investigated acid rain and its effects, so in 1980 President Jimmy Carter commissioned the National Acid Precipitation Assessment Program (NAPAP) to examine the damage being caused by acid rain and to recommend solutions.

President Ronald Reagan expanded the NAPAP program in 1982 from a $10 million to $100 million annual budget, and within a year, evidence emerged that contradicted the "mineral titration" theory then prevalent. The mineral titration theory assumed that acidic soils could not buffer acid rain, which then ran directly into lakes and streams and acidified them. But Krug and Frink (1983) disputed the connection between lake acidity and acid precipitation, claiming that the composition of soil in the watersheds of lakes and streams had a greater impact upon surface water acidification than did rainfall.

Krug later wrote (1990, 1991) that core samples taken from acidic Adirondack lakes showed that those waters had been acidic even before the Industrial Revolution. Furthermore, other NAPAP studies found that the region with the highest concentration of acidic lakes was not upstate New York but rather Florida, where acid rain did not fall (NAPAP 1988). This finding was especially important because the news that many of Florida's lakes were acidic originally had been reported as an example of the peril of acid rain. Furthermore, researchers found numerous lakes in the mountains of Australia and New Zealand to be acidic despite the fact that rainfall in that region had a normal pH (Anderson 1992).

NAPAP scientists (1989) also failed to turn up evidence that acid rain was destroying U.S. forests, including sugar maples. Researchers did document acid precipitation damage to approximately one-tenth of 1 percent of the high-altitude eastern red spruce forests, but found no other evidence of harm to trees and crops from acid rain (NAPAP 1989).

Much of this news was included in the 1987 *Interim Report* that NAPAP presented to Congress. While one might suppose that members of Congress would have been pleased to find that an acid rain environmental disaster was *not* in the making, that is not what happened.

Instead, most in Congress gave the researchers a hostile reception. Representative James Scheuer (D-NY), then chairman of the House Subcommittee on Natural Resources, Agriculture Research, and the Environment, attacked both the report and NAPAP director J. Laurence Kulp, who resigned his position a week later. Environmental groups described the *Interim Report* as "political propaganda" from the Reagan administration.

NAPAP scientists were not unanimous about the findings, even though Roberts wrote: "The quality of NAPAP's research effort is generally considered to be quite good, perhaps first rate, and there is little quarrel with the individual facts." The disagreement among scientists, she adds, involved "the way the facts are presented—which tends to minimize the extent of the problem." (Roberts 1987: 1404).

Although scientists generally respected the NAPAP study, congressional leaders sided with environmentalists. When President George H. W. Bush pursued stringent new rules for SO_2 reduction (50 percent), he and William Reilly, Bush's appointed head of the Environmental Protection Agency, never publicly acknowledged any of the NAPAP findings. In the fall of 1990, Congress passed the new rules, which required a 10-ton reduction from daily point sources in SO_2 emissions, and a 2-ton reduction in N_2O from 1980 levels, along with an 8.9-ton cap on SO_2 by the year 2000. The EPA permitted the NAPAP report to be released only after the bill became law.

The earliest discussions of acid rain occurred within a worldview that accepted at face value that rainfall more acidic than "normal" automatically was dangerous. That modern industrial society has produced environmental degradation is well accepted and, like many paradigms, is rooted in observable facts.

IV

Acid Rain Coverage by Six Newspapers

THE EMPIRICAL EVIDENCE covers acid rain articles from six daily U.S. newspapers:

Atlanta Constitution — *Boston Globe*
Chicago Tribune — *New York Times*
Los Angeles Times — *Washington Post*

These papers were chosen because of the markets they serve, and because acid rain allegedly did not affect their cities in the same way, thus potentially influencing the nature of their coverage of that issue. For example, the *New York Times* represents readership in New York City, but it also is read nationwide. It is also the most influential newspaper in the Northeast, where the country's most acidic rainfall existed. The *Washington Post*, while read in other cities, influences the direction of government policy in the nation's capital. The problem of acid rain, however, was not seen to be as great in Washington, D.C., as it was in New York and New England.

The *Tribune* and the *Constitution* both serve markets where acid rain was not as significant as in the Northeast. However, their circulation markets were also served by electric utilities that generate most of their electricity by coal, and residents of Chicago and Atlanta found themselves potentially facing higher electric bills under pending acid rain legislation. The *Boston Globe* is the most influential newspaper in New England, where residents especially perceived danger from acid rain. Electric utilities in New England also use higher-cost fuels than coal, the fuel of choice for utilities in the Midwest and Southeast. While expensive, these fuels do not emit as great an amount of pollutants blamed for acid rain as do coal-burning power plants.

The *Los Angeles Times* is the major newspaper in the western United States, and no acidic rainfall occurred in Los Angeles. However, the *Times* historically has long favored strict pollution measures, as air pollution in Los Angeles has always been an important topic.

V

How the Press Covered Acid Rain

JOURNALISTS MOSTLY DISREGARDED the NAPAP study, choosing to stay with the original paradigm of acid rain as an immediate peril. All six papers carried the story of the September 1987 announcement, but discounted the information. The *Boston Globe* carried an "objective" story that said lake acidification in the United States "is not a new phenomenon," but still implicated acid rain. The *New York Times* centered its story upon the controversial nature of the report, interviewing opponents of the summary.

The *Chicago Tribune* interpreted the report as saying that acid rain did not pose a "broad threat to the environment," affecting "only small areas." The *Washington Post*, not surprisingly, dwelt upon the report's political aspects instead of the science. The *Los Angeles Times* emphasized the controversy and environmentalist criticisms. Only the *Atlanta Constitution* noted that the report had its supporters, reprinting a favorable editorial from the *Wall Street Journal*.

Newspapers in the United States were selective in quoting "experts." The treatment given to Gene Likens as opposed to Edward Krug is a telling point. Likens was an ecology professor at Cornell University, and Krug was employed by the Illinois State Water Survey Division. According to Abdullah (1989), in 1985 Likens was ranked seventh in mean citations in the *scientific* acid rain literature, while Krug ranked third.

Yet, the mainstream press liberally quoted Likens, while Krug was nearly invisible, according to a LexisNexis search of acid rain articles from 1980 to 1993. The search reveals that Likens was cited 39 times in the press, most coming from papers like the *New York Times, Boston Globe, Washington Post*, and the Associated Press and United Press International.

Krug, however, was cited only nine times and then usually on editorial pages of conservative newspapers such as the *Washington Times, Daily Oklahoman*, and *St. Louis Post-Dispatch*. His only citation in a major newspaper was in the *Washington Post*, and that was only because an EPA administrator attacked his views. The wire services cited him once, a UPI dispatch in 1983 in which other scientists attacked his article in *Science*.

In December 1990, CBS's *60 Minutes* questioned why the government ignored NAPAP, interviewing Krug and others. Howard Kurtz of the *Washington Post* then asked why environmental reporters had ignored the other side. He wrote that although the NAPAP report contained good news, it "was virtually ignored by the *Washington Post* and given scant attention by most major news organizations last year, even while Congress debated and approved new acid rain controls that will cost as much as $4 billion a year." Wrote Kurtz: "Some reporters say privately that it is difficult to write stories that debunk the conventional wisdom of environmental activists, whom the press treats more deferentially than industry spokesmen and other lobbyists" (Kurtz 1991: A3).

For example, the *New York Times* during the measurement period never quoted a scientist who was skeptical of the apocalyptic acid rain claims. Skeptics quoted by the *Times* usually were electric or coal industry representatives, or elected officials representing districts in which coal-burning power plants were located.

VI

Testing the "Statist Quo" and Other Hypotheses

As Sutter (2001) has pointed out, there exists no accurate way statistically to test for "liberal bias," or for any other real ideological persuasion, for that matter. Although Lichter, Rothman, and Lichter (1986) report a "liberal bias" on behalf of reporters, making the transition that viewpoints of reporters automatically mean their news dispatches will be biased toward the "left" can be problematic, according to Sutter, who notes that the direction of "liberal" causality is difficult to determine.

Given Sutter's warnings, we employ an alternative hypothesis to explain the direction and "slant" of newspaper coverage, using reporting on acid rain as our subject. We believe that Brookes's "statist quo" claim presents a testable hypothesis. Our question regarding acid rain coverage is this: Did journalists pay the same amount of attention to this issue *after* it became clear that the Clean Air Act Amendments would pass as before?

If the answer is "yes," we can conclude that government action had no effect upon newspaper coverage of the issue. However, if the certain prospect of government regulation were followed by a significant decrease in acid rain stories, one could conclude that once the government acted against acid rain, journalists would lose interest in the story. Another way to put it is that one of the goals of acid rain coverage *was* the imposition of new laws to deal with the source of the problem.

However, this goes to motivation and allegations of bias, which cannot be proven. While that is true, we do have the patterns of daily news coverage here, and although we do not attempt to be psychologists, we do believe that the numbers tell an interesting story.

We use descriptive statistics and statistical tests, our data being monthly time series from 1980 through 1993 for the *Boston Globe, Chicago Tribune, Los Angeles Times, New York Times*, and *Washington Post.* (Monthly data for the *Atlanta Journal-Constitution* could be found only for 1983–1993.) The information on the newspapers comes from bound newspaper indices published by the *New York Times* (*Boston Globe, New York Times*, and *Atlanta Journal-Constitution*) and by Bell and Howell (*Chicago Tribune, Los Angeles Times*, and *Washington Post*). Information on acid rain stories in the *Globe* from 1980–1982 was found in the archives of the *Globe*'s site on the Internet. While most publications are news stories, we also include editorials, op-ed columns, and letters to the editor.

We first means by employing pooled-variance *t*-tests to examine changes in newspaper coverage of acid rain from January 1980 through December 1988 and from January 1989 through December 1993. We also compare the means of monthly acid rain stories from January 1984 through December 1988 with the mean of monthly stories from January 1989 to December 1993, using the same tests.

In our second set of tests, we regress the number of acid rain stories against variables representing time trends, a "legislative effect," and two different measurements of U.S. presidential campaigns. We also conducted a large number of tests that we do not use but will mention later.

The first test of the "statist quo" hypothesis is to find whether or not the average number of stories on acid rain decreased significantly after it became clear the government would "solve" the problem. Both presidential candidates—and especially Bush, who declared that he wanted to be known as the "environmental president"—had campaigned on pushing tougher legislation to stop acid rain. Thus, the election of Bush in November 1988 is seen as a strong sign that new legislation would be forthcoming.

In Figure 1, we show the pattern of acid rain stories that appeared in the previously listed newspapers (excluding the *Atlanta Journal-Constitution*). As one can see, the largest numbers of stories are grouped around the election campaign of 1984, the release of the 1987 *Interim Report* of NAPAP, and the 1988 elections. It is also obvious from Figure 1 that the number of stories tails off rapidly after 1990, when Congress passed the Clean Air Act Amendments.

(We use 1989 for the benchmark instead of 1990 because of the certainty of new clean air legislation coming with Bush's election. While we do not doubt that journalists would have tried to influence the pending legislation in the year before the laws were passed, there could have been no doubt that Congress and Bush were going to "do something" about acid rain.)

For the first set of tests, we grouped the stories into two groups: the number of monthly stories in each newspaper (again, excluding the *Atlanta Journal-Constitution*) from 1980 through 1988, and the second group of monthly stories published from 1989 to 1993. The null hypothesis states that there is no difference in the average

Figure 1

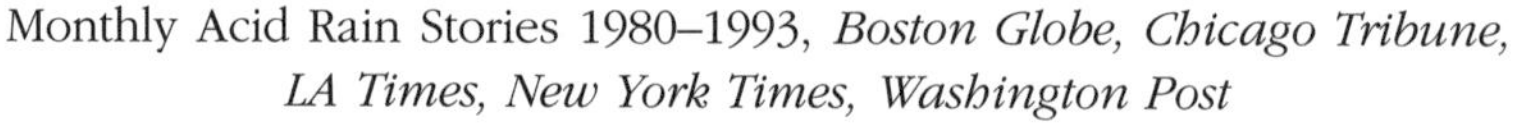

Monthly Acid Rain Stories 1980–1993, *Boston Globe, Chicago Tribune, LA Times, New York Times, Washington Post*

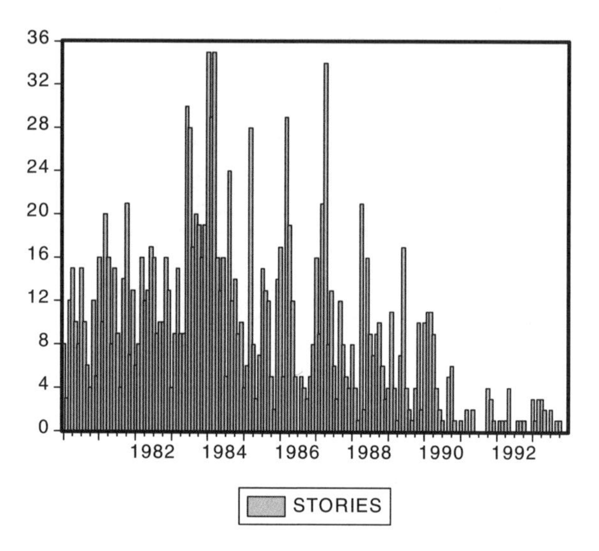

number of stories per month from one period to the next. In the last test in Table 1, we combine all of the monthly stories to see how the papers performed in the aggregate.

For each case, the *t*-test statistics are significant at the 1 percent level, meaning that the coverage of acid rain fell drastically after it was clear that with the election of George H. W. Bush, no real political obstacles stood in the way of Congress passing new clean air laws—even after the NAPAP scientists had published credible studies questioning the need to radically curtail acid precipitation. While Bush's opponent, Massachusetts Gov. Michael Dukkakis, also campaigned as an environmentalist, Bush aggressively campaigned on an environmental quality program, including an endorsement of acid rain amendments to the Clean Air Act.

However, there could also be another force at work, as readers would tire of similar stories over a long period of time. For acid rain coverage (or coverage of any hot-button issue, for that matter), is

Table 1

t-Test Comparison of Means 1980–1988/1989–1993

Newspaper	Monthly Mean 1980–1988	Monthly Mean 1989–1993	*t*-Statistic
Boston Globe	4.49	0.75	7.943***
Chi. Tribune	1.72	0.35	5.375***
LA Times	0.944	0.15	4.144***
NY Times	4.12	1.267	6.605***
Washington Post	1.56	0.4	4.787***
All newspapers	11.89	2.77	8.783***
	n = 108	n = 60	

***Significant at the 1% level.
Source: Newspaper Indexes.

there a natural progression over time for stories on *any* subject to decline due to reader fatigue? That hypothesis is tested in two different sets of tests, one being another series of means tests seen in Table 2, and the other being regressions that look at the "fatigue factor" and other variables that we hypothesize would have had an effect upon news coverage of acid rain.

In Table 2, we look at three coverage periods—1980–1983, 1984–1988, and 1989–1993—again employing the statistical technique of testing the equality of the monthly mean of acid rain stories to see if a "fatigue factor" was in place before the Bush election. We compare the mean monthly coverage of the newspapers (except the *Atlanta Journal-Constitution*) of 1980–1983 to the period covering 1984–1988. We then compare the mean monthly coverage of all the newspapers from 1984–1988 to the period 1989–1993. The results are shown in Table 2.

Only in the case of the *Boston Globe* is there a significant decrease in acid rain coverage from the first period to the second period, and in the case of the *Chicago Tribune* and *Washington Post*, the monthly average of stories is shown to have a statistically significant increase. (The *New York Times, Los Angeles Times*, and the sum of all stories

Table 2

t-Test Comparison of Means: 1980–1983/1984–1988 1984–1988/1989–1993

Newspaper	1980–1983 Monthly Mean	1984–1988 Monthly Mean	*t*-Statistic	1989–1993 Monthly Mean	*t*-Statistic
Atlanta Journal-Constitution	NA	2.678	NA	0.3898	–8.205***
Boston Globe	6.25	2.864	–5.632***	0.7627	–5.813***
Chi. Tribune	1.188	2.136	2.633**	0.356	–6.265***
LA Times	0.75	1.068	1.144	0.153	–4.393***
NY Times	4.063	4.051	0.019	1.237	–6.271***
Washington Post	1.042	1.915	2.621**	0.3898	–5.636***
All newspapers	12.542	10.966	1.073	2.746	–7.545***
	n = 48	n = 60		n = 60	

**Significant at the 5% level; ***significant at the 1% level.
Source: Newspaper Indexes.

do not show statistically significant changes from the first to the second period.)

That is not the case with the second set of measurements shown in Table 2, however. For each newspaper, and for the monthly mean number of stories, the change from the pre-1989 period to the post election period is negative and statistically significant.

In order to gain a better measure of the alleged "fatigue factor," as well as other factors that might have influenced acid rain newspaper stories, we construct three sets of regressions that contain not only a "fatigue" variable but also measure the effects of the acid rain legislation and U.S. presidential campaigns upon acid rain newspaper coverage.

The regressions are measurements of the following models:

$\text{Coverage}_i = f(\text{Time, Legislation})$,
$\text{Coverage}_i = g(\text{Time, Legislation, Campaign}_1)$,
$\text{Coverage}_i = h(\text{Time, Legislation, Campaign}_2)$,

where

Coverage_i = monthly acid rain coverage of newspaper *i* from 1980–1993 (for the *Atlanta Journal-Constitution*, we measure monthly coverage from 1983–1993),

Time = a time trend beginning with 0 and progressing to 168,

Legislation = a dummy variable that assigns a value of 1 from January 1989–December 1993, 0 otherwise,

Campaign_1 = U.S. presidential campaigns of 1984 and 1988, which assigns a value of 1 from January to November of those years, 0 otherwise, and

Campaign_2 = U.S. presidential campaigns of 1980, 1984, and 1988, which assigns a value of 1 from January to November of those years, 0 otherwise.

The results are demonstrated in Tables 3–5.

In Table 3, the only individual newspaper to demonstrate a statistically significant "fatigue effect" is the *Boston Globe*, although the estimator for the variable *All* is negative and statistically significant, albeit at a weaker level than the coverage for the *Globe*. At the same time, one must remember that the *Globe* was the first newspaper to begin beating the acid rain drum, even ahead of the *New York Times*.

Table 3

Coverage of Acid Rain 1980–1993 Regression #1 $N = 168$

Dependent Variables / Independent Variables	*Atl. Journal-Const.*[+]	*Boston Globe*	*Chicago Tribune*	*LA Times*	*NY Times*	*Wash. Post*	All[++]
C	2.70803	7.0684	1.33574	0.6887	4.7735	0.95929	14.137
	(4.932)***	(11.231)***	(3.111)***	(2.693)***	(6.97)***	(3.047)***	(9.073)***
Time	−0.00454	−0.0482	0.0076	0.00499	−0.0116	0.01135	−0.04085
	(−0.485)	(−5.956)***	(1.222)	(1.335)	(−1.077)	(2.585)***	(−1.888)*
Legislate	−1.87292	0.30484	−2.0321	−1.2251	−1.9389	−2.10416	−5.77032
	(−3.49)***	(0.398)	(−3.391)***	(−3.34)***	(−1.802)*	(−4.696)***	(−2.624)***
R-Squared	0.289	0.421	0.168	0.109	0.229	0.158	0.345
F-Statistic	26.249***	60.029***	16.701***	10.066***	24.533***	15.47***	43.471***

[+]Data available only from 1983–1993. [++]Excluding *Atlanta Journal-Constitution*. ***Significant at the 1% level. *Significant at the 10% level.

The estimators for the variable *Legislate* are all negative and statistically significant (except for the *Globe*) and with the exception of the *New York Times*, are significant at the 1 percent level. Given the presence of the time variable, this provides stronger evidence of the "statist quo" than do the results of Table 1 and Table 2.

When creating a regression, however, one looks to include as many explanatory variables as possible, so we have added dummy variables in Tables 4 and 5, which include U.S. presidential campaigns. Table 4 includes a dummy variable that covers the presidential campaigns of 1984 and 1988, while Table 5 also includes the campaign of 1980. The reason we include these variables is that we wanted to see if a national campaign during that period would have significant effects upon newspaper coverage of acid rain, given that presidential contenders in all three elections broached the subject.

The inclusion of $Campaign_1$ affects the estimators for the *Legislate* variable, decreasing the level of significance in all but one case (*Washington Post*). But while the estimators for the *Legislate* variable see their statistical significance reduced, we also see that, except for the *Atlanta Journal-Constitution*, the $Campaign_1$ estimator is not significant for any newspaper.

When we include the 1980 presidential campaign in our campaign dummy variable, the picture changes again. The estimator for *Legislate* is statistically significant in five of the six regressions, while the estimator for $Campaign_2$ is not statistically significant at all.

In each of the regressions from Tables 3–5, the regression *F*-statistic is significant to the 1 percent level, although the R^2 measurements are much more diverse, ranging from 0.436 (*Boston Globe*, Table 4) to 0.111 (*Los Angeles Times*, Table 5). We also tested for unit roots and, using the augmented Dickey-Fuller test, the alternative hypothesis (that there is no unit root) held each time.

Because of inevitable problems of autocorrelation that generally accompany the use of time series data, we often employed corrective matrices as outlined by Newey and West (1987). In converting the ordinary least squares regressions to general least squares, we were able to deal with the problems of bias, although in some cases the estimators lost some efficiency. For our purposes, however, we believed that eliminating the problem of bias was more important.

Table 4

Coverage of Acid Rain 1980–1993 Regression #2 $N = 168$

Dependent Variables / Independent Variables	*Atl. Journal-Const.*[+]	*Boston Globe*	*Chicago Tribune*	*LA Times*	*NY Times*	*Wash. Post*	All[++]
C	2.43947	7.10414	1.36578	0.69612	4.79316	0.97397	14.2371
	(7.071)***	(11.168)***	(3.242)***	(2.728)***	(6.995)***	(3.097)***	(9.171)***
Time	−0.00792	−0.05428	0.00249	0.00373	−0.01494	0.00885	−0.05787
	(−1.03)	(−5.878)***	(0.337)	(0.917)	(−1.178)	(1.619)	(−2.077)**
Legislate	−1.263	1.10203	−1.36284	−1.05944	−1.5004	−1.77713	−3.53834
	(−2.043)**	(1.238)	(−1.856)*	(−2.451)**	(−1.131)	(−3.026)***	(−1.224)
Campaign1	1.2488	1.39322	1.16965	0.28948	0.76636	0.57154	3.90077
	(2.868)***	(1.395)	(1.497)	(0.804)	(0.695)	(0.951)	(1.324)
R-Squared	0.332	0.436	0.211	0.114	0.235	0.17	0.368
F-Statistic	21.221***	42.294***	14.595***	7.009***	16.804***	11.164***	31.818***

[+]Data available only from 1983–1993. [++]Excluding *Atlanta Journal-Constitution.* ***Significant at the 1% level. **Significant at the 5% level. *Significant at the 10% level.

Table 5

Coverage of Acid Rain 1980–1993 Regression #3 $N = 168$

Dependent Variables						
Independent Variables	*Boston Globe*	*Chicago Tribune*	*LA Times*	*NY Times*	*Wash. Post*	All
C	6.8922	1.2866	0.7354	4.8849	0.9526	14.0162
	(10.797)	(3.30)***	(2.757)***	(7.608)***	(2.958)***	(10.344)***
Time	−0.0482	0.0076	0.00499	−0.0116	0.01135	−0.0409
	(−5.964)***	(1.173)	(1.378)	(−1.132)	(2.566)**	(−1.839)*
Legislate	0.48102	−1.9830	−1.2718	−2.0503	−2.0975	−5.6497
	(0.598)	(−3.045)***	(−3.303)***	(−1.867)*	(−4.411)***	(−2.309)**
Campaign 2	0.5713	0.1593	−0.1515	−0.3612	0.0217	0.39107
	(0.78)	(0.247)	(−0.476)	(−0.421)	(0.049)	(0.174)
R-Squared	0.425	0.170	0.111	0.231	0.16	0.345
F-Statistic	40.4***	11.16***	6.809***	16.438***	10.253***	28.849***

***Significant at the 1% level. **Significant at the 5% level. *Significant at the 10% level.

VII

Interpreting the Results

ALTHOUGH THE EMPIRICAL RESULTS ARE MIXED, we believe that they do support our "statist quo" hypothesis. First, the estimators for the *Time* for the *Boston Globe* and for *All* are statistically significant and negative, at the same time as the *Time* estimator for the *Washington Post* is *positive* and statistically significant.

On the other hand, the estimator for *Legislate* is negative and statistically significant in several instances in all three sets of tests. (The *Boston Globe* has the only *Legislate* estimator with a positive sign.) Furthermore, the estimator for *Legislate* is negative and significant at the 1 percent level for all of the regressions involving the *Washington Post*, a newspaper that more than the others listed depends heavily upon the daily goings-on of the federal government.

As for the two variables measuring coverage during U.S. presidential campaigns, the only estimator that is statistically significant is found in the regression in which the *Atlanta Journal-Constitution* is the dependent variable. In other words, the most powerful set of estimators in the series of regressions used in our paper belong to the *Legislate* variable, even when accounting for other factors like time and presidential campaigns.

Furthermore, the *Legislate* estimator is statistically significant despite the diversity of the markets for the newspapers that were used. The only newspaper whose coverage did not follow the patterns of the others was the *Boston Globe*. Yet, one should remember that the *Globe* began beating the acid rain drum before the others, including the *New York Times*. Given the extensive coverage that the *Globe* gave to acid rain in the early 1980s, it would seem logical that the "fatigue factor" might be stronger for that paper's readers than for the others.

We also looked at other theories, including the possibility that an influential newspaper like the *New York Times* or *Washington Post* would be able to drive the coverage. To test for such an effect, we employed a number of Granger causality tests. However, the results were inconclusive at best and were not able to establish any patterns of influence by one or more newspapers.

VIII

Conclusion

THIS PAPER HAS EXAMINED the acid rain controversy, looking both at the scientific arguments and the press coverage of that phenomenon in an effort to model journalists' behavior. The analysis covers six major U.S. newspapers, measuring how journalists from those papers covered acid rain.

Examination of the news coverage reveals that, while it reflected the scientific consensus in the early days of research, journalists failed to report changes in the scientific literature that challenged the original theories of lake acidification and harm to forests and vegetation. In fact, the mainstream press seemed to be in a time warp on this subject and continued to repeat the old theories even after they had been successfully rebutted in the scientific literature.

The empirical evidence consists of simple difference-of-means tests, descriptive statistics, OLS and GLS regressions, and some anecdotal evidence. It indicates that at least some of the acid rain coverage can be explained by Brookes's "statist quo" hypothesis, even when controlling for other factors.

We would like to emphasize that we do not see our study as explaining the *only* source of motivation for acid rain stories in particular and news stories in general. While we believe that the "statist quo" hypothesis has explanatory power, we also understand that to attribute all journalistic motivation to this hypothesis would be an example of overreaching our conclusions. It remains to other researchers to hypothesize and test for explanations of the nature and tone of press coverage on important issues.

References

Abdullah, Szarina B. (1989). *The Contributions of Scientists to Popular Literature, Their Role as Expert Witnesses and Their Influence Among Peers: A Case Study in the Field of Acid Rain.* Doctoral Dissertation, University of Illinois, Urbana–Chamapign, IL.

American Society of Newspaper Editors. (1975). *Statement of Principles.* Adopted October 23 1975, supplanting the *Cannons of Journalism 1922.*

Anderson, William L. (1992). "Acid Test." *Reason* 23(8): 20–27.
Brookes, Warren T. (1991). "Sense and Nonsense on the Environment." *Quill* 79(1): 14–18.
Crain, W. Mark, and Robert D. Tollison. (1997). "Expansive Government and the Media." Unpublished paper. Presented at the Media Symposium, Center for the Study of Market Processes, November 1997.
Efron, Edith. (1971). *The News Twisters*. Los Angeles: Nash Publications.
Goldberg, Bernard. (2003). *Bias: A CBS Insider Exposes How the Media Distorts the News*. New York: Harperperennial Library.
Krug, Edward C. (1992). "Acids and Bases in Watersheds." In *Encyclopedia of Earth System Science*, Vol. 1. Academic Press.
——. (1989). *Assessment of the Theory and Hypotheses of the Acidification of Watersheds*. Prepared for the U.S. Department of Energy.
——. (1990). "Fish Story." *Policy Review* Spring: 44–48.
——. (1991). "Review of the Acid-Deposition-Catchment Interaction and Comments of Future Research Needs." *Journal of Hydrology* 128: 1–27.
Krug, Edward C., and Charles R. Frink (1983). "Acid Rain on Acid Soil: A New Perspective." *Science* 221: 520–525.
Kurtz, Howard. (1991). "Is Acid Rain a Tempest in News Media Teapot?" *Washington Post* January 14: A3.
LaBastille, Anne. (1981). "How Menacing is Acid Rain?" *National Geographic* 160(5): 652–681.
Lichter, S. Robert, Stanley Rothman, and Linda S. Lichter. (1986). *The Media Elite*. Bethesda, MD: Adler & Adler.
Likens, G.E., and F. H. Bormann (1978). "Acid Rain: A Serious Regional Environmental Problem." *Science* 184: 1176–1179.
McCormick, Robert E., and Robert D. Tollison. (1981). *Politicians, Legislation, and the Economy*. The Hague: Martinus Nijhoff Publishing.
National Acid Precipitation Assessment Project. (1988). *Annual Report*.
——. (1989). *Annual Report*.
——. (1992). *Report to Congress*.
National News Council. (1981). *After "Jimmy's World": Tightening Up in Editing*. New York: National News Council.
Newey, W. K., and K. D. West. (1987). "A Simple, Positive Semi-Definite, Heteroskedasticity and Autocorrelation Consistent Covariance Matrix." *Econometrica* 55: 703–708.
Paulos, John Allen. (1995). *A Mathematician Reads the Newspaper*. New York: Basic Books.
Roberts, Leslie. (1987). "Federal Report on Acid Rain Draws Criticism." *Science* 237(4821): 1404–1406.
Rosenqvist, I. T. (1978). "Alternative Sources for Acidification of River Water in Norway." *Science Total Environment* 10: 39–49.
——. (1980). "Influence of Forest Vegetation and Agriculture of the Acidity

of Fresh Water." *Advances in Environmental Science and Engineering* 3: 56–79.

Shabecoff, Philip. (1987). "Study Discounts Immediate Peril from Acid Rain." *New York Times* September 18: A1.

Stigler, George J. (1971). "The Theory of Economic Regulation." *Bell Journal of Economics and Management Science* Spring: 3–21.

Sutter, Daniel. (2001). "Can the Media be so Liberal? The Economics of Media Bias." *Cato Journal* 20(3): 431–451.

A Comparative Political Economy Approach to Farming Interest Groups in Australia and the United States

By SEAN ALLEY and JOHN MARANGOS*

ABSTRACT. Commons ([1950] 1970: 34) insisted that "economics should be the science of activity." In this tradition, the aim of this paper is to investigate the impact of farming interest groups on natural resource policy by using a comparative political economy approach. Special attention will be given to farming interest groups in Australia and the United States. Curiously, each group takes a very different ideological approach to promoting farming interests. Our contention is that each group tends to display values that were prominent during its formation. The ideology and thus behavior of interest groups cannot be isolated from the history, the economic conditions, and the changing alternatives open to individuals. It is very reasonable to argue that two groups with similar goals might pursue different means to the same ends; the different means simply reflect values that were important in the formation of the groups. As such, there might be a concerted effort of the farming interest group, based on history, economic conditions, and custom, to either encourage a higher degree of competition or protect against the degree of competition.

*Sean Alley is a graduate student in the Department of Economics, Colorado State University. John Marangos is an Associate Professor in the Department of Economics, Colorado State University. The authors are grateful to the comments received by Scott Findley, John Loomis, Ronnie Phillips, and Laura Taylor.

American Journal of Economics and Sociology, Vol. 65, No. 3 (July, 2006).

I

Introduction: Farming Interest Groups and Natural Resource Management

IT IS UNQUESTIONABLE THAT farm practices impact the quantity and quality of natural resources. For example, forests can be inefficiently converted to cropland by artificially high crop prices. Domestic farm subsidy programs not only impact domestic agriculture markets but also have international ripple effects that disrupt farm production around the world. Local surpluses generated by artificially high prices put downward pressure on world prices and disrupt farm operations and renewable natural resource management decisions around the globe (Simon 1996: 125). Consequently, it is also undeniable that farming interest groups influence and impact on natural resource management. Partly as a result of interest group activity, governments have encouraged sustainable soil-use practices, not by telling farmers what they must do, but by combining cash with alternative land-use practices so that the farmers have financial incentives to use the sustainable alternative (Parsons 1941: 756). The political economy of natural resource economics, that is, the continual conflict between future and present appropriation (Gray 1913: 505), cannot escape the indispensable analysis of the political economy of farming collective action. As such, an examination of farming interest groups—formation, evolution, institutional decision-making processes, and goals—would reveal the direct link between farming applications, government policies as a result of lobbying, and natural resource management. Farming applications/practices are not simply the result of government policies, though for our purposes we will only concentrate on the values and ideology of farming interest groups, which naturally determine lobbying that influences government policy.

Commons ([1950] 1970: 34) insisted that "economics should be the science of activity." In this tradition, the aim of this paper is to investigate the structure, development, and evolution of farming interest groups using a comparative political economy approach. The analysis in this paper is not only restricted to the American context and, as such, is a contribution to the literature. Moe indicates that there is little research of a comparative nature in the analysis of farming interest groups (1980: 181), and to our knowledge this argument is still valid today. Special attention will be given to farming interest groups in Australia and the United States. Our research proposes to appeal to theories of collective action by Commons ([1950] 1970), Olson ([1965] 1971), and others in order to perform a comparative study of each country's largest farming interest group. The aim is to determine why two farming groups with supposedly similar interests have evolved such different means of achieving those interests. Our contention is that each group tends to display values that were prominent during its formation. The ideology and thus behavior of interest groups cannot be isolated from the history, the economic conditions, and the changing alternatives open to individuals. Knowledge of history is necessary for an understanding of the ideology of an interest group, the relative importance of the different factors, and the different proposals and demands for the future.

II

The Political Economy of Collective Action

GENERALLY, INTEREST GROUPS FORM because agents feel there is an opportunity to accomplish something collectively they cannot do alone. If members of a group have a common interest or objective, and if they would be better off by achieving that objective, it has been thought to follow logically that the rational, self-interested individuals would form a group to act collectively to achieve the objective (Olson [1965] 1971: 1; Knoke 1990: 6). This objective, or purpose, of group formation is important in understanding how groups evolve. A group's purpose is defined as that desirable state of affairs that members intend to bring about through joint action. A group's purpose is thus tied closely to the group's value set and ideology. With an understanding of the origin of a group, we can be equipped to explain how members become involved, make plans, and take collective action (Zander 1985: ix).

Interest groups arise spontaneously in response to a feeling of common interests among individuals; emotional or ideological elements are often present in group formation. Economic, political, or environmental changes disturb the "normal" behavior of potential group members, prompting them to interact and become increasingly aware of their shared interests. As this awareness grows, they form an interest group to serve as their representative. For example, U.S. farmers acted cooperatively in the 1870s to control railway corporations through politics and through the so-called Granger laws, and in 1890 the farmers joined with others to dissolve manufacturing corporations through antitrust laws. In Australia, the political system has enabled rural representatives to be elected to federal and state parliaments in disproportion to the votes received and to occupy powerful government positions such as deputy prime minister and trade minister.

Olson ([1965] 1971: 17–19) was very critical about the assumption that collective action is an inherent human instinct or propensity; this assumption does not add anything to our knowledge of group formation. The specific conditions that provoke a group's formation affect the way a group's membership and purpose change (or resist change) over time (Zander 1985: ix). History and political climate play a large part in what types of groups form and when they form. Insight into the historical context of group and purpose formation is critical if we are to study not only what groups do but also why they do what they do (Castles 1967: 17). Economic conditions certainly affect the presence of "trigger states" for group formation. Results from a historical survey confirmed that groups tend to form in waves concurrent with times of political or social unrest (Hrebenar 1982: 17).

Parsons (1941: 751–760) suggested that group power is real. Free, self-regulating markets are an ideal. Markets were, in fact, used initially for conflict resolution. They grew out of societies iteratively forming rules to ease the burden of continually conflicting interests. Markets today are the sum of these rules that were established to grease the wheels of social progress so that people would have a reliable way to settle disputes. Markets are not a natural order (Parsons 1941: 751–760). Commons ([1950] 1970: 15) accepts conflicts of interest as natural and indispensable ingredients of social processes: "Conflict and power are real." This is not conflict in the narrow sense of friction; the political economy of conflict between interest groups is inherently conjoined with mutual dependency that evolves into the achievement of some form of order. Thus, social processes intrinsically embody conflict, mutual dependency, and order in a decentralized but consensual democratic polity. Nevertheless, it is not order in the static equilibrium sense of individualist economics; rather, it is order in the dynamic sense of continuously and eternally resurfacing problems. Collective action, which is essentially an "institution" in modern capitalism, provides a tool of analysis and investigation for comprehending the similarities and differences in the varied forms of order as the result of collective action. In order to be relevant, social (economic) analysis must account for the effect of power on relations between economic agents, whether they are individuals or groups (Galbraith 1973: 6).

The formation and evolution of an interest group is path-dependent. During this process, the wills of individuals are brought together into a created collective will. Zander (1985: 5) states that initial conditions under which a group forms determine its evolution. One reason that purposes of groups are path-dependent is that individuals in groups develop a common point of view with regard to the group's interests. Beliefs of group members are influenced by the individual's association with the group (Zander 1985: 11). Perhaps relatedly, group leaders can impact members' involvement as well as their desired involvement. The fact that leaders affect members' opinions means that shifts in member interests are likely to be muted. As mentioned above, group membership in terms of population of a group has a direct effect on group ideologies. Having a large membership means that a shift in beliefs on the part of a few individuals will not affect the direction of the group (Zander 1985: xiv). Large group values are also path-dependent because they often have a large staff. Staff members influence the ideology of the group due to the mere fact that they are the ones dealing with the day-to-day operations of the group. Further, groups whose membership is based primarily on provision of selective incentives tend to display more path-dependent behavior. Constituents are not associated on the basis of common political beliefs, so they are less likely to be concerned about group activity not directly related to them. Such groups tend to persist over time, as membership does not grow and decline around social issues

(Hrebenar 1982: 26). These characteristics indicate that large, persistent, diverse groups such as the farm lobbies studied here should display even less ideology change over time than most groups.

Thus, group purposes tend to stay unchanged over time. Zander (1985: 40) states that groups tend to hold on to original value sets indefinitely. Zander further contends that some purposes themselves resist change. Goals are less likely to be changed if they are immeasurable. Purposes that concern outcomes for the group as a whole should be less changeable than goals that affect the separate fates of individual members (Zander 1985: 142). Thus, group purposes tend to be more path-dependent than individual purposes. The organization tends to "insist" on the original value set. Parsons sums it up nicely: "social arrangements have something approaching careers of their own" (Parsons 1941: 763). In addition, as the organized form of collective action is evolved, custom comes into play. Customary behavior—the mere repetition, duplication, and variability of practices and transactions—is stabilized as social behavior, affording to the group the expectation that the usual successful ways of doing things must be carried on (Commons [1950] 1970: 354).

The group/government interaction takes the form of lobbying, or the attempt to secure specific policy by government. Among other things, interest groups either defend themselves against intrusions by government or demand government assistance. Lobbying frequently involves an adversarial and competitive process between different interest groups (Mack 1989: 2). So both agricultural interest groups in our study are promoting path-dependent and customary values that originated when the groups formed. It is very reasonable to argue that two groups with similar goals might pursue different means to the same ends; the different means simply reflect values that were important in the formation of the groups. During a credit collapse, manufacturers respond to falling prices through closing factories and laying off workers. Farmers cannot shut down their farms, nor lay off themselves and families. They must tolerate the adverse economic conditions and go on producing a surplus at falling prices while manufacturers can maintain prices by unemployment. Similarly, farmers often increase their output even when the demand is inelastic and this is contrary to their common interests. Farmers had always to deal with these adversarial environmental conditions, encouraging them to form interest groups. As such, there might be a concerted effort of a farming interest group, based on history, economic conditions, and custom, to either encourage a higher degree of competition or to protect against the degree of competition. As stated before, the action of collective groups embodies conflict, mutual dependency, and order in a decentralized but consensual democratic polity. An investigation into farming groups provides reasons to comprehend the varied forms of "order" as the result of collective action.

III

Farming Collective Action in Australia and the United States

THE NATIONAL FARMERS FEDERATION (NFF) and the American Farm Bureau Federation (AFBF) are the largest agricultural interest groups in Australia and the United States, respectively. Both maintain that a main function is to serve the interests of domestic farmers. Curiously, as revealed from the vision, mission, and goals of each interest group in Table 1, each takes a very different ideological approach to promoting these interests. NFF's dominant ideology promotes free competition and the elimination of agricultural subsidies worldwide. AFBF's recent agenda has centered on supporting the 2002 Farm Bill, under which one-half of net farm income comes from various levels of government subsidies.

Table 1 provides the first indication of the values and ideology of each farming group. At first glance, many of the purported values and goals are quite similar. The legislative agenda of each group tells a very different story, however. In 2002, the NFF's legislative agenda focused on lobbying for improved competition. Government aid was focused on initiative programs designed to improve infrastructure for efficient farms. NFF preferred programs that discourage the continuing survival of inefficient farms. Specifically, it argued for abandonment of unenforceable price regulation in the transportation industry and the establishment of a competitive licensing system for transporters of agricultural goods. Additionally, it lobbied for increased penalties for collusion generally, and promotion of competition in banking and energy industries. Internationally, the NFF suggested that

Table 1

Vision, Mission, and Goals of NFF and AFBF

	National Farmers Federation (NFF)	American Farm Bureau Federation (AFBF)
Vision	Australian farmers operating profitable and sustainable farm businesses	A growing agriculture industry that depends less on government payments and more on returns from the marketplace
Mission Statement	To influence the Commonwealth government, Parliament, and agencies to deliver policy outcomes to the advantage of Australian farmers	To implement policies that are developed by members and provide programs that will improve the financial well-being and quality of life for farmers and ranchers
Goals	• Increase profitability by reducing costs • Increase markets and revenue • Improve sustainability	• Implement policies that grow markets • Eliminate trade sanctions • Increase farm income by increasing farmer investment

Industrial Relations	• Reduce labor and employment costs • Reduce labor administration costs • Increase utilization of enterprise agreements • Improve farm labor supply	
Taxation	• Lower tax burden and compliance costs • Increase access, use, and benefit of farm tax programs	• Capital gains taxes should be repealed • Death taxes should be eliminated • Farmers and ranchers should not have to pay self-employment taxes on unearned income
Telecommunications	• Increased access to voice, Internet, and mobile services • Reduced cost of voice, Internet, and mobile services • Promote increased Internet usage	
Macroeconomics	• Ensure macroeconomic policies are to the benefit of farmers (i.e., low interest rate, low inflation regime)	• Use market-based incentive programs to reduce compliance costs of regulation • Oppose Kyoto Protocol

Table 1 *Continued*

	National Farmers Federation (NFF)	American Farm Bureau Federation (AFBF)
Farm Costs	• Reduce costs of farm inputs • Increase competition in farm inputs	• Reduce costs of farm inputs • Promote low-interest loan programs
Trade	• Reduce market barriers to Australian farm exports • Encourage international trade reform • Maintain Australia's science-based quarantine and SPS rules	• Eliminate unilateral trade sanctions • Eliminate foreign trade barriers • Enforce agreements with EU and China to open markets
Competition	• Improve competition in market chain • Ensure market power is not abused	• Resist consolidation by corporate farmers
Food Industry	• Increase access, use, and benefit from government food programs	
Environment	• Ensure water resource security • Ensure land resource security	• Increased role for renewable fuels in natural resource policy

	• Increase access, use, and benefit from environmental stewardship programs • Limit costs of greenhouse impacts, adaptation, and abatement	• Farmers respond to economic incentives to improve sustainability (voluntary, incentive-based conservation) • Reduce compliance costs
Agricultural Programs	• Agricultural programs improve farmers' risk management and self-reliance • Increase access, use, and benefit from these agricultural programs	• Need continued income support • Reduced use of ad hoc support; increase in long-term, reliable income support • Provide safety net for farmers so they can participate fairly in markets • Oppose supply management programs • Increase funding for agricultural research
Regional Development	• Ensure government regional programs boost the profitability of the rural economy • Increase access, use, and benefit from these regional programs	• Increased farm income from investment provides rural growth opportunity • Increase rural economic development via access to competitive grants

the best way to help efficient Australian farms is through world trade liberalization and elimination of price and income supports (National Farmers Federation 2002).

On the other hand, AFBF's legislative agenda focused on maintaining farms. Notably, the agenda put a priority on maintaining the 2002 farm bill. This bill provides countercyclical payments to farms in the form of both direct supports and marketing loans. This bill provides insurance against low returns from the marketplace. Elsewhere, AFBF supported using public funds to encourage alternative energy sources that come from farms (ethanol, for example). Internationally, AFBF favored reduction in tariffs, but not income supports explicitly (American Farm Bureau 2003). As such, we believe that each group's legislative agenda is more telling of the dominant ideology in each group. These ideologies and values were present in the earliest days of these groups' formation. The evolution of these values requires the investigation of the long process of development and evolution of each farming group, which takes place in the following section. This will help us to determine the behavior and the influence of each group.

IV

The National Farmers Federation in Australia

In July 1979, the National Farmers Federation was established. The achievement of farmer unity in Australia and the establishment of "one voice" for Australian farmers were long sought. It took almost 90 years after the establishment of the first federal organization, the Pastoralists Federal Council (1890), for the divisions between primary producers to be bridged. The rift between primary producers, known throughout Australia's agricultural history as "farmers" and "graziers," was deep-seated. The reasons why people producing the same commodities chose to organize in distinct associations were historical, ideological, social, class-related, and even due to snobbery. Primary producers were victims of their own diversity. Not only were farm organizations spilt along commodity lines, but it was possible to find two or more organizations representing producers of one commodity. The fact that primary producers thought of themselves as producers of specific commodities rather than as general primary producers was a major factor to overcome in order for unity to be achieved.

The roots of the division can be traced back to the 1800s, when pastoralists used their political influence to control large tracts of arable land for sheep, denying small farmers land for cropping. They first fought over land (Halpin and Martin 1999: 35). This left bitter scars that were aggravated by differences in wealth, property size, social status, lifestyles, and methods of farming. Graziers usually employed full-time staff as well as casual shearers, and farmers were far more reliant on family labor. Many graziers belonged to the employer class and had off-farm investments and social links to the city. The wealthier graziers sent their children to expensive private schools—the high value placed on private school education by grazing families was not blemished by hard times and falling farm income—and shopped for a range of household items in the city in preference to county towns. Some had city residences. Farmers generally had more modest lifestyles, only employing outside labor during busy seasonal periods, and making use of local schools and stores. There were also extensive differences in ideology. While graziers supported the free market and had close ties with wool brokers (grazier associations accepted broker membership), farmers had an inbuilt fear of middlemen and wanted the marketing of farm produce taken out of outsiders' hands and placed in a government-sponsored organization.

In 1893, Australian farmers organized in New South Wales under the banner of the Farmers and Settlers' Association in order to change the colonial land laws and remove land from the pastoralists. A few years earlier, the grazers had united as the Pastoralists' Federal Council (1890) to oppose the demand of shearers that they employ only members of the Amalgamated Shearers' Union. For many decades, the pastoralists had comprised the economic and social elite of Australia, but had lost much of their power in the colonial parliaments through the extension of suffrage and democratic reforms. They were now being threatened by their own "servants": the shearers, farm laborers, and wool carriers. They had little choice other than to organize against organized labor. The Pastoralists' Federal Council was originally an industrial relations body with the purpose of fighting the closed-shop principle, establishing "freedom of contract," and stopping what they considered the excessive pay demands of the Amalgamated Shearers' Union. In August 1891, a meeting between the two opposing groups in Sydney saw the shearers accept freedom of contract as a precondition for an industrial agreement to cover shearing throughout the colonies. This was a moral boost for the newly formed Pastoralists' Federal Council. This success established industrial relations as a forefront issue for the pastoralists, even in subsequent organizations. The pastoralists later took upon themselves all primary producers' organizations in arguing against national wage increases. It became customary that industrial relations was the responsibility of the subsequent pastoralist-grazer interest groups.

There were deep and traditional divisions between wheat farmers, who demanded government intervention and distrusted the grain merchants, and wool growers, who supported the free market and had strong links with pastoral houses. However, the divisions became blurred when many small wool growers demanded price guarantees for wool and a monopoly marketing board for wheat. The Victorian Wheatgrowers' Association changed its name to the Victorian Wheat and Woolgrowers Association in 1929 for two major reasons. One was to cater to the increasing number of its members who had diversified into wool to lessen their dependence on wheat. The second was to attract small wool growers who were unhappy with the free market philosophy of the grazier organizations. In New South Wales, the Farmers and the Settlers' Association was predominantly a wheat growers' body, but it attracted many wool growers through its support for organized marketing. Many Farmers and Settlers' Association members combined wool production with wheat farming while others diversified into fat lambs, which produced wool as a byproduct.

The radical wheat grower organizations—the Australian Wheatgrowers Federation (1931) and the Australian Wool and Meat Producers' Federation (1939)—came into prominence between World Wars I and II to demand government intervention in their industry to protect the livelihoods of farm families. In contrast, pastoralists, who adhered to the free market and had close ties with the wool brokers, thought the wheat growers demanded government-backed marketing schemes that would provide price guarantees and eliminate the industry of the middlemen they distrusted. However, by the fourth decade of the 20th century, there were only two strands. The pastoralists, now calling themselves graziers, had remained true to their freemarket philosophy; the small wool growers of the farmers and settlers' movement had joined the radical wheat growers in the quest for "orderly marketing." This consisted of grower-controlled marketing boards with monopoly trading powers and government-backed price guarantees. Farmer organizations became affiliates of the Australian Wheatgrowers Federation and argued that orderly marketing should be extended beyond wheat to wool. Unlike their counterparts in North America, Australian wheat growers did not seek answers to marketing problems through cooperatives (Connors 1996: 31).

The year 1948 brought the first of the unbroken run of five-year wheat stabilization schemes that continued virtually unchanged until 1968. There were many factors behind the eventual achievement of wheat stabilization, including two world wars that saw the introduction of controlled marketing under wartime defense powers. Of greater significance were the increased stature of the Australian Wheatgrowers Federation by the wars' end, the growing importance of the wheat vote, and the wide acceptance of the fact that the government had the right to intervene in the Australian wheat industry. The Australian Wheatgrowers Federation had not only achieved its aim of a wheat stabilization plan but had also dominated the Australian Wheat Board, the sole seller of the annual harvest. When many graziers diversified into wheat in the 1960s, however, the Australian Wheatgrowers Federation firmly resisted their demand to affiliate and gain seats in the Australian Wheat Board. The Australian Wheatgrowers Federation perceived that the graziers, once inside, could attempt to undermine the wheat stabilization program.

In 1968, the Graziers' Association of New South Wales produced a unity study report that suggested a number of reasons why farmer-grazier unity would be difficult to achieve. Among them was the superiority of graziers as men belonging to the employer class, having higher education standards, and being the type of people who (unlike United Farmers and Woolgrowers' Association members) were persuaded by reasoned argument (Connors 1996: 10). Yet unity remained on the agenda, and a major impediment was removed within two years of the report's completion when graziers succumbed to low wool prices, drought, rising costs, and political pressure. They accepted what farmers had long desired: a reserve price scheme for wool. As well, they entertained the belief that compromises should be made in the interests of unity. There was also a shift in the Graziers' Association's power base toward a reserve price. More members had received a taste of orderly marketing and price guarantees following the big swing to wheat in the 1960s. At the same time as the wool industry was in dire trouble, governments were about to change, the exchange rate was being managed and manipulated upward, the mining industry was growing in importance, the Greens were emerging, and the manufacturing lobby was reaching its peak. All these factors hastened the development of the National Farmers Federation (Connors 1996: 156). In 1970, the graziers had accepted a single marketing authority for wool, which had the power to operate flexible reserve prices to ease the ups and downs of the market. By 1974, they gave way on fixed floor prices and were securely trapped in the protectionist web.

The election in December 1972 of the Whitlam Labor government, the first Labor government in 23 years, also encouraged unity. Now, there was a government in control that owned its existence to the trade union movement with very little support within the farming community. The Labor government transformed the Tariff Board into the Industries Assistance Commission to advise government on all forms of assistance to all sectors of the economy (Halpin and Martin 1999: 36). The emergence of the Industries Assistance Commission was another factor encouraging farm organizations to amalgamate. Its establishment meant that farm organizations had to present detailed submissions for the continuation of assistance activity. Farmers' organizations required more resources and more skilled staff to present farmers' cases in public submissions with comprehensive and economically literate submissions before the Industries Assistance Commission. Unity would cut costs and acquire those resources. However, few farm organizations had the staff and resources to develop detailed policy submissions for their industry. Only the Australian Woolgrowers and Graziers Council and its close affiliate, the Graziers' Association of New South Wales, were probably capable of sophisticated economic and social analysis in those days. The Australian Woolgrowers and Graziers Council, as already mentioned, had customarily taken up the burden for all farm organizations in arguing against national wage increases before the Conciliations and Arbitration Commission. It employed economists and industrial advocates and took the view that what was good for the economy, such as lower wage costs, lower interest rates, and lower tariffs, was also good for farmers. The remaining federal commodity organizations operated on a very tight budget. In the fight for new members and retaining existing members, some farm organizations kept their fees very low. This meant a lack of money to employ staff to prepare submissions of quality. Farm organizations found the task burdensome and called on the federal government for assistance in making claims for assistance! Only the graziers had the wealth and resources, the result of higher membership fees applied to their larger and wealthier producers and property investment, to take on the broad issues. Apart from the resources, the graziers, as primary producers operating in the free market, also had the inclination to tackle such issues. The economists employed by the farm organizations in the early 1970s had been trained by university economists, who tended to the side of freemarket economic rationalism. The new breed of young economists in farm organizations played a role in shifting farm leadership away from a handout mentality and toward concentrating on reducing costs, including lower tariffs, structural reform, and industrial relations.

In 1975, annual conferences of both the United Farmers and Woolgrowers' Association and the Graziers' Association voted by large majorities in favor of amalgamation and setting up a joint working group. After a successful vote, it was announced on August 1, 1977 that unity was established (Connors 1996: 188). The National Farmers Federation was so named without public protests from "graziers." The term "farmers" now meant all primary producers. The positions on the NFF Council and its various commodity councils were hotly contested, with "farmers" competing against "graziers." The Western Australian dairy farmer Don Eckersley defeated grazier Ian McLachlan in the ballot of July 20–21, 1979 to become the first president. Eckersley won not just on his popularity, but also because the farmers outnumbered the graziers in positions of power on the state organizations that nominated delegates to NFF. The graziers may have been beaten by the farmers for key positions such as NFF president and commodity council presidents, but they got their staff into top positions of the NFF secretariat. Thus, while farmers were the victors in the ballots, senior staff positions were filled by the economists and other professionals from the now-defunct grazier organizations. This had a major impact on the style and direction of the NFF. The main people preparing the policy documents, writing speeches for President Don Eckersley and commodity council heads, and writing submissions to the government and the Industries Assistance Commission were former employees of the now-superceded grazier organizations with an inherited free market ideology. In the second NFF Council meeting, in October 1979, a set of policy resolutions was passed that were basically prepared by the staff with a free market ideology. In regard to economic policy, the resolutions were what the graziers had been arguing for a long time: lower protection levels and greater competition throughout the economy (Connors 1996: 218).

An increasing focus in agricultural policy on "self-reliance" and "individual risk management" are the principles on which the NFF operates. State intervention should aim ultimately toward restructuring and making the market work more "efficiently" in the interest of larger-scale farmers and agribusiness. The increasing emphasis is on farming as a business, rather than as a lifestyle (Higgins 1999: 141). With regard to natural resource management, the NFF has always strongly supported voluntary, incentive-based approaches toward sustainable management of farms. The NFF prefers the devolution of management responsibility to discrete regions because the best people to make decisions about natural resources are those who live in these regions. This argument highlights the need to access to good-quality baseline data. Lovett (1998: 16), from the NFF, states: "The bottom line for farming is that it must be both profitable and environmentally sustainable. The two are naturally complementary and neither can exist without the other. Protecting biodiversity protects farms, and protecting farms from environmental degradation means more profitable farms."

V

The American Farm Bureau Federation

The American Farm Bureau Federation arose as a national organization in 1919 after numerous, short-lived attempts to establish a national organization. The idea of organizing to exert collective economic power had begun gathering momentum half a century earlier, around 1870. Farmers of this era percieved the railroads, farm machinery manufacturers, and bankers who held farm mortgages to be monopolies that should be regulated. Immediately following the Civil War, the frontier expanded rapidly, and labor-saving machinery enabled many farmers to expand their operations. National railroads made long-distance marketing feasible for the first time, and heavy overproduction resulted in plummeting prices for many agricultural commodities. Currency was unstable, and tariffs often worked against farmers, both as exporters of food products and as consumers of manufactured articles. They were at a disadvantage in procuring favorable legislation along these lines, compared with the influence of the bankers. The farm organizations were demanding price increases, but the bankers were demanding and obtaining price decreases through retirement of the greenbacks (1867) and demonetization of silver in 1879 and 1900 (Commons [1950] 1970: 212). On top of this, wealth was being transferred to urban populations, and the social status of the farmer was eroding. From all appearances, rural farmers were being left behind in the economic scheme of things, and they blamed the city dweller for the farm predicament. Farmers, and then wage earners, began to organize politically in order to legislatively and administratively constrain corporations and employers, whom they began to distinguish as "capitalist" class (Commons [1950] 1970: 265).

In 1867, Oliver Kelly (Kile 1921: 47) decided that this plight was largely the result of the independence of farmers, blindly following what their fathers had done. Kelly organized the National Grange to be a secret agricultural society, modeled after the Masonic Order, of which he was also a member. The purpose of the Grange was intended to be social and educational. The Grange grew quickly after it began to promote itself as a means of protection against corporations and of opportunities for cooperative buying and selling. The Order spread quickly after 1873, when a panic caused creditors to press farmers for loan payments. The Grange actively played to farmers' discontent to encourage membership. The organization was able to garner enough political support to cause legislators to regulate railroads so that farmers could ship their produce without handing over all profits to the transportation conglomerates. After that, the power of the Grange unraveled, as some thought the organization should be more political while others thought it should stick to its educational roots and be less actively involved in politics. Grangers were able, however, to establish "Grange stores," which were the first coordinated efforts at community-wide cooperative buying and selling of farm products. Some of these were quite successful, but when a couple of larger commercial cooperatives failed, local Granges began disbanding for fear they might be held responsible for the debts. The organization, as a national entity, splintered.

The discontent that fostered the success of the group was still around, however. The benefits to be derived from an organization were too alluring to be forgotten. A string of attempts at organizing farmers followed. The Farmers' Alliance was built on the successes of the American industrial laborer and sought to draw attention to cooperative possibilities. This organization (along with organized labor) was able to move its members to political action, but it had no core business. Lack of funds caused the power of the Alliance to dissipate several years later, when political disagreements between state units arose. In 1902, the Farmers' Union was built on the ruins of the Alliance by Newt Gresham. A national organization was set up, and former Alliance organizations were invited to join. Membership grew rapidly for the Union, which emphasized cooperative and other economic features of the organization, rather than the social and educational features prominent in earlier organization efforts. A great variety of cooperative enterprises were started, and many were successful. Partly because of this, the Union was not able to support a truly national organization. Cooperatives, naturally segregated to areas of particular crop interest, resulted in an inability to sustain a national organization. Other, smaller organizations came and went before the

American Farm Bureau Federation grew to prominence. Around this time, many commodities were establishing, or attempting to establish, cooperative selling markets. These met with varying degrees of success but, importantly, they brought previously independent farmers together, taught organizing principles, and furthered the idea of gain through cooperative enterprise.

The U.S. government showed initial signs of getting involved in agricultural life in 1908 when President Theodore Roosevelt established the Commission on Country Life. This was to be a survey and investigation to see what could be done to improve the quality of rural life and inject a little enthusiasm in the "country life movement." The seeds for a national organization were planted. Simultaneously, state agricultural colleges were becoming established. Through these colleges, the government began to organize and pay for farmers' institutes, where progressive farmers could go and receive instruction on the newest farming techniques. It was soon realized that, especially in the poorer areas, a very small percentage of the rural population was being reached. Information needed to be taken out to the farmers. Extension work was born, where agriculture professors were paid to disseminate their findings by visiting farms in their area.

The next development was the coming of the county agent. A group of agricultural experts was sent to Texas to spread the latest information about combating the boll weevil. The idea was to use demonstrations on farms to prove to the rural public that science could conquer problems that farmers alone could not. Rural farmers could see firsthand that the new information worked, and the demonstrators gained their confidence. Farmers in Smith County, Texas, became the first to contribute toward the salary of their own full-time demonstrator, or county agent. Others quickly followed, as it became clear that having someone assigned to keep up with agricultural advances was good for the business of farming. The Smith-Lever Act in 1914 made federal funds available to hire a county agent for each county in the United States.

There were soon county agents everywhere, spreading agricultural awareness to the masses. They found that one agent could not, because of time constraints, advise every farm in the county. In fact, the agent was often only seeing a few of the more progressive farmers. Agents decided to enlist the help of these progressive farmers. These farmers were encouraged to organize into a "county farm bureau," which could collect dues to help pay the agent's salary and help the agent disseminate information and demonstrations to the farmers further out in the county. Since more members meant more money and greater influence, the county agent often took part in membership drives. The influence of the farm bureau was extended throughout nearly every farm in the country, and the county agent became jointly answerable to the state college of agriculture and to the local farm bureau. This intertwining of the government and farm bureau operations from the earliest days is important to the group's evolving ideology. Farm bureaus were first organized under the aegis and with the assistance of the federal government.

The relationship between the county agent and the farm bureau was the source of much consternation and criticism on the part of competing interest groups and free market organizations. Because the agent's salary and continued employment was tied to bureau membership fees, the agent had an incentive to make a showing on a cash basis. The agent began to assist the local farm bureaus with cooperative buying organizing for buying common farm supplies in bulk for the entire county. The farmers who joined were normally put on the mailing list for technical publications; the farmers who did not join were not. The farmers who joined had first call on the county agent's services: the farmers who did not join normally had last call or no call at all. Farmers thus had a specific incentive to join the farm bureau. The dues they had to pay were an investment (and probably a good investment) in agricultural education and cooperation (Olson [1965] 1971: 150).

Producers of farm machinery and other farm implements were forced to sell at very narrow profit margins in order to secure the patronage of these cooperative buyers. Those that did not were unable to sell anything at all, as the business of farmers in each county went to only one supplier. Because county agents were involved in the administration of these cooperatives, the suppliers complained loudly that federal funds from the Smith-Lever Act were being used to favor farmers' interests. The U.S. Department of Agriculture was forced to regulate the extent to which the county agent could participate in cooperative buying. The agent could help organize the group, but could go no further. Farm bureaus had to hire purchasing agents to oversee cooperative buying. Cooperating with the county agent was now only one of the lines of work of the farm bureau. From this point forward, the farm bureau grew in power as it operated more independently of the county agent.

Local farm bureaus crystallized into state units to take advantage of the organization that had been provided for educational means by the county agents. Almost immediately, the state units came together to form a national organization, the American Farm Bureau Federation. The size and the relative stability of the American Farm Bureau Federation has resulted from two factors. One is that, for a long time, it was the natural channel through which farmers could get technical aid and education from the government; the other is that it controls a vast variety of business institutions that normally provide special benefits to AFBF members (Olson [1965] 1971: 157). The primary motivation for nationwide organization was to keep control of food products until they reached consumers, thereby reducing what were seen as excessive profits on the part of railroads and other middlemen. After intense argument between leaders who wanted to keep up the educational focus of the farm bureau and others who wanted to take advantage of the political and economic possibilities of organization, a middle ground was reached. Education would still be on the agenda, but exploiting the organization to improve the lot of the farmer was a higher priority. National dues of 50 cents per member were established in 1921, giving the AFBF a financial base that previous attempts at farm organization had not enjoyed. Economic goals were as follows: to extend cooperative marketing of farm crops to the point that maximum benefits were secured for the producer and consumer; to limit the profits and reduce the costs of distribution in all lines not handled cooperatively; to estimate world supply of farm products so as to regulate flow and stabilize prices; to establish new foreign markets for surplus American farm products; and to provide cheaper sources of farm inputs. So, from the very beginning, cooperative buying and selling were at the heart of AFBF's agenda. Also, the AFBF was notably in favor of using legislation and regulation to improve the farmer's competitive position, control output and price of farm commodities, and reduce tariffs on agricultural exports.

As the group evolved, a close relationship with the federal government continued. In 1933, the Roosevelt administration began a vast program of aid to agriculture under the Agricultural Adjustment Act. To get the program off to a rapid start, the administration had to rely on the only nationwide administrative system that had any experience with agriculture—the Agricultural Extension Service, with county agents in every county (Olson [1965] 1971: 152). Battles were fought with varying degrees of success over price controls of inputs and outputs. Cooperative organizations first established by the AFBF were quite successful; by this time, however, they were organized along commodity lines and were managed regionally, rather than nationally.

Today, the legislative agenda addresses many of the same concerns. Although management of many of the remaining buying and selling cooperatives has long been turned over to specific commodity interests, the ideology that promoted government help with export and input markets remains a key part of AFBF's strategy. Concerning export markets, AFBF has asked for federal help with eliminating export licensing for previously sanctioned markets, continuing to seek new markets globally, providing funding for international market development programs, and continuing to monitor and enforce existing trade agreements. AFBF supports temporary immigrant worker programs that are agriculture specific, and supports regulation of packing and shipping companies. These and other programs seek to keep prices of agricultural inputs low. The dominant ideology of the AFBF, then, displays considerable evidence of path-dependence. Major support of cooperative enterprises (which proved untenable on a national scale) notwithstanding, the AFBF of today embodies many aspects of the same ideology that it showed in the early 1900s when it was forming. While these things certainly impact natural resource management, the AFBF advocates a natural resource policy very similar to that of the NFF. That is, it supports sustainable development in all its forms, as long as it does not inhibit productivity of agricultural business. Specifically, it encourages development of water resources and reformation of the Endangered Species Act and other legislation to reduce regulatory compliance costs for farmers.

VI

Conclusion: A Comparative Approach

The NFF, while established relatively recently, inherited the ideology of the free market as a result of a concerted effort by the graziers' organizations to install the free market ideology in the newly formed organization. The graziers' organizations had the funds to support professional staff (with a free market ideology) and were able to provide the necessary infrastructure for the newly formed organization to be operative, independently of the election results to the Executive Board. In this way, the employment of the previous staff of the graziers' organizations guaranteed that the values of the free market would be installed in the NFF. The circumstances under which the group was formed are important. Large, previously existing groups were joined into a national organization. The NFF began as a national organization with an already-developed set of values taken from the graziers and large farm interests. In contrast to the AFBF, government involvement and cooperative ideology were inhibited in the advancement of the NFF.

The AFBF enjoyed very strong support by the government. Its historical development is linked with government support, a connection that is maintained even today. The AFBF originated in a much more decentralized way, as a collection of small farmers seeking a voice in a large economy. With government agents helping to provide structure, individuals gathered power until they formed a national organization. This bottom-up development structure tended to encourage the interventionist ideals of the small, powerless, and independent farmer. While the AFBF is in favor of free competition, this competition should not be to the detriment of the stability of farmer's income. In this context, natural resource policy for both organizations should not impact negatively on the profitability of farms.

In its 2002 *Annual Review*, the NFF's position on environmental management is based on four principles: information, property rights, incentives, and partnerships. For the NFF, managing the balance between natural resource management and property rights is a major issue. Restoring to farmers security of their use of natural resources would encourage private investment to manage farm business in an environmentally sustainable way. However, the NFF is concerned that the government will adopt a top-down approach to the issues regarding environment management systems, which would prohibit farmers' involvement in the decision-making process and in the process of change (Nationl Farmers Federation 2002: 27). In general, farm businesses should be based on profitability, sustainability, competitiveness, risk management, and self-reliance; the role of the government is to provide the supporting public goods of information and the appropriate institutional environment.

The Organisation for Economic Cooperation and Development (OECD) reports show a wide range of farm income support across OECD countries. They estimate U.S. subsidy levels to be around 20 percent. More notably, Australian farmers receive only 4 percent subsidy, second-lowest only to New Zealand (OECD 2003). This is in keeping with the lobbying agendas of the two large groups analyzed here. The AFBF, as revealed in the *Legislative Agenda 2003*, strongly supports the 2002 Farm Bill. The Farm Bill offers a combination of countercyclical income support and direct support for marketing and loans. While the market should be competitive, these same markets should offer "fair" prices for agricultural products. These policies support "broad national goals" and "the associated benefits for our entire nation, and the futures of the families and rural communities involved" (American Farm Bureau 2003: 4). Although these policies have not resulted in significant increases in subsidies to U.S. farmers as yet, they have the potential to do so to a significant degree if farm prices drop (NSW Farmers Association 2003: 3). With regard to natural resource management, the Farm Bill has provided increased resources to assist farmers in the improvement of resource management. However, AFBF stipulates that this increased funding should be provided for voluntary, incentive-based programs because these are the most effective and efficient ways of improving and protecting the environment. Thus, both organizations have similar positions regarding natural resource management. There is a major difference, however: the AFBF lobbies for and promotes the passage of the Farm Bill.

Implicitly at least, it is willing to tolerate and support inefficient farms through the provision of government income support and establishment of a "fair" price. This would be totally unacceptable for the NFF. The NFF has been demanding the elimination of domestic and foreign subsidies for agricultural products. If farm businesses cannot meet the market test, they should close down, whether the shutdown is the result of price fluctuations or of lagging behind voluntary environmental sustainability programs.

It is not by accident that the positions of the NFF and the AFBF are so different. The dominant ideology of the NFF is free market, while the AFBF favors government intervention, government income support, and a "fair" price. Current world conditions and domestic political settings certainly play some role in the contemporary behavior of these organizations, but the evolution of each organization has played the key role in the positions they now take regarding how best to improve domestic agriculture. The historical development and evolution of both organizations demonstrated that conflict, mutual dependency, economic conditions, and custom have shaped the ideology of each organization, even though both organizations have the same goal: to protect the interests of farmers.

References

American Farm Bureau. (2003). *Legislative Agenda 2003: 108th Congress.* http://www.fb.org.

Castles, F. G. (1967). *Pressure Groups and Political Culture: A Comparative Study.* London: Routledge and Keegan Paul.

Commons, J. R. ([1950] 1970). *The Economics of Collective Action.* New York: MacMillan.

Connors, T. (1996). *To Speak with One Voice: The Quest by Australian Farmer for Federal Unity.* Kingston, Australia: Lamb Printers.

Galbraith, J. K. (1973). "Power and the Useful Economist." *American Economic Review* 63(1): 1–11.

Gray, L. C. (1913). "The Economic Possibilities of Conservation." *Quarterly Journal of Economics* 27.

Halpin D., and P. Martin. (1999). "Farmer Representation in Australia: Avenues for Changing the Political Environment." *Australian Journal of Public Administration* 58(2): 33–46.

Higgins, V. (1999). "Economic Restructuring and Neo-Liberalism in Australian

Rural Adjustment Policy." In D. Burch, J. Goss, and G. Lawrence (eds.), *Restructuring Global and Regional Agricultures: Transformations in Australasian Agri-Food Economies and Spaces.* Aldershot, UK: Ashgate.

Hrebenar, R. (1982). *Interest Group Politics in America.* Armonk, NY: M.E. Sharpe.

Kile, O. M. (1921). *The Farm Bureau Movement.* New York: MacMillan.

Knoke, D. (1990). *Organizing for Collective Action: The Political Economies of Associations.* Hawthorne, NY: A. de Gruyter.

Lovett, A. (1998). Farmers at the Forefront of Sustainable Agriculture. *Issues.* March.

Mack, C. (1989). *Lobbying and Government: A Guide for Executives.* New York: Quorum Books.

Moe, T. (1980). *The Organization of Interests: Incentives and the Internal Dynamics of Political Interest Groups.* Chicago: University of Chicago Press.

National Farmers Federation. (2002). *Annual Review.* Barton, Australia: National Farmers Federation Limited.

NSW Farmers Association. (2003). *The Primary Report.* http://www.nswfarmers.org.au.

OECD. (2003). "Estimates of Support to Agriculture." http://www.oecd.org/agr/policy.

Olson, M. (1971 [1965]). *The Logic of Collective Action: Public Goods and the Theory of Groups.* Cambridge: Harvard University Press.

Parsons, K. (1941). Social Conflicts and Agricultural Programs. *Journal of Farm Economics* 23(3): 743–764.

Simon, J. (1996). *The Ultimate Resource 2.* Princeton: Princeton University Press.

Zander, A. (1985). *The Purposes of Groups and Organizations.* San Francisco: Jossey-Bass.

THE MANAGEMENT AND UTILIZATION OF LAND AND OTHER RESOURCES: OLD AND NEW

Valuing Nature

Economic Analysis and Public Land Management, 1975–2000

By ROBERT H. NELSON*

ABSTRACT. During the 1970s, Congress created a new statutory foundation for public land management by the U.S. Forest Service and Bureau of Land Management. The stated goal was to establish a rational administrative process for resolving the demands of competing users. Economists argued that public land decisions therefore must be made through comprehensive application of benefit-cost and other economic methods. The hopes to ground public land management in economic analysis, however, were not realized. It would have required a radical change in the politics of the public lands, including a large loss of influence among historically dominant groups, and there was no powerful constituency to make that happen. By the 1980s, moreover, the environmental movement was promoting ecosystem management as a replacement for traditional multiple-use management. In place of economic benefits, ecosystem management substituted biological goals that could not effectively be captured by economic methods. This article offers a case study of the failure of professional economic analysis to have much impact in many real-world government settings.

*The author is a Professor in the School of Public Policy of the University of Maryland, College Park, MD 20742; e-mail: nelsonr@umd.edu. From 1975 to 1993, Prof. Nelson worked in the Office of Policy Analysis in the Office of the U.S. Secretary of the Interior. Trained as an economist, he is the author of three books on public land management: *The Making of Federal Coal Policy* (1983), *Public Lands and Private Rights: The Failure of Scientific Management* (1995), and *A Burning Issue: A Case for Abolishing the U.S. Forest Service* (2000).

American Journal of Economics and Sociology, Vol. 65, No. 3 (July, 2006).

More federal environmental laws were passed in the 1970s than in any previous decade of American history. It amounted to a complete rewrite of the statutory foundations of U.S. environmental management and policy. The Clean Air Act (1970) and the Clean Water Act (1972) are the best known, but many important laws also were enacted in other environmental areas. For the public lands, there were five key laws—the National Environmental Policy Act (NEPA, signed on New Year Day, 1970); the Endangered Species Act (1973); the National Forest Management Act (1976); the Federal Land Policy and Management Act (1976); and the Alaska National Interest Lands Conservation Act (technically not a 1970s law, since it was enacted in 1980, but largely written in the 1970s). This body of legislation set new goals and decision-making procedures for public land institutions that in many cases had been operating under statutory guidance dating back to the progressive era.

The federal government owns about 30 percent of the land area of the United States. The largest amounts are managed by two agencies, the U.S. Forest Service in the Agriculture Department, created in 1905, and the Bureau of Land Management (BLM) in the Interior Department, created in 1946. Significant federal lands are also managed by the National Park Service, the Fish and Wildlife Service, and the Defense Department (the Bureau of Indian Affairs is a trustee for tribal lands that are not a part of the federal lands). Historically, however, common usage limits the term "public lands" to those areas managed by the Forest Service and the BLM (Gates 1968; Dana and Fairfax 1980). The former include 192 million acres and the latter 264 million acres, mostly in the West, and in total equal to 20 percent of the U.S. land area.

About half the coal reserves of the United States and large amounts of oil and gas lie under these public lands. Indeed, total mineral royalties from federally owned resources equaled $9.4 billion in FY 2005. The national forests, managed by the Forest Service, alone contain 46 percent of the softwood timber inventory of the United States (Pacific Northwest Research Station 2003). Since the 1960s, the traditional commercial uses of these resources have increasingly come into conflict with the protection of environmental assets. The use of public lands for hiking, skiing, backpacking, mountain climbing, picnicking, and other recreational activities has soared since World War II. The public lands also contain many of the less populated mountain and desert areas of the nation, including many places now designated formally as wilderness areas.

According to the law, the uses of the public lands are mainly to be decided as administrative decisions of the federal land agencies. The Forest Service and the BLM are legally guided by a management philosophy of "multiple use and sustained yield." In practice, however, the lands historically have been allocated to the groups with the greatest political influence (Anderson and Hill 1994). Cattle and sheep ranchers dominated the management of grazing lands; the most valuable forests were made available to timber companies; mining companies acquired mineral-rich lands; and so forth (Foss 1960; Wilkinson 1992). As they sought to challenge these arrangements, environmentalists faced a large political hurdle: How could they prevail against the traditional groups in order to win greater access for recreational and other environmentally-oriented uses? Moreover, environmentalists in some cases were not seeking a particular use of the lands but to minimize any kind of use at all. In a wilderness area, even recreational activities are to some extent a compromise with the goal of keeping the land as little touched by human hand as possible.

When confronted in the 1960s and 1970s with growing environmental pressures, the Forest Service and BLM were forced to defend their decision-making processes in newly explicit terms. How had they decided which uses would be chosen over others? What were the decision criteria? How was the overarching goal of multiple-use management being put into practice? On the public lands, issues like these had not been raised in such a principled fashion since the progressive era. Forced now to respond, the federal land agencies often lacked an adequate intellectual framework. For many decades, a public land status quo had simply been ratified. During the go-go years of the 1920s, the Depression decade of the 1930s, the war and its aftermath in the 1940s, and the focus on economic prosperity of the 1950s, the United States had not paid a great deal of attention to public land management. This finally changed with the rise of the environmental movement in the second half of the 1960s.

It fell, therefore, to the new 1970s laws to revisit the basic principles of public land management and policy (Clawson 1983). As is often the case, Congress blended together diverse and sometimes conflicting elements. One impulse was to revive the formal management philosophy inherited—if not much practiced—from the progressive era. The progressives had sought the "scientific management" of American society. As Eliza Lee writes of such progressive-era aspirations, "scientific management redefines what had hitherto been political problems as management problems, the solution of which is governed by the logic of science." Thus, the progressive proponents sought "the establishment of science as the institution of governance and the centralization of power in the hands of scientists." This was seen as possible and desirable because the very processes of government administration were regarded as "objective, universal, natural, altogether devoid of historical and cultural contexts, and dictated only by scientific laws" (Lee 1995: 543).

Harking back to these themes, NEPA now required in the 1970s that a comprehensive factual and scientific analysis be made publicly available in an "environmental impact statement." The goal of federal decision making, according to NEPA, should be to "utilize a systematic, interdisciplinary approach which will insure the integrated use of the natural and social sciences and the environmental design arts in planning and in decisionmaking." Similar requirements were incorporated into the National Forest Management Act (NFMA) and the Federal Land Policy and Management Act (FLPMA) (Le Master 1984). The Forest Service and the BLM were directed to prepare formal land-use plans as a guide to future land allocations, environmental protection, and other key management decisions. NFMA, for example, declares that "resource plans and permits, contracts, and other instruments for the use and occupancy of National Forest System lands should be consistent with the land use plans." The Forest Service is directed to "formulate and implement, as soon as practicable, a process for estimating long-term costs and benefits to support the program evaluation requirements of this Act." Overall, NEPA, NFMA, and FLPMA indicated a congressional intent that the land management process should be "rational," based on a systematic analysis of all the alternatives, a comparison of their relative impacts, and the assembly of the relevant information to achieve a "scientific" resolution.[1] Rather than by the interest-group politics of the past, these laws declared, the allocation of the public lands should be determined by objective neutral criteria that reflected the widest public interest.

The environmental movement, however, also included new elements in American public life skeptical of science and progress. This outlook had more influence on laws such as the Endangered Species Act. If an endangered species was present, this very fact would trump any other considerations—there would be no "multiple-use" balancing. The Alaska National Interest Lands Conservation Act (ANILCA) similarly removed much of Alaska from the NFMA and FLPMA framework of the Forest Service and the BLM. Instead, the main part of the vast federal land holdings in Alaska were transferred to the jurisdiction of the National Park Service and the Fish and Wildlife Service, where the preservation of wild nature would be given the highest priority.

This was possible in Alaska because the lands were largely free of existing uses (although the large oil and gas deposits of the Arctic National Wildlife Refuge would eventually come into basic conflict with preservationist goals). Although FLMPA, by contrast, was largely a "multiple-use" statute, it did include a requirement that the BLM must comprehensively review its lands for possible future wilderness designations. A wilderness was another place on the public lands where utilitarian calculations, as emphasized in the progressive foundations for public land management, should have little weight.

I

Being Economically "Rational"

For the great majority of the Forest Service and BLM lands, however, the "neo-progressive" goal of a rational scientific decision process was mandated by NFMA and FLPMA. Because Congress did not specify any final objectives, it would fall to federal administrators in the public land agencies—and increasingly also to the courts—to address the difficult question of the actual "objective function" of public land management. Initially, administrators again looked back to the progressive heritage of the public lands (Hays 1959). The first Chief of the Forest Service, Gifford Pinchot, had proclaimed that the management goal would be to maximize the net long-run benefits to society—the "greatest good of the greatest number [of people] in the long run," as he famously put it (Pinchot 1947: 261). Thus, in the late 1970s, the public land agencies were confronted with the issue of how to maximize their total contribution to national social—human—welfare. The professions most closely involved in the past with public land management—the foresters, rangeland scientists, wildlife biologists, and so forth—had little to offer. To the extent that they had even considered final objectives, they tended to think in physical terms: to maximize long-run timber outputs, rangeland forage produced, numbers of game animals, and so forth. In contrast, the one professional group that could lay a claim to understanding the technical methods of aggregating all public land outputs to achieve a maximum contribution to social well-being was the economics profession.

Until the 1960s, the public land agencies had almost no economists working for them. As the agencies came under increasing pressure to justify their land allocations, they began to bring a few on board. With the new legislative mandates of the 1970s, some economists now also advertised their services to the public land agencies. The Forest Service for many years supported the work of the economist John Krutilla of Resources for the Future, eventually culminating in Krutilla's grand treatise, *Multiple-Use Management: The Economics of Public Forestlands* (Krutilla 1989). After the passage of NFMA, Krutilla argued that this law (together with the 1974 enactment of the Forest and Rangeland Renewable Resources Planning Act, or RPA) should be interpreted as a mandate for a system of comprehensive economic management of the national forests.

If the laws meant what they said, as Krutilla now explained, there could be no other interpretation. Public land management must be put on a scientific basis, and this would have to mean defining the final objective, developing a full range of alternatives, and finally choosing the alternative that maximized total social benefits minus costs. Only economics could provide the technical tools to accomplish this task. A public land agency should be run like a business corporation, but in place of private profit, it would pursue a wider objective of maximizing its contribution to social welfare.

In 1979, Krutilla criticized the Forest Service for its past failures in this regard. Rather than applying rigorous methods of analysis, "up until the present the instincts and proper impulses of the profession expressed themselves somewhat more as high motives and sincere exhortations." It would be necessary in the future, Krutilla now argued, to bring economic analysis to bear "to provide the 'correct' level and mix of the various resource services which the national forests are capable of providing." The Forest Service for too long had appeared confused and beholden to private interests. Unable to provide a convincing scientific justification for its actions, it had been left to suffer "indignities at the hand of one or another group insisting that the national forests satisfy their mutually incompatible demands," each group thinking that a denial of its own favored use represented a management failure on the part of the Forest Service (Krutilla 1979: 5–6).

As Krutilla concluded, both as a matter of law and practical necessity, the Forest Service now had to recognize the "need to bring into its management frame of reference the developments which have advanced in other scientific or management disciplines which, while not required of the forester during the 'golden age of forestry,' have elements of direct applicability to national forest management today" (Krutilla 1979: 7). In practice, this would mean the following. Where a public land output was sold in the market, a private price could be observed and the total public land outputs valued at this price. If there was no observable market price, economists could estimate the equivalent—the marginal dollar benefit to society of one more unit of the output. Thus, Forest Service economists would be put to work developing dollar estimates of the value of additional hiking, fishing, hunting, trail biking, and a host of other nonmarket activities that occur on the national forests (Krutilla and Haigh 1978). In other writings, Krutilla had also argued that economists could provide dollar estimates even of public land "non-use" benefits that were derived without any direct acts of consumption (Krutilla 1967). Thus, the designation of a wilderness in Colorado might have a significant monetary "existence value" to a person living in New York City, even if this person had no intention of ever visiting the wilderness area (the person might be elderly and in poor health).

Finally, given all the dollar values, market based and otherwise, economists would estimate the total social value of each alternative set of uses of a particular area of the national forests. In practice, it would be necessary to simplify somewhat in order to reduce the range of alternatives to a manageable number. Then, the final alternative, the one the Forest Service should choose, would maximize the total dollar social benefits minus costs. Any other approach would necessarily involve large subjective elements. Only in this way, as Krutilla explained to Forest Service leaders, could the agency comply with the mandates of 1970s legislation for a scientific process of decision making on the public lands. Lacking such a procedure, Forest Service decisions would merely further inflame the growing numbers of critics who now saw it as acting in an unprincipled manner, often contrary to the national interest.

Krutilla was not alone. Indeed, he was expressing accurately the views of much of the American economics profession. In a 2000 brief to the U.S. Supreme Court, a group of distinguished American economists, including four Nobel prize winners, asserted the necessity of doing comprehensive benefit-cost analyses for government environmental decisions. Their brief stated that "economists and other policy experts . . . have produced a large literature on the methods and applications of benefit-cost analysis. There are, and always will be, many uncertainties and disagreements about those methods and their applications in particular cases. Nevertheless, a wide consensus exists on certain fundamental matters (Arrow 2000: 9)." Admittedly, some management considerations were not amenable to a formal economic analysis, but in most areas doing a "benefit-cost analysis can help the decision maker better understand the implications of a decision" and should "play an important role" in environmental decision making (Arrow 2000: 9).[2] Indeed, government administrators of environmental and natural resource agencies ideally should begin by estimating total benefits and costs, and then adjust these estimates to reflect those other factors that had necessarily been omitted from the economic calculations.

II

A Report from the Ground

In June 1975, having earned a Ph.D. in economics in 1971 from Princeton University, I joined the Office of Policy Analysis in the Office of the Secretary of the Interior in Washington, D.C. (where I ended up working until 1993). My first assignment was to develop a system for estimating the benefits and costs of public rangeland investments, as were being made by the Bureau of Land Management. NEPA had already mandated a scientific decision process, and many of the BLM programs were under court order to develop comprehensive environmental impact statements and otherwise to comply with legal requirements for a newly rational administrative procedure. Although FLMPA was not enacted until the next year, the core provisions were already under discussion, and their inclusion in a forthcoming law could be anticipated. At the same time, within the executive branch, the Office of Management and Budget (OMB) was pressing the BLM to provide clearer economic justifications for its rangeland investments.

I was assigned to work with senior officials of the BLM to develop estimates of the values of rangeland outputs, to write an instruction manual, to give training to field employees, and in general to introduce benefit-cost analysis to the BLM livestock grazing program. The main use of the BLM rangelands was for the grazing of cattle and sheep. Under the provisions of the Taylor Grazing Act of 1934, individual ranchers (or sometimes groups of ranchers) had long been assigned exclusive grazing "privileges" to particular areas of BLM rangeland ("allotments"). For many years, the BLM had been installing fences and watering facilities on these lands, designed in part to give greater control over livestock movements and thus to improve grazing practices and to increase total forage yields. There could also be environmental benefits, such as greater forage availability for wildlife or the exclusion of cattle from sensitive riparian areas. OMB and others were asking: Is this worth it? Does the public achieve an adequate return on these BLM expenditures of taxpayer dollars?

To that point, the BLM had few economists and little experience with doing economic analysis. Nevertheless, when required by powerful outside forces, the BLM sought to comply. With my assistance, specific dollar estimates were developed of the values of additional hunting days, fishing days, and other forms of nonmarket benefits on BLM grazing lands. These estimates were necessarily crude, partly because a single value for each output was usually applied to all BLM lands in the same state; individual BLM district offices lacked the resources or capacity to develop valuations specifically tailored to varying local circumstances. A single nationwide estimate of the value of increased production of rangeland forage was developed. This was a sensitive matter, partly because the BLM charged a grazing fee that included a significant element of public subsidy, despite the requirement at the time that the BLM charge fair market value. The BLM was caught between a desire to show a high value of rangeland forage, in order to justify proposed investments, and its continuing need to defend its low grazing fees.

A discount rate also had to be selected, equal to 6.125 percent in this case, following the recommendation of the Water Resources Council at the time. Since costs were largely incurred up front, and the benefits accrued slowly over time, a higher discount rate could significantly reduce the number of viable investment projects.

About half a year of work effort on my part went into this project, and the BLM also committed perhaps a person-year to the development stage. By 1976, an operational system of benefit-cost analysis was ready for field testing (Nelson 1976). Visits were made to field offices where the new benefit-cost system was applied to selected fencing and water investments. The resulting estimates showed that some rangeland investments offered a reasonable rate of return, while others did not. In the latter cases, the BLM might modify the investment proposal to increase benefits and decrease costs. Although the calculations were crude, they could often be used to weed out the worst projects. Perhaps the most useful feature of the process was the requirement that the BLM be more explicit about expected benefits and costs. Once BLM assumptions were put on the table for examination, it might not be necessary to do any formal economic evaluation at all—some of the projects were obvious "losers."

In the end, however, all this effort came to very little. The Office of Policy Analysis reported to the Secretary of the Interior, not the

BLM. It had many other concerns, and I was shifted by mid-1976 to work on new issues, such as the management of the large federal coal reserves in the West (Nelson 1983). The Carter administration replaced the Ford administration in early 1977 and had its own priorities. Many of the incoming Carter officials in the Interior Department came from the environmental movement and had little experience with or even interest in economic analysis. Once the external pressures subsided, few BLM managers showed much continued commitment to benefit-cost analysis as a basis for internal decision making. The BLM had hired a few economists to help with rangeland economic studies, but they were gradually shifted to other tasks. All in all, the momentum for economic analysis of 1976 soon dissipated, and by 1980 there were very few benefit-cost analyses of BLM rangeland investments being done.

III

Alternative BLM Objective Functions

THIS CASE ILLUSTRATES how a variety of powerful obstacles work against the use of benefit-cost analyses in public land management, as well as in many other areas of government. Krutilla and other economists assumed that the management goal of a government agency such as the BLM was well defined: to maximize the contribution of its resources to national income. Government officials in agencies such as the BLM, however, did not share this understanding. In part, this was revealed by their lack of interest in doing economic analysis except when required by outside forces. If circumstances changed, as for the BLM in the late 1970s, economic analysis would no longer be done. As a matter, therefore, of "revealed agency preference," one might say that other agency purposes trumped an "economic objective function." Admittedly, this will not be a great surprise to most political scientists, but economists have traditionally paid less professional attention to the political and administrative circumstances in which their analytical products will be applied.

The BLM's indifference to economic objectives was further illustrated by the reception to another study of mine at the Interior Department (Nelson and Joseph 1982). In 1982, I developed nationwide revenue and cost estimates for public land management, including a separate breakout for the BLM livestock grazing program. The main public revenues from livestock grazing were obtained from the grazing fee, equal in 1981 to $25 million in total for all BLM lands. By my detailed calculations, as made over the course of several months, the total administrative costs of the BLM grazing program were about $125 million per year.[3] It was possible, admittedly, that the BLM grazing program was producing large nonmarket benefits that could justify such high costs. However, livestock grazing competes with wildlife for forage, often damages stream fisheries, sometimes threatens endangered species, and otherwise can cause environmental problems. Environmentalists in the 1970s and 1980s were expressing strong objections to the past environmental impacts of livestock grazing on BLM lands.

So why did the BLM grazing program exist at all? The economically rational answer was that it should be abolished, or at a minimum much transformed, and only maintained in those places where rangeland revenues (plus any nonmarket benefits) exceeded the management costs. As a whole, BLM livestock grazing was a large money loser; a business corporation with a division showing such poor financial results would have closed it down long ago. Of course, the BLM leadership, top Interior officials, and members of Congress had essentially no interest in such a step. As a matter again of "revealed preference," the BLM grazing program clearly had an objective that was well removed from maximizing its contribution to net national income.

What were the agency objectives? Not surprisingly, they were multiple. The BLM relied on the political support of various user groups and had strong incentives in managing grazing to try to accommodate their needs. Members of Congress found numerous opportunities to provide constituency service in the workings of the BLM livestock grazing program. The grazing program also had internal constituencies; the BLM provided thousands of government jobs in the rural West that were often among the highest paying in attractive settings to raise a family. Indeed, from the 1960s on, such political and private objectives of government agencies were receiving increasing attention among professional economists in the writings of the public choice school, including James Buchanan, who won the Nobel prize in economics in 1986 for his work in this area (Mueller 1997).

Many economists had long argued that their methods were value-neutral. The systematic application of benefit-cost analysis, however, would have set a new management objective with which BLM officials were unfamiliar. The reality was that doing—and following the results of—a benefit-cost analysis for BLM rangeland investments would have amounted to a political revolution, carried out in the name of a technical and scientific improvement. The conclusions of economic analysis, as determined by the specific economists who did the calculations, would have superseded traditional BLM decision-makers and the pursuit of longstanding BLM goals. Whether they knew it or not, economists were engaged in a struggle for political power rather than a mere analytical exercise. Their own clout, and that of their patrons, however, was limited. The ultimate rejection of economic analysis closely reflected the real distribution of power in the management of BLM livestock grazing.

It should be noted that besides the BLM leadership, another important political loser would have been many members of Congress. When the 1970s environmental laws required rational analysis as the basis for administrative actions, Congress was voting to sharply reduce its own prerogatives. Most members, however, were unaware of this and would have opposed any such action if they had understood clearly the implication. Whether they realized it or not, the members of congress were actually voting for administrative rationality as a symbolic or "feel-good" measure. When push came to shove, there was little congressional support for the depoliticization of BLM (and Forest Service) decision making through the application of benefit-cost or other technical methods. According to the requirements of the law, formal economic analyses had to be done. But the resulting analyses could then be ignored, as I soon learned from my own experiences with the BLM program of livestock grazing investments.

In short, the 1970s environmental laws prescribed a ritual that mostly worked to affirm the public land status quo as a "scientific" and "rational" outcome, even as no large changes in land management occured. Top BLM officials understood this at least implicitly, although they could not be very direct about it with the public. But many economists were less familiar with the ways of Washington and thus were disappointed when the requirements for benefit-cost and other professional methods of analysis eventually proved to be mainly window dressing.

IV

Uneconomic Economics

ALTHOUGH SUCH CONSIDERATIONS largely explained the rapid demise of benefit-cost analyses in the BLM grazing program—and there were similar experiences in other areas of Forest Service and BLM management—there were also more technical problems. One factor, ironically, was that in some cases it might be uneconomic to do an economic analysis. BLM rangeland investments frequently involved small amounts of money, such as $20,000 to build a new fence in order to enclose a pasture or $25,000 to build a new water tank where cattle could congregate, rather than seeking out nearby streams. Doing a benefit-analysis itself involves a commitment of taxpayer resources; it might cost $5,000—or even more—to do a good job for one grazing investment. Admittedly, a "quick and dirty" analysis will often be better than no analysis at all (Leman and Nelson 1981). The pressures in the public sector, however, work in the opposite direction. An environmental group, OMB, or another party may well question a proposed BLM rangeland investment and demand that the assumptions of a formal analysis be fully documented and supported. In the 1970s and 1980s, the environmental impact statements written by many government agencies even for limited projects were often the size of big-city telephone books.

For many years, a standard analytical assumption of the economics profession was that a state of "perfect information" existed. As a result, economists had long neglected the administrative costs of collecting information and of processing it. It was not until George Stigler's 1961 article "The Economics of Information" that this issue began to receive wider professional attention (Stigler 1961). Once the subject began to be studied, however, it raised the possibility that a benefit-cost analysis might itself be uneconomic. It might cost more to collect the necessary data and to conduct the economic analysis than the total social benefits of doing the analysis in the first place. Indeed, given the small size and the local character of BLM rangeland investments and the relatively high costs of doing benefit-cost studies in such circumstances, this was a distinct possibility. The valid economic answer might be to rely instead on "professional judgment."

The BLM did not do a formal economic analysis of this issue. It would have required a second tier of benefit-cost analysis, making estimates of the total benefits and costs of the first tier of benefit-cost analyses. But, then, a third tier of analysis of the second tier might have been required. Indeed, this line of reasoning could go on indefinitely. Obviously, at some point it is necessary to make a subjective judgment that no further analysis will be done. Economists have no logical way within the technical framework of their analytical methods to determine when this "stopping point" might be reached. In concept, and it might well have been the case for many of the smaller investments being made on the BLM rangelands, even the first step of benefit-cost analysis might not be worth it. Paradoxically, an effort to be completely "rational" might itself be "irrational," and there was no strictly rational way to say when.

The 1970s attempts to create a rational administrative process thus failed for the BLM rangeland program, as happened in most other areas of public land management. George Hoberg comments that "the role of science and expertise in the policy process is . . . exceptionally complex." The reality is that economic analyses and other "science can rarely answer with an adequate degree of certainty the questions policy makers pose. As a result, science becomes politicized, as interest groups adopt whatever factual claims support their views." The progressive ideals of scientific management presumed the existence of a "politics-administration dichotomy, in which politics provided the values and administrators provided the expertise," but this political model by the end of the 20th century had "long since faded" (Hobert 2003: 3). Economic analysis seldom provided definitive answers to social benefits and costs but was often put to use by private groups to defend their own claims and interests on the public lands.

V

A Property Rights Analysis

ACCORDING TO ANOTHER and older tradition, dating back at least to Adam Smith, economic analysis should focus on the structure of property rights and the workings of markets. Although property right issues were neglected in the economics profession for much of the 20th century, the writings of University of Chicago economist Ronald Coase helped to stimulate a "new institutional economics" that has revived interest in the past few decades (Furubotn and Richter 1997). Indeed, it is possible to give a property rights interpretation of the 20th-century management history of the public lands. This may be the most persuasive economic analysis of all.

After the Taylor Grazing Act was enacted in 1934, ranchers received what amounted to "grazing rights" to run their cattle and sheep on particular areas of BLM land (actually, it was the Grazing Service that managed the lands until the BLM was created in 1946). The rangelands remained public and were freely open to hiking, camping, and other recreational uses that did not directly conflict with livestock grazing, but the use of the land for grazing purposes was effectively privatized. Because the rights to graze were attached to particular nearby ranches, the capital values of these ranches were increased by the amount of their "permit value." Banks and other financial institutions explicitly recognized and took account of this permit value in appraising the ranch property. Local ranchers were also informally allowed by the BLM to transfer grazing rights among each another in exchange for payment. Thus, a market price—typically around $50 to $100 per "animal unit month" (AUM) of grazing—could be directly observed for the rights.

Until the 1960s, this public/private system functioned with low administrative costs. The most important BLM role in the grazing program, which was similar on Forest Service grazing lands as well, was to prevent the trespass of one rancher's livestock on the rangeland allotments of other ranchers. The BLM, in effect, provided a public rangeland "police force" and served as a "court of law" for resolving grazing disputes among ranchers.

The rise of the environmental movement, however, created a new property rights claimant. Changes in property rights are seldom planned in a deliberative process. Historically, they often simply reflect a new reality of power; a newly arriving army might, for example, evict the previous owners, as happened to many Native Americans. On the public lands of the West, environmentalists did not wage a military campaign, but they did seek to wield the coercive powers of the state to assert new rights to BLM and Forest Service rangeland forage (Nelson 1980). To the degree they were successful, traditional rancher grazing rights would be curtailed or even eliminated.

The new environmental laws of the 1970s proved a useful tool for environmentalists in this political struggle. However much ignored over the years in practice, the public land system nominally still functioned under the scientific management principles espoused by the progressive-era founders (Hess 1994). Environmentalists sought to challenge the rancher rights on the public lands by exposing the longstanding gaps between the high progressive ideals and the prevailing practices (National Academy of Sciences 1984). They brought NEPA suits asserting that the BLM had failed to analyze adequately the environmental impacts of grazing and other uses as required by law. In other cases, it was the failure to address a full range of alternatives in a land-use plan that was legally challenged. Still other suits alleged that the Forest Service or BLM had ignored existing plans or had failed in other ways to demonstrate a sufficient rational basis for their actions.

Environmental groups won many of these legal contests. There was little increase in the overall administrative rationality of public land decision making. But the ultimate decision results were often changed; there was a significant increase in the number of decisions favorable to environmental objectives. Bringing successful lawsuits required extensive preparation and was expensive. The ability to assemble a high-powered legal team became a new proxy for political power. Ranchers and other traditional public land users could often be outgunned in this respect. They retained much of their traditional political leverage in the Congress, but there was erosion there as well. At a time when decision making throughout American society was shifting to the courts, environmental legal skills and growing political influence succeeded in asserting new environmental rights on the public lands (Nelson 1995).

In short, demanding more benefit-cost studies, environmental impact statements, and other tools of "rational" analysis and seeking to influence the contents of the analysis were not the main ends. Rather, these were an environmental tactic in the pursuit of other purposes relating to political power and control over the use of the public lands—to create new de facto property rights. Few environmentalists cared, for instance, that the costs of BLM rangeland decision making were rising well above the total economic value of the BLM resources involved. Not many people were committed to the quality of the final analyses, except insofar as political leverage could be derived thereby. Perhaps OMB was the only important player in the public land system that showed any real commitment to abide by the conclusions of economic analysis—or a rational administrative process—as a matter of basic principle.

As economists observed these developments, a property rights perspective suggested a new economic approach. A struggle to control rights to property was nothing new—that is what a private market is all about. In market exchanges, however, transaction costs are minimized by the rules of the game; potential buyers make money offers to acquire new rights, and sellers either accept or reject the offers. If two parties seek the same property, the winner normally is the one that bids the highest. In a market contest for rights, it is not necessary to prepare elaborate legal documents and otherwise to incur large strategic costs in order to make a "bid."

As early as 1963, the natural resource economist Delworth Gardner offered a novel policy recommendation for public grazing lands based on this understanding. He suggested that grazing on public rangelands should be recognized legally as a private right that could then be bought and sold in the market (Gardner 1963). If environmental groups wanted to limit livestock grazing, they could simply buy up the grazing rights in willing buyer/willing seller exchanges. They could then retire the public rangeland from grazing altogether or perhaps in some cases allow continued grazing under tight new rules that reflected their environmental concerns. In the end, this might be a much less expensive strategy for the environmental groups, as compared with long legal battles and other longstanding efforts to manipulate the administrative process. It would also be better for the rancher, because he or she would receive compensation for any rights lost. Finally, it would probably be less expensive for the public land agencies, which would be relieved of their large existing burdens in preparing environmental impact statements and otherwise enacting the official rituals of administrative rationality. Society might benefit economically in that the grazing rights would be allocated to the highest value use—as shown by the willingness of one party to pay the most for the grazing forage.

Until recently, this economic analysis was received with minimal enthusiasm outside a small band of economists.[4] Ranchers saw the potential sale and retirement of grazing rights as a symbolic repudiation of the business of livestock grazing and a challenge to their "way of life" in small western communities. Environmentalists were offended by the prospect of paying for grazing rights, which they did not regard as legitimately belonging to the ranchers. According to the strict requirements of the law, the allocation of public rangelands was supposed to be an administrative decision guided by objective criteria in the public interest. The sale of administrative decisions to the highest bidder would be tantamount to bribery. When ranchers saw their historic place on the public rangelands in similar terms, the longstanding political struggles for control of use rights intensified, along with the very high management and other costs of overseeing these bitter struggles.

Economists in their attempts to establish a system of administrative rationality might have neglected yet another important factor. It seems that some people simply like to fight. The public land struggles were not a cost for some participants but, economically speaking, a form of consumptive activity. The economic way of thinking has always suffered from its limited "objective function." When economists press for economic solutions to management and policy issues on the public lands and elsewhere, they are seldom being value-neutral, despite their usual claims to this effect. The reality is that they are asserting—if symbolically, by focusing on some elements of the problem and neglecting others—a particular vision of a correct human nature and a corresponding desirable future human condition (Nelson 1991, 2001).

VI

Wild Nature for Its Own Sake

In some important ways, the goals of environmentalism are fully compatible with the economic way of thinking. Lower levels of pollution reduce adverse health effects and yield other direct human benefits that are in principle measurable in dollar terms. Much the same can be said of additional outdoor recreation and other environmentally-oriented activities on public lands. Such environmental benefits are in the same general category economically as having more meat on the table or owning a larger house. There is another type of environmental objective, however, that fits much less comfortably with standard economic methods. A main goal of the environmental movement is to preserve nature for its own sake, independent of any human visitation or other direct benefits experienced. As environmental law professor Richard Lazarus commented recently, "many of the modern resource conservation and pollution control laws are . . . premised on a theory of ecosystem equilibrium and seek to protect the natural environment from change" that would cause a departure from a desired "natural"—equilibrium—state (Lazarus 2004: 15).

This side of environmentalism is seen, for example, in the Endangered Species Act, among the most important of the 1970s laws. Protecting a species is not justified because more people will then see members of the species and thus experience a good feeling (a higher level of utility). Rather, a species should be protected as an end in itself. It reflects a value that the continued existence of each and every species on earth is a worthwhile social goal—a secular version of God's command to Noah in the Bible. Indeed, if the language of the Endangered Species Act is to be taken literally—and of course things do not work out precisely this way in practice—the willingness of society to sacrifice to protect each species should be virtually limitless. In deciding whether or not to list a species as endangered or designing a recovery plan and other government protective actions, the Fish and Wildlife Service is supposed to give little if any weight to economic factors.

Protecting an endangered species thus fits awkwardly within a conventional framework of economic analysis. The real objective is not a "use" by some group of people but the well-being of the species. Economic methods have long been designed, however, to estimate values where direct acts of use and consumption are involved. In essence, environmentalists were bringing to the management of the public lands a moral imperative. Because it was difficult to square economic methods with this moral—to some extent, even religious—dimension of modern environmentalism, economists and environmentalists often found themselves on opposite sides of public debates (Dunlop 2004; Nelson 2004). As Stanford law professor Barton Thompson explains, "environmental moralists firmly reject the idea of determining environmental goals by trying to maximize economic value" (Thompson 2003: 179). Indeed, the ethical disagreements sometimes escalated to the point that a leading environmental philosopher would seek in a professional environmental journal article to explain "Why Environmentalists Hate Mainstream Economists" (Norton 1991).

By the 1980s, moreover, protection of species and other preservationist goals were becoming increasingly important to the environmental movement. Environmentalism proved a formidable political force, and by the 1990s a new value system was being established in public land management. Although it has never officially been blessed by Congress, a philosophy of "ecosystem management" in agency practice and regulation began to replace the old management philosophy of multiple use. Ecosystem management substituted biological objectives such as species preservation and other forms of "biodiversity" in place of progressive-era objectives of the highest and best human uses of the land (Worster 1993).

The decisive political struggle was fought between 1987 and 1993 over the old-growth public forests in northern California, Oregon, and Washington, much of which are found on Forest Service and BLM lands. These virgin forests—"old growth" means they have never been cut—are the habitat of the northern spotted owl, a species whose future existence was threatened by longstanding agency plans for continued logging. Environmentalists pressed to have the owl designated officially as an endangered species. If that occurred, it would mean a halt to future timber harvesting over much of the public forests in the region.

After several years of legal struggle and environmental pressures, the spotted owl was in fact designated a threatened species in 1990 by the Fish and Wildlife Service under the Endangered Species Act. Environmentalists had also filed lawsuits that involved alleged failures to comply with the requirements of the public land-use planning process. As Hoberg comments, in the Pacific Northwest, the "NFMA planning rules went through an extraordinary evolution . . . which changed the [Forest Service's] mission from multiple use to giving priority to ecosystem protection, then spilled over into other regions and filtered its way up to the agency leadership and finally was adopted as regulation" nationwide for the forests (Hobert 2003: 6). A recovery plan for the spotted owl was written in the early 1990s that eliminated timber harvesting on large parts of the federally owned forests in the Pacific Northwest. In 1995, the Interagency Ecosystem Management Task Force declared that "as a matter of policy, the federal government should provide leadership in and cooperate with activities that foster the ecosystem approach to natural resources management, regulation, and assistance" (*Ecosystem Approach* 1995: 8).

For the Forest Service, there might still be grounds for the harvesting of timber, but no longer as an economic goal in itself. Any continued timber harvesting should now be a management tool for achieving desired ecological conditions in the forests. The new management goal should be a "healthy" ecosystem; it should be a forest that displays ecological "integrity" and maintains its "natural" qualities. It was, admittedly, difficult to specify such terms in an operational way. In practice, partly because there was no other obvious alternative, Forest Service managers came in many cases to interpret "natural" as the condition of the public lands prior to European settlement. As a group of ecologists thus wrote of national forest goals for the Sierra Nevada region, "ecosystem management is an attempt to maintain the historical structural complexity and suite of processes that occurred in these ecosystems before Euro-American influence" (Elliot-Fisk et al. 1997: 307).

In concept, then, under ecosystem management, a first task would be to identify the "natural" conditions of the lands—in practice in the West, the conditions around 1870 to 1890 when European settlement first reached this region. Then, the ecological workings of the lands should be restored to this earlier historical functioning as far as possible. For this purpose, it might be useful to do a cost-effectiveness analysis. If two ways had been identified to restore a public land area to a natural condition, the public land agency should choose the least expensive—the cost-effective solution. A full-scale benefit-cost analysis, however, was unnecessary: a national forest was no longer to be regarded as a "natural resource" to be developed. Under ecosystem management, the final goal would be a natural state of affairs, defined in biological and ecological terms. Lazarus notes, reflecting such thinking, that "many environmentalists are . . . hostile to application of cost-benefit analysis in environmental law," and their views increasingly prevailed (Lazarus 2004: 28).

As public forests were newly managed for biodiversity and other ecological objectives, the total timber harvests on Forest Service and BLM lands plummeted. Nationwide, Forest Service timber harvests declined from around 12 billion board feet in 1989 to 3 billion in 1997. It might be argued that total economic values could still be rising because equally or more valuable nontimber uses might be increasing even as timber harvests were declining. The Forest Service, however, calculates an "All Resources Net Present Value" from the national forests, including the economic value of recreational uses and other nonmarket outputs. From 1991 to 1997, by Forest Service calculations, the net economic value of all the diverse outputs of the national forests declined from more than $1 billion to $345 million (U.S. Forest Service 1998). Ecological purposes, as a matter of a new national policy, were trumping economic goals.

VII

Economics as the Measure of Everything

THIS WAS A DISTURBING OUTCOME for many economists, especially those working in the environmental field (Sedjo 2000). There might be one method, however, for sustaining a central role of economic analysis in a new environmental era focused on the protection of nature. What if economists could put a dollar value on the existence of a species, forest, or other environmental object in its natural condition? Such a value would not be derived from any tangible uses or other direct outputs. Rather, the economic benefit would be obtained from the very fact of the existence of a state of wild nature. By the 1980s, a number of environmental economists were advocating this approach (Mitchell and Carson 1989). Using novel methods of economic analysis, they proposed to calculate the "existence value" of a wilderness area or natural condition in the world. Such estimates could then be entered into economic calculations, and public land management could once again be determined by the results of a—newly—comprehensive benefit-cost analysis.

Since no direct act of using nature for productive or consumptive purposes was involved, existence values could not be obtained by traditional economic methods involving measurements of observed demands for goods or services. Indeed, it would be necessary to turn to the use of survey instruments. An economic researcher should ask an interviewee, for example, how much he or she would be willing to pay to know that the spotted owl would continue to exist as a species in the Pacific Northwest. In 1992, Walter Mead surveyed a number of early existence value estimates, finding in one case an estimated total benefit among all Americans of $6.8 billion per year from the continued existence of the spotted owl. In another study, the existence value of the whooping crane was estimated to be $32 billion per year nationwide. Yet another existence-value study estimated that each U.S. household would be willing to pay $1.90 to preserve visibility in the Grand Canyon, adding up to a long-run discounted value of $6.8 billion for all Americans (Mead 1992).

There were a host of problems, however, as a critical literature on existence value soon pointed out (Hausman 1993; Nelson 1997a). Changes in the questions asked, or in the manner of their administration, could yield large differences in results. Seeming to violate normal economic assumptions, survey respondents often gave similar dollar numbers when asked how much they might be willing to pay to preserve, say, 20 ducks versus 2,000 ducks in a given area. Another problem was that environmental economists studied almost exclusively the values associated with protection of wild nature. There were many other elements of American life, however, in which significant existence values might be found. In the Pacific Northwest, many people might have put a significant value on the continued existence of logging jobs or the continued existence of the small communities that depended on public timber. These kinds of nonenvironmental values were seldom calculated in economic studies that sought to incorporate existence values in environmental decision making.

Moreover, there could be both large positive and negative existence values for the same state of nature.[5] Many environmentalists, for example, want to protect the Arctic National Wildlife Refuge as a "cathedral" untouched by human hand, but other people experience a strong distaste for the very idea of "wasting" so much oil and gas—and this aside from any direct production values of the ANWR oil and gas resource. Indeed, given the large number of objects that can have powerful symbolic meanings in American society, the potential number of existence values could be extraordinarily large. If economists had to calculate the existence value of every item that potentially has some symbolic importance to Americans, the total volume of economic calculations in society might soon become overwhelming. Finally, because existence values are in essence a form of public good and are not reflected in private market incentives, the required domain of government decision making might grow correspondingly.

In practice, there have been few public land decisions that were much influenced by existence-value calculations. Indeed, estimates of existence values have been mainly developed in support of environmental decisions already reached for other reasons—as noted above, typically a sense of moral imperative relating to retaining wild nature. The leading advocates of existence value have been environmental economists who were often happy enough to produce high estimates. Thus, the environmental philosopher Mark Sagoff finds the whole effort misguided. As he says, "many of us recognize an obligation to places and objects that reflects a moral judgment about what society should do, not a subjective expectation about what may benefit us" and therefore can be assigned a dollar value in the manner of a typical item of consumption. We do not consume beliefs or, as Sagoff says, "beliefs are not benefits." Hence, it is meaningless to ask what a person would be willing to pay to hold a belief. The benefit to society of behaving according to a moral judgment cannot be assigned a dollar value because "there would seem to be no relevant benefits to measure" in any understandable economic sense (Sagoff 2004: 46–47).

VIII

Conclusion

As this article has reviewed, economic analyses done by professional economists played a very limited role in public land management during the period 1975–2000. Few public land decisions were made as a result of a benefit-cost analyses. This does not mean, however, that economic factors were insignificant. Indeed, whether anyone likes it or not, considerations of costs have a large impact on decisions in almost every area of life. Even in matters supposedly decided by ethical principles, if the costs are too high, it will be necessary in practice to make compromises. Manhattan could never be a wilderness area, even if most Americans were somehow to come to believe that God Himself had so commanded it.

How, then, is the limited role of professional economic analysis to be reconciled with the large significance of economic factors in public land decision making? The answer reflects in part the character of the professional methods used by the economists themselves. As the case of public land management illustrates, they are often poorly suited to the circumstances of government decision making. Economic considerations are important, if seldom exclusively controlling, but they enter into the decision calculus in ways that economic professionals find difficult to serve.

Part of the problem is that the formal methods of professional economic analysis reflect a vision of comprehensive administrative rationality. The ambitions of public land managers, however, are more modest. They operate in a world of large uncertainties with respect to scientific facts, future social and economic trends, likely political pressures, and many other factors. The economic tradeoffs that have to be made are necessarily crude and typically not much illuminated by the false formal precision of professional economic studies. When economists seek to incorporate political, ideological, and other noneconomic considerations into their analyses, using technical methods such as existence value, their credibility suffers further. Even in narrowly economic domains, it is mainly the simplest of economic calculations that are the most useful to Forest Service and BLM managers (Nelson 1987). Indeed, people without a great deal of formal economic training may be able to make these calculations as well or better than professional economists. A newly-minted Ph.D. in economics might be insistent on applying refined technical methods that he or she has recently mastered in graduate school. But these methods will often require such heroic assumptions that they are of little applicability in any complex real-world setting.

It is also difficult to incorporate formal economic analysis into the pervasively political framework of public land management. Although they may not realize it, economists' real goal often is to depoliticize land management. However laudable this might be, it is a radical political objective in light of existing agency practices, and few economists have a plausible political strategy for achieving such radical change. Indeed, if the maximization of national product were to become the dominant decision criterion for public land management, much of the public lands should simply be privatized (Nelson 2000). But there is little public support at present for privatization. The progressive-era retention of the public lands in federal ownership means that land management decisions will continue to be collective—that is, a product of political choice. Determining the proper role of professional economists within a political system in itself requires a political theory. But economists have seldom sought to develop any such theory. Instead, they have commonly portrayed a utopian world in which the members of the economics profession in effect become the politically dominant group.

Ideology—or secular religion—also plays a large role in the management of public lands. The political strength of the American progressive "gospel of efficiency"—the founding set of values of public land management—was evident in 1913, when Gifford Pinchot prevailed over John Muir in seeking the construction of the Hetch Hetchy dam in Yosemite National Park (Fox 1985: 138–147). As recently as the 1980s, Forest Service lands still provided as much as 20 percent of the softwood timber harvests of the United States. The public land agencies were staffed by foresters, rangeland scientists, wildlife biologists, soil scientists, and members of other professional disciplines whose world views were formally grounded in progressive utilitarian values.

From the 1960s, however, environmentalism offered a new gospel that was often skeptical of, and sometimes outright antagonistic to, progressive aspirations for rapid national economic growth. Indeed, ecosystem management by the 1990s had shifted the focus of public land management to preserving "natural" conditions. John Muir was finally winning out over Gifford Pinchot. But it was difficult if not impossible to put an economic value on the existence of a particular desired state of wild nature. Indeed, there was little point in doing an economic analysis.

Professional economists think of their role in society as detailed problem solvers. But in that capacity they have much less influence than many economists assume. This does not mean, however, that the efforts of professional economists have no practical significance. Although it is not a role much advertised by economists themselves, the members of the profession do function importantly as dispensers of moral legitimacy in American society. Economists are much like a new priesthood in an era when the older priesthoods have lost much of their traditional authority in matters of governance and public affairs. The concept of social legitimacy does not have much place in professional economic thinking, but political scientists and sociologists know better. Indeed, a socially illegitimate regime can erode from within and may eventually collapse economically, as happened to the former Soviet Union when its Marxist tenets were no longer believable to its own people.

The economics profession was a product of the progressive era, and it continues to express fundamental progressive values. By asserting its own scientific status, the actual message of professional economics is that the scientific management of American society is still a feasible goal. On the public lands, to the extent that they have any real influence, economic arguments mainly work in practice to legitimize and sustain a public land status quo. The technical apparatus of professional economics dresses the current realities of land-use allocation as a scientific decision process. This puts public land decisions in a higher moral and even quasi-religious framework of advancing American progressive values.

When the environmental movement initially sought in the 1970s to change the uses of the public lands, environmentalists at first argued for a "higher progressivism." Public land management in practice, as it was not difficult for environmentalists to show, had failed by the standard of its own high ideals. By the 1990s, however, environmental goals had become more radical; many environmentalists now asserted the ethical failure of the progressive ideals themselves. Instead of upholding the values of scientifically managed growth and development—of economic "progress"—leading environmental prophets more and more questioned the foundational moral claims for such progress. The public lands should not be managed to maximize their contribution to total national economic outputs, but as a refuge in American society from the very consequences of rapid economic growth. This new set of public land values—a "religion" for the public lands—achieved its first great legislative success in 1964 with the enactment of the Wilderness Act. By the 1990s, a "wilderness ethic" was triumphing widely across all the public lands. Maintaining or restoring future natural conditions as a moral imperative in itself, rather than maximizing total social benefits minus costs, became the ethos of public land management, as expressed in the tenets of "ecosystem management."

American economists have often been confused about their professional role in society. In understanding the place of economic analysis, what is needed today is a new sociology of the economics profession. Economists do in fact have important sociological roles to play in the practical workings of American society. It is just that these roles have little to do with the "problem-solving" image of most professional economists and much more to do with public symbolism—with maintaining the rituals of social legitimacy. That is to say, and however much contrary to their own self-image, in the real world professional economists have been better priests than economic scientists (Nelson 2001).

Notes

1. Besides the general public land laws, Congress also enacted new laws to revise specifically the management of federally owned mineral resources. These laws included most importantly the Federal Coal Leasing Amendments Act of 1976 and the Outer Continental Shelf Lands Act of 1978 (the latter addressing the leasing of the valuable oil and gas resources of the Gulf of Mexico and elsewhere on the outer continental shelf). Much like surface land management, these laws emphasized formal land-use planning and other elements of progressive scientific management as the required decision-making framework for federal leasing of minerals (see Nelson 1983).

2. Besides Arrow, the authors included Nobel prize–winning economists Milton Friedman, Robert Solow, and Joseph Stiglitz, as well as Jagdish Bagwati, William Baumol, Alfred Kahn, Charles Schultze, and a number of other distinguished American economists.

3. In 2005, the Government Accountability Office published similar estimates, showing that little had changed (see U.S. Government Accountability Office 2005).

4. By the last half of the 1990s, environmentalists were showing a greater interest in buyouts of rancher grazing rights. Their longstanding efforts to reduce livestock grazing on public lands by command-and-control means had proved frustratingly slow, and buyouts offered the prospect of a more rapid and less contentious process to achieve this end. As of 2005, however, it was still not officially possible to buy and sell grazing rights directly for non-livestock uses (see Nelson 1997b).

5. The core concept of ecosystem management was in itself offensive and had a negative existence value for some critics (see Fitzsimmons 1999).

References

Anderson, Terry L., and Peter J. Hill, eds. (1994). *The Political Economy of the American West*. Lanham, MD: Rowman & Littlefield.

Arrow, Kenneth J. et al. (2000). *Brief to the United States Supreme Court in the Case of American Trucking Association v. Browner,* July 21, 2000.

Clawson, Marion. (1983). *The Federal Lands Revisited*. Washington, DC: Resources for the Future.

Dana, Samuel Trask, and Sally K. Fairfax. (1980). *Forest and Range Policy: Its Development in the United States,* 2nd ed. New York: McGraw-Hill Publishing Co.

Dunlop, Thomas R. (2004). *Faith In Nature: Environmentalism as a Religious Quest*. Seattle, WA: University of Washington Press.

The Ecosystem Approach: Healthy Ecosystems and Sustainable Economies, Report of the Interagency Ecosystem Management Task Force. (1995). Washington, DC: National Technical Information Service.

Elliot-Fisk, Deborah et al. (1997). "Mediated Settlement Agreement for Sequoia National Forest, Section B. Giant Sequoia Groves: An Evaluation." In *Status of the Sierra Nevada: Addendum, Sierra Nevada Ecosystem Project, Final Report to Congress.* Wildland Resources Center Report No. 40, March: 217–277.

Fitzsimmons, Allan K. (1999). *Defending Illusions: Federal Protection of Ecosystems.* Lanham, MD: Rowman & Littlefield.

Foss, Phillip O. (1960). *Politics and Grass: The Administration of Grazing on the Public Domain.* Seattle: University of Washington Press.

Fox, Stephen. (1985). *The American Conservation Movement: John Muir and His Legacy.* Madison, WI: University of Wisconsin Press.

Furubotn, Eirik G., and Rudolf, Richter. (1997). *Institutions and Economic Theory: The Contributions of the New Institutional Economics.* Ann Arbor: University of Michigan Press.

Gardner, Delworth. (1963). "A Proposal to Reduce Misallocation of Livestock Grazing Permits." *Journal of Farm Economics* 45(1): 109–120.

Gates, Paul W. (1968). *History of Public Land Law Development.* Washington, DC: Government Printing Office.

Hausman, Jerry A., ed. (1993). *Contingent Evaluation: A Critical Assessment.* Amsterdam: North-Holland.

Hays, Samuel P. (1959). *Conservation and the Gospel of Efficiency: The Progressive Conservation Movement, 1890–1920.* Cambridge: Harvard University Press.

Hess, Karl, Jr. (1994). *Visions Upon the Land: Man and Nature on the Western Range.* Washington, DC: Island Press.

Hobert, George. (2003). *Science, Politics, and U.S. Forest Service Law: The Battle Over the Forest Service Planning Rule.* Discussion Paper 03-19. Washington, DC: Resources for the Future.

Krutilla, John V. (1967). "Conservation Reconsidered." *American Economic Review* 57(4): 779–786.

——. (1979). "Adaptive Responses to Forces for Change." Paper presented at the Annual Meetings of the Society of American Foresters, Boston, MA, October 16.

——. (1989). *Multiple-Use Management: The Economics of Public Forestlands.* Washington, DC: Resources for the Future.

Krutilla, John V., and John A. Haigh. (1978). "An Integrated Approach to National Forest Management." *Environmental Law* 8(2): 373–416.

Lazarus, Richard J. (2004). *The Making of Environmental Law.* Chicago: University of Chicago Press.

Lee, Eliza Wing-yee. (1995). "Political Science, Public Administration, and the

Rise of the American Administrative State." *Public Administration Review* 55(6): 538–546.

Leman, Christopher K., and Robert H. Nelson. (1981). "Ten Commandments for Policy Economists." *Journal of Policy Analysis and Management* 1(1): 97–117.

Le Master, Dennis C. (1984). *Decade of Change: Remaking the Forest Service Statutory Authority During the 1970s.* Westport, CT: Greenwood Press.

Mead, Walter J. (1992). "Review and Analysis of Recent State-of-the-Art Contingent Valuation Studies." In *Contingent Evaluation: A Critical Assessment.* Ed. J. A. Hausman. Amsterdam: North-Holland.

Mitchell, Robert Cameron, and Richard T. Carson. (1989). *Using Surveys to Value Public Goods: The Contingent Valuation Method.* Washington, DC: Resources for the Future.

Mueller, Dennis C., ed. (1997). *Perspectives on Public Choice: A Handbook.* New York: Cambridge University Press.

National Academy of Sciences, National Research Council. (1984). *Developing Strategies for Rangeland Management: A Report Prepared by the Committee on Developing Strategies for Rangeland Management.* Boulder, CO: Westview Press.

Nelson, Robert H. (1976). *Benefit-Cost Analysis of Public Range Investments: A Case Study.* Washington, DC: Office of Policy Analysis, U.S. Department of the Interior.

——. (1980). *The New Range Wars: Environmentalists versus Cattlemen for the Public Rangelands.* Washington, DC: Office of Policy Analysis, U.S. Department of the Interior.

——. (1983). *The Making of Federal Coal Policy.* Durham, NC: Duke University Press.

——. (1987). "The Economics Profession and the Making of Public Policy." *Journal of Economic Literature* 25(March): 49–91.

——. (1991). *Reaching for Heaven on Earth: The Theological Meaning of Economics.* Lanham, MD: Rowman & Littlefield.

——. (1995). *Public Lands and Private Rights: The Failure of Scientific Management.* Lanham, MD: Rowman & Littlefield.

——. (1997a). "Does 'Existence Value' Exist?: An Essay on Religions, Old and New." *Independent Review* 1(March): 499–522. Rpt. in Robert Higgs and Carl P. Close, eds. (2005). *Re-Thinking Green: Alternatives to Environmental Bureaucracy.* Oakland, CA: Independent Institute.

——. (1997b). "How to Reform Grazing Policy: Creating Forage Rights on Federal Rangelands." *Fordham Environmental Law Journal* VIII(3): 645–690.

——. (2000). *A Burning Issue: A Case for Abolishing the U.S. Forest Service.* Lanham, MD: Rowman & Littlefield.

——. (2001). *Economics as Religion: From Samuelson to Chicago and Beyond*. University Park, PA: Pennsylvania State University Press.

——. (2004). "Environmental Religion: A Theological Critique." *Case Western Reserve Law Review* 55(Fall): 51–80.

Nelson, Robert H., and Gabriel, Joseph. (1982). *An Analysis of Revenues and Costs of Public Land Management by the Interior Department in 13 Western States—Update to 1981*. Washington, DC: Office of Policy Analysis, U.S. Department of the Interior.

Norton, Bryan G. (1991). "Thoreau's Insect Analogies: Or, Why Environmentalists Hate Mainstream Economists." *Environmental Ethics* Fall: 235–251.

Pacific Northwest Research Station, USDA Forest Service. (2003). *An Analysis of the Timber Situation of the United States, 1952–2050*. General Technical Report PNW-GTR-560. Portland, OR.

Pinchot, Gifford. (1947). *Breaking New Ground*. New York: Harcourt Brace.

Sagoff, Mark. (2004). *Price, Principle, and the Environment*. New York: Cambridge University Press.

Sedjo, Roger A. (2000). "Does the Forest Service Have a Future?: A Thought-Provoking View." In *A Vision for the U.S. Forest Service: Goals for its Next Century*. Washington, DC: Resources for the Future.

Stigler, George J. (1961). "The Economics of Information." *Journal of Political Economy* 69(9): 213–225.

Thompson, Barton H., Jr. (2003). "What Good Is Economics?" *Environs* Fall: 175–201.

U.S. Forest Service, Department of Agriculture. (1998). *National Summary: Forest Management Program Annual Report, Fiscal Year 1997*. Washington, DC: U.S. Forest Service.

U.S. Government Accountability Office. (2005). *Livestock Grazing: Federal Expenditures and Receipts Vary, Depending on the Agency and the Purpose of the Fee Charged*. Report No. GAO-05-869. Washington, DC: GAO.

Wilkinson, Charles F. (1992). *Crossing the Next Meridian: Land, Water, and the Future of the West*. Washington, DC: Island Press.

Worster, Donald. (1993). *The Wealth of Nature: Environmental History and the Ecological Imagination*. New York: Oxford University Press.

The Role of Ethnicity and Language in Contingent Valuation Analysis

A Fire Prevention Policy Application

By John Loomis, Lindsey Ellingson, Armando Gonzalez-Caban, and Andy Seidl*

Abstract. In order to satisfy legal requirements, many federal agencies must assess the potential effects of their policies on the public. This is often done through surveys, but frequently those surveys are only administered in English. This paper tests whether there are differences in survey response rates, refusals to pay, and willingness to pay (WTP) across different ethnicities and language for forest fire reduction in the State of California. The ethnicities studied were Caucasian, African American, and Hispanic (half in Spanish, half in English).

There was a statistical difference in survey response rates across all ethnicities, and no statistical difference among ethnicities for reasons of refusing to pay. The influence of ethnicity and language was tested using a logit model with ethnicity intercepts and bid slope interaction terms. The Hispanic-Spanish intercept shifter and the Hispanic-English dollar bid amount interaction terms were statistically significant and positive. There was a significant difference in the logit willingness to pay coefficients between Hispanics surveyed in Spanish with each of the other ethnicities. The annual willingness to pay of Hispanics taking the survey in Spanish was twice that of Caucasians, but no statistical difference in mean and median WTP between these

*John Loomis is a Professor of Agricultural and Resource Economics at Colorado State University, Fort Collins, CO 80523: e-mail: jloomis@lamar.colostate.edu. Lindsey Ellingson is a Research Assistant in the Department of Agricultural and Resource Economics at Colorado State University, Fort Collins, CO. Armando Gonzalaz-Caban is an economist at the USDA Forest Service, Riverside, CA. Andy Seidl is an Associate Professor of Agricultural and Resource Economics at Colorado State University, Fort Collins, CO. The authors would like to acknowledge Hayley Hesseln, University of Saskatchewan, Doug Rideout, Colorado State University, and the staff at the Survey Research Center at California State University Chico.

American Journal of Economics and Sociology, Vol. 65, No. 3 (July, 2006).

two groups was found, due to large confidence intervals around each estimate. Nonetheless, the WTP of both Hispanics and Caucasians for the forest thinning program is substantial, and statistically different from zero, suggesting there may be broad support for this program in California.

I

Introduction

FEDERAL AGENCIES ARE REQUIRED under several laws to assess the potential effects of their proposed actions on the public in general, distinct groups of the public, and the human environment. In particular, the National Environmental Policy Act (NEPA) requires federal agencies to evaluate the effects of proposed policies on the environment via an Environmental Impact Statement. The agencies are required to use this information to "consider" the effect on the human environment when selecting the preferred alternative. The State of California has a parallel regulation in the California Environment Quality Act, which also requires an Environment Impact Report on any relevant state policy under consideration. In 1994, then-President Clinton issued Executive Order 12898, which requires agencies to address the concern for environmental justice among ethnic minorities and low-income households in the United States. The Executive Order states: "each Federal agency shall make achieving environmental justice part of its mission by identifying and addressing, as appropriate, disproportionately high and adverse human health or environmental effects of its programs, policies, and activities on minority populations and low-income populations in the United States and its territories and possessions" (Clinton 1994). Under this Executive Order, federal agencies are required to evaluate the effects of policy and its alternatives on ethnic minorities and low-income households.

Surveys are one popular and accepted means for government agencies to gauge support or opposition to proposed policy alternatives or to estimate whether the public would benefit from or be harmed by proposed environmental policies. Survey information helps agencies refine policy proposals and choose a policy alternative that is responsive to public concerns. For example, the U.S. Forest Service (USFS) has conducted a series of surveys over the last several years to gain an understanding of public preferences for forest and recreation management. However, historically, nearly all surveys dealing with natural resource issues have been administered only in English, leaving out a substantial proportion of the citizenry that either does not speak English or is more comfortable interacting in another language. Since language is an indicator of culture, and since culture can influence attitudes and preferences for the natural environment, conducting surveys only in English may create bias in the data collected and subsequent policy decisions. Clearly, then, environmental impact assessments of public policies should take into account any differences due to language and ethnicity. If there are little or no differences by language and ethnicity, then current survey administration practices may be acceptable. If there are systematic differences, this suggests that government agencies need to expand their sampling strategies and perform surveys in the primary languages of the citizens in order to formulate policies that reflect the wants of all citizens.

This issue has become of great concern, as Hispanics have now become the nation's largest minority group, surpassing African Americans, a transition occurring well sooner than anticipated (Cohn 2003). Of the 18 percent of the U.S. population that does not speak English at home, 60 percent speak Spanish (U.S. Census Bureau 2003). Hispanics may evaluate environmental policy options differently than non-Hispanics. In addition, it is not sufficient to assume that Hispanics who are bilingual adequately represent Hispanics who only speak Spanish. Biased survey results may lead to inefficient natural resource policies in which the actual or realized benefits are less than the costs.

Of interest for this paper is the intersection of ethnicity and language with forest fire policy. During 2000 and 2002, the United States experienced its two worst fire seasons in a half-century. Throughout the 2002 fire season, forest fires burned a total acreage that was larger than the states of Maryland and Rhode Island combined (White House 2003). The Healthy Forests Initiative was developed by President

George W. Bush to restore the health of the forests and rangelands in the western United States and reduce the severity of wildfires. The two main methods of fuel reduction are prescribed burning and mechanical fuel reduction. In this paper we focus on the mechanical fuel reduction method, which consists of cutting and/or chipping smaller trees and brushy vegetation. This creates mulch that acts as a barrier to the growth of vegetation on the forest floor. While this reduces fire potential, it also reduces vegetation for wildlife. The mechanical fuel reduction method prevents wildfires from reaching the tops of the mature trees. Unlike prescribed burning, mechanical fuel reduction does not produce any fire smoke, a potential negative external effect of a prescribed burning program.

California is among the states experiencing some of its worst fire seasons in the past few years (Bush 2002). Over the past decade, California has also experienced a 43 percent increase in Hispanics and a 14 percent increase in African Americans. Hispanics, of any race, comprise 32 percent of California's population, while African Americans are 7.5 percent and Caucasians are 64 percent of the California's population. In addition, nearly half of the California residents who speak a language other than English at home speak Spanish (U.S. Census Bureau 2003). Due to the recent increasing prominence of both ethnic diversity and wildfires in California, this study will examine responses across different ethnicities with regard to willingness to pay for the mechanical fire fuel reduction program option to prevent wildfires in California.

Smedley defines ethnicity as "all those traditions, customs, activities, beliefs, and practices that pertain to a particular group of people who see themselves and are seen by others as having distinct cultural features, a separate history, and a specific socio-cultural identity" (Smedley 1993; Schelhas 2002). This definition will be used when referring to ethnicity throughout this study. The prominent ethnicities to be examined are Hispanic, African American, and Caucasian. Hispanic responses will be examined further when the survey is administered in English and Spanish. There have been no studies to date that research differences in ethnicity to the mechanical fire fuel reduction program for fire prevention in the State of California or any other states in the western United States.

II

Addressing Ethnicity in Surveys

ALTHOUGH MORE FREQUENT IN the sociology literature, only a few economic studies have dealt with the distinctions among ethnicities in the United States. For example, Turner et al. (1996) conducted a study that targeted Hispanics through a health-related survey. They found that the geographic dispersion of linguistic minorities, such as Spanish-speaking individuals, presents barriers to their response to nationally representative surveys.

Few studies have focused on the interaction of survey administration mode in evaluating ethnic differences among interviewers and respondents. Lotade-Manje (2002) conducted a face-to-face interview in a grocery store using an African and a Caucasian interviewer to ask about WTP for "fair share" coffee. The study found that willingness to pay differed across products and interviewers, and the willingness to pay responses were higher when asked by the African interviewer due to social desirability bias. Lenski and Leggett (1960) also used the face-to-face interview approach and found that the interview brought into existence a social relationship between interviewer and respondent, such that the responses to the interview involve perceptions of the interviewer and social norms. While these two studies found differences in responses through face-to-face interviews, this present study is evaluating a difference in responses to phone interviews conducted in English and Spanish.

Welch et al. (1973) and Weeks and Moore (1981) conducted two separate face-to-face interviews using English and Spanish. Welch et al. (1973) used three different interview groups, a Caucasian interviewer, a Hispanic interviewer, and a team of one Caucasian and one Hispanic interviewer to interview Hispanics in Nebraska about different health and political issues. No statistically significant differences among responses across the different interviewers were found. Weeks and Moore (1981) focused their face-to-face interviews on elementary school children in Miami, El Paso, northeast Arizona, and San Francisco. The survey was originally administered in English and then translated into Spanish at the request of the respondent. The study found no statistically significant differences among responses across the different locations and languages. These two studies did not find any statistical significance across their responses when they allowed for differences in language during face-to-face interviews. Our study also allows differences in language but it uses the phone interview approach, which may generate distinct responses.

Two studies used phone interviews to compare differences in responses across respondents and interviewers. Cotter et al. (1982) used phone interviews among African-American and Caucasian respondents and interviewers. The study found no statistical difference in responses when the respondents were asked nonracially sensitive questions, but did find a difference when the respondents were asked racially sensitive questions. Reese et al. (1986) used phone interviews to examine cultural questions across Caucasians and Hispanics in English and Spanish. The study concluded that respondents answered the questions in an effort to sound more like the interviewer. However, our study does not touch on any overtly racial or cultural questions and, therefore, is unlikely to generate responses in which the respondent would want to sound more like the interviewer.

III

Valuation of Public Natural Resources

FORESTS PROVIDE MANY BENEFITS to society. In addition to their aesthetic and recreational value, forests provide an ecosystem for plants and animals. The many social benefits of the forest cannot be valued by simple market prices. For example, a market price does not exist on the aesthetic value of public forest or on the ecosystem services it provides. Further, wildfire prevention programs on public lands, such as the mechanical fire fuel reduction program, cannot usually be purchased in a market. Therefore, the market value of timber on public forests protected by these programs would be an incomplete measure of the benefits of fire prevention. A nonmarket valuation technique needs to be used to measure social benefits from a nonmarket public good, like wildfire prevention programs.

The contingent valuation method (CVM) is a direct interview (survey) nonmarket valuation approach that can be used to provide acceptable measures of the economic value of the preservation of natural resources (Loomis and Walsh 1997). Contingent valuation asks the respondent how he or she would value a particular resource, contingent upon a qualitative or quantitative change in that resource. Previous studies that used the CVM approach to value forest resources include Champ et al. (2002), Loomis et al. (2002), VanRensburg et al. (2002), Loomis and Gonzalez-Caban (1998), and Hagen et al. (1992), to name just a few. The contingent valuation method requires the construction of a hypothetical market or simulated referendum for a good or service that is not normally allocated via the market. Implementing a constructed market for the nonmarket good being valued requires an appropriate valuation question format. Among the several types of valuation question formats, the dichotomous choice or referendum format follows the recommendation of the U.S. Department of Commerce's National Oceanic and Atmospheric Administration (NOAA) report on CVM (Arrow et al. 1993). In the referendum format, individuals are asked if they would vote for a particular tax increase, the level of which varies across the sample.

Champ et al. (2002) found the referendum question format to be superior to other donation type formats in a contingent valuation setting. The referendum question format creates a voter-like scenario that most respondents are familiar with from elections. It asks respondents if they would vote for or against the contingent scenario pertaining to the nonmarket resource if it were on the next ballot. While some respondents may vote against the program, they may still hold a positive but lower value than the bid amount they are asked to pay in the survey. However, other respondents may protest the program completely. In order to differentiate between the two types of "no" responses, a follow-up question needs to be asked on why they voted against the program. From this question, we can evaluate whether the respondent held some value of the program or protested the program completely (Halstead et al. 1992; Jorgensen et al. 1999). Mitchell and Carson (1989) identify protest responses as those that show the respondent does not accept the contingent valuation scenario.

One of the only published CVM studies to examine language-related differences in response to both prescribed burning and mechanical fuel reduction fire prevention programs in Florida is Loomis et al. (2002). Their study primarily examined only the difference in responses on implementing the survey in English for Caucasians and Spanish for Hispanics. However, it did not take into account potential response differences by ethnicity without language differences. The Florida study found a statistical difference among responses based on language with the protest responses excluded. However, they did not find any statistical differences between languages in the mean willingness to pay for either fire prevention program. However, it was noted by Loomis et al. (2002) that Hispanics in the Florida study were primarily of Caribbean descent. Therefore, additional studies in other states, such as California or Texas, where Hispanics have a greater tendency to be of Mexican descent, needed to be implemented to determine if the contingent valuation method is useful in evaluating benefits for nonmarket valuation with Hispanics of Mexican descent (Loomis et al. 2002).

IV

Contingent Valuation Method Empirical Specification

As recommended by the NOAA panel on contingent valuation, we elicited willingness to pay using a dichotomous choice CVM question format (Arrow et al. 1993). The binomial logit regression is an appropriate statistical analysis tool to analyze the dichotomous choice (yes/no) responses to the tax increase (Hanemann 1984). In our case, the dependent variable is 1 if the respondent votes for the mechanical fuel reduction program and 0 if the respondent votes against the program.

The cumulative logistic distribution function for the yes/no response is as follows:

$$\text{Prob}(Y = 1) = [\exp(\beta X_i)]/[1 + \exp(\beta X_i)]. \tag{1}$$

β is the set of parameters that reflect the impact of changes in the independent variables, X_i, on the probability of choosing the dependent variable, Y (Greene 1993). From the cumulative distribution function, we can develop the odds ratio of voting for (Y = 1) or against

(Y = 0) the mechanical fuel program, and then taking the log of the odds ratio, we have the logit model:

$$L = \ln\{[\text{Prob}(Y = 1)]/[1 - \text{Prob}(Y = 1)]\} = \beta_0 + \beta_i X_i. \tag{2}$$

The log of the odds ratio is linear in the coefficients and the independent variables (Gujarati 2003). From the coefficients in the logit model, mean and median WTP can be calculated. When estimating the logit model, the maximum likelihood estimation is used.

Maddala (1996) states three goodness of fit measures for the logit model. The three measures are the McFadden R-squared, predicted-expected percentage, and the likelihood ratio (LR) statistic. The McFadden R-squared is:

$$1 - [\text{ULL}/\text{RLL}], \tag{3}$$

where ULL is the unrestricted log likelihood and RLL is the restricted log likelihood. The unrestricted log likelihood is the maximum of the likelihood function with respect to all the parameters in the regression. The restricted log likelihood is the maximum of the likelihood function when the parameters, excluding the constant, are set equal to zero (Maddala 1996). The closer the McFadden R-squared is to 1, the better the fit of the regression.

The predicted-expected percentage shows how well the model can forecast responses. The percentage is calculated between the estimated probabilities and the observed response frequencies (Maddala 1996). Finally, the LR statistic tests to see if the overall fit of the estimated regression is statistically significant. The three goodness of fit measures were used throughout this study to assess the fit of the logit regression.

V

Hypotheses

THE OBJECTIVE OF THIS STUDY is to determine whether there are differences in responses by Caucasians, African Americans, and Hispanics to a contingent valuation of mechanical fuel reduction in the State of California. In addition, the Hispanic responses are compared in relation to the language the survey was administered to them, in English or in Spanish. Administering the survey in both English and Spanish facilitates understanding of how language may shape ethnic minorities' participation and responses to a nonmarket valuation survey as well as their values for the mechanical fire fuel reduction program. This study tests for ethnicity and language differences in overall survey response rates (RR), logit equation coefficient equality (β), willingness to pay (WTP), and reasons for refusing to pay for the mechanical fuel reduction program.

The language of the initial interview determined which language was used for the Hispanic respondents, with half receiving the survey in English and half in Spanish. The surveyor for the initial phone interview identified himself or herself as being from a California university. It is hypothesized that the three different ethnic groups may have different responses to taking a survey when asked by a university regarding a government program. The null hypothesis is that the overall response rate (RR) is independent of the language and ethnicity:

$$H_0: RR_{\text{Caucasian}} = RR_{\text{African American}} = RR_{\text{Hispanic-English}} = RR_{\text{Hispanic-Spanish}}.$$

The overall response rate is estimated by comparing the number of respondents who participated in the first interview and the number of respondents who completed the lengthier, second follow-up interview. The chi-squared statistic is used to estimate the difference in response rates across ethnicities.

Because different ethnicities may have different views on government programs, it can be hypothesized that there may be systematic differences among ethnic groups in their votes for and against the mechanical fuel reduction program, holding other variables constant. The first null hypothesis is tested using ethnicity slope and intercept shifter variables on ethnicity in the logit willingness-to-pay equation.

$$\begin{aligned} \text{Vote} = \beta_0 &- \beta_1\text{Bid Amount} + \beta_2\text{Age} + \beta_3\text{Education} \\ &- \beta_4\text{Exp Smoke} + \beta_5\text{Gender} + \beta_6\text{Income} \\ &+ \beta_7\text{Income}^2 + \beta_8\text{Own Home} + \beta_9\text{People} \\ &- \beta_{10}\text{Resp Prob} - \beta_{11}\text{AA} - \beta_{12}\text{AA Bid} \\ &- \beta_{13}\text{HE} - \beta_{14}\text{HE Bid} - \beta_{15}\text{HS} - \beta_{16}\text{HS Bid} \end{aligned} \tag{4}$$

The null hypothesis for Equation (4) is that there is no statistical difference between the intercept and slope shifters among ethnicities. The null hypotheses can be represented as:

$$H_o: \beta_{11}AA = \beta_{13}HE = \beta_{15}HS = 0$$

$$H_o: \beta_{12}AA\ Bid = \beta_{14}HE\ Bid = \beta_{16}HS\ Bid = 0.$$

These two null hypotheses will be tested using *t*-statistics on each coefficient. Table 1 gives a detailed description of the different variables used in the logit willingness-to-pay equation.

A less restrictive statistical test of differences across ethnicities is to create separate logit models for each ethnicity and compare coefficient equality across each model. This allows all the coefficients to vary by ethnicity. The second null hypothesis will be:

$$H_o: \beta_{\text{Caucasian}} = \beta_{\text{African American}} = \beta_{\text{Hispanic-English}} = \beta_{\text{Hispanic-Spanish}}.$$

The vector of coefficients for each ethnicity is denoted as β. In order to test for the coefficient equality, a likelihood ratio (LR) test is performed. The LR test involves taking twice the difference of the restricted log likelihood function (log likelihood of the pooled model) from the unrestricted log likelihood function (summation of log likelihood for each ethnicity model). The result is the calculated chi-square, and if it is greater than the critical chi-square statistic, then the coefficients across the ethnicities are statistically different from each other. The logit regression to be evaluated through the LR test is:

$$\begin{aligned}\text{Vote} = \beta_0 - \beta_1\text{Bid Amount} + \beta_2\text{Education} + \beta_3\text{Gender} \\ + \beta_4\text{Income} - \beta_5\text{Income}^2.\end{aligned} \tag{5}$$

The hypothesized sign of the bid amount variable is negative because as the amount a respondent is asked to pay for the program increases, the probability of voting "yes" decreases. Education is expected to be positively related to voting "yes" because the more educated a respondent, the more he or she may know about recent changes in forest management and the greater the probability of voting for the program. The predicted sign of the gender coefficient is indeterminate. The combined effect of income and income squared is expected to be

Table 1

Explanation of Variables

Variable	Explanation
African American (AA)	Dummy variable: 1 if respondent is African American, 0 otherwise
AA Bid	Interaction term: African American dummy * Bid Amount
Age	Age of the respondent
Bid Amount	The dollar amount respondent is asked to pay, with the following values: $15, $25, $45, $65, $95, $125, $175, $260, $360, and $480
Education	Education level of the respondent
Exp Smoke	Dummy variable: 1 if the respondent experienced smoke from a wildfire or prescribed burn, 0 otherwise
Gender	Dummy variable: 1 if respondent is male, 0 if respondent is female
Hispanic-English (HE)	Dummy variable: 1 if respondent is Hispanic and received the survey in English, 0 otherwise
HE Bid	Interaction term: Hispanic-English dummy * Bid Amount
Hispanic-Spanish (HS)	Dummy variable: 1 if respondent is Hispanic and received the survey in Spanish, 0 otherwise
HS Bid	Interaction term: Hispanic-Spanish dummy * Bid Amount
Income	Household income of the respondent
$Income^2$	Household income of the respondent, squared

Table 1 *Continued*

Variable	Explanation
Own Home	Dummy variable: 1 if respondent owns a home, 0 if respondent rents
People	Number of people living in the household, including the respondent
Respiratory Problem (Resp Prob)	Dummy variable: 1 if respondent suffers from respiratory or breathing problems, 0 otherwise
Voting for Mechanical Program (Vote)	Dummy variable: 1 if respondent is in favor of the mechanical fire fuel reduction program, 0 otherwise

positive because as income increases, the probability of voting for the program would increase, but possibly at a decreasing rate (a negative sign on income squared).

If it is hypothesized that different ethnicities may respond differently to the overall survey, it can additionally be hypothesized that their willingness to pay for the mechanical fire fuel reduction program may vary across ethnicities. Lotade-Manje (2002) found that willingness to pay is statistically different across the race of the interviewers in a face-to-face interview. While we did not conduct face-to-face interviews, it can be hypothesized that the difference in language would cause the willingness to pay to differ across responses. The null hypothesis is that the willingness to pay per household for the California mechanical fire fuel reduction program is independent of language and ethnicity:

$$H_o: WTP_{\text{Caucasian}} = WTP_{\text{African American}} = WTP_{\text{Hispanic-English}} = WTP_{\text{Hispanic-Spanish}}.$$

The mean willingness to pay per household will be evaluated for the purpose of benefit cost analysis. The median willingness to pay will be evaluated to assist in determining the dollar amount for which a majority would vote in favor of the mechanical fuel reduction program. To test for differences in willingness to pay, confidence intervals on WTP as calculated using Park et al. (1991) are used.

If respondents voted against the fire fuel reduction program with the given bid amount, they were asked a follow-up question. The follow-up question was also a dichotomous choice referendum question that asked respondents if they would be willing to pay $1 for the mechanical fire fuel reduction program. If they stated they would not pay $1, they were asked why they voted this way. The responses on why they voted no to the dollar question were postcoded and determined as either protest or nonprotest responses. The null hypothesis is that the protest response is independent of language or ethnicity:

$$H_o\text{: Protest}_{\text{Caucasian}} = \text{Protest}_{\text{African American}} = \text{Protest}_{\text{Hispanic-English}} = \text{Protest}_{\text{Hispanic-Spanish}}.$$

The reasons for protesting the mechanical fire fuel reduction program are hypothesized as varying across ethnicities. If different ethnicities view the government and the fuel reduction program differently, their reasons for not wanting the program implemented will also be different.

VI

Survey Design and Methodology

THE SURVEY WAS IMPLEMENTED through a phone-mail-phone method. The initial phone contact was obtained through random digit dialing with a short 5-minute interview. During the phone call, the respondents' language was verified, their name and address were requested to mail a survey booklet, and the 20-minute in-depth interview was scheduled. The survey booklet was sent in English to Caucasians and African Americans and half of the Hispanic households. The other half of the Hispanic households received the survey booklet in Spanish.

The basic format of the survey booklet and phone script (including the Spanish language version) had previously been through six focus groups in two different states, so it was only necessary to pretest the booklet and script on the four ethnic subgroups in English and

Spanish prior to survey implementation (Loomis et al. 2002). The survey booklet that was mailed to participants contained a brief description of the California Fire Management Program with definitions of common fire management terms. It provided colored pictures of the current forest fire problem and two potential solutions, the method of prescribed burning and the alternative method of mechanical fire fuel reduction. For each method, the booklet explained the features of the program, the results of the program, and the funding of the program. The mechanical fire fuel reduction program was defined in the booklet as the following:

> Another approach to reducing the buildup of fuels in the forest is to "mow" or mechanically chip the low- and medium-height trees and bushes into mulch. This is especially effective at lowering the height of the vegetation, which reduces the ability of fire to climb from the ground to the top or crown of the tress. In addition, mechanical "mowing" slows the growth of new vegetation with the layer of mulch acting as a barrier.

The mechanical fire fuel reduction program was stated as not producing any smoke, unlike the prescribed burning method, so as not to deteriorate the air quality. The survey booklet explained the number of acres and houses that are destroyed each year in California due to wildfires, and it stated that only one of the two programs would be implemented.

The mechanical fire fuel reduction dichotomous choice willingness-to-pay question was stated as follows:

> If the Mechanical Fire Fuel Reduction Program was undertaken instead of the Expanded Prescribed Burning Program, it is expected to reduce the number of acres of wildfires from the current average of approximately 362,000 acres each year to about 272,500 acres, for a 25% reduction. The number of houses destroyed by wildfires is expected to be reduced from 30 a year to about 12. Your share of this Mechanical Fire Fuel Reduction Program would cost your household $X a year. If the Mechanical Fire Fuel Reduction program were the ONLY program on the next ballot would you vote?
>
> ______ In favor
>
> ______ Against

The funding of the program was explained as being on a county-by-county basis, where if a majority of the county residents voted for the program, the State of California would match funds for the approved counties and everyone in the county would be required to pay the additional stated amount for their county. The bid amount, denoted by $X, varied across respondents and had the following values: $15, $25, $45, $65, $95, $125, $175, $260, $360, and $480. The final portion of the interview consisted of demographic questions, including the question pertaining to the respondent's ethnic background.

VII

Results

A. Response Rate Analysis

There are two survey response rates to be examined. The first-wave screener response rate is the percent of respondents from the total initial sample that was contacted and that completed the initial interview. The second-wave in-depth interview response rate is the percent of the net sample that completed the in-depth interview. The net sample is the completed initial sample less the respondents who refused to give their address, were not available, or were not contacted by the end of the data collection period. Table 2 breaks down the different response rates across ethnicities. The chi-square statistic for the overall difference in response rates by ethnicity was significant at the 1 percent level for both the initial interview and the follow-up interview. Therefore, we can infer that there is a statistically significant difference among response rates across ethnicities. Further, there is a statistically significant difference between response rates at the 1 percent level for Hispanics given the initial interview in English and Hispanics given the initial interview in Spanish. For the in-depth interview, there was a statistically significant difference in response rates between African Americans and Caucasians (1 percent level) and between Hispanics given the second interview in English and Hispanics given the second interview in Spanish (5 percent level).

Hispanics given the interview in Spanish had the highest screener response rate (75.5 percent), while Caucasians had the lowest screener response rate (41.3 percent). The opposite effect occurred during the in-depth interview. Hispanics given the interview in

Table 2

Survey Response Rates Across Ethnicities

First Wave: Screener	Caucasian	African American	Hispanic English	Hispanic Spanish	Total
Total initial sample contacted	794	708	733	620	2,855
Completed initial	328	308	421	468	1,525
1st wave response rate	41.3%	43.5%	57.4%	75.5%	53.4%
Chi-square total					**58.61******
Chi-square (AA vs. C)					0.298
Chi-square (HE vs. HS)					**9.98****

Second Wave: In-Depth Interview	Caucasian	African American	Hispanic English	Hispanic Spanish	Total
Refused to give address	4	4	9	1	18
Phone disconnected, moved, not available	16	25	37	47	125
Not called by end	51	3	0	0	54
Net sample for 2nd wave	257	276	375	420	1,328
Total surveys completed	187	126	170	139	622
2nd wave response rate	72.8%	45.7%	45.3%	33.1%	46.8%
Chi-square total					**34.25****
Chi-square (AA vs. C)					**10.51****
Chi-square (HE vs. HS)					**5.48***

*Significant at the 5 percent level.
**Significant at the 1 percent level.

Spanish had the lowest in-depth response rate (33.1 percent) and Caucasians had the highest response rate (72.8 percent). Potentially, the drop in the in-depth interview response rate for Hispanics given the interview in Spanish could be because Hispanics were unaccustomed to having the initial interview in their native language and were more accommodating. However, once they received the booklet, they may have realized the time commitment necessary to complete the second interview and chose not to do so. Potentially, Hispanics may not have had the same average facility in written English as Caucasians and African Americans, increasing the required time commitment for completing the written survey relative to the other groups and providing sufficient disincentive to reduce completion rates measurably.

In order to analyze protest responses, if the respondents were not willing to pay $1 for the mechanical fire fuel reduction program, they were asked why. The responses to the question are in Table 3. The upper half of the table reflects nonprotest refusals to pay, while the lower half reflects protests or rejections of the premise of the CVM-constructed market. Hispanics had the least amount of protest responses (5), while African Americans protested the most, with 14 responses. A chi-square test was used to determine if there were differences in reasons for refusals to pay across ethnicities. With three degrees of freedom, the calculated chi-square was 1.68 and the critical chi-square value was 7.815. Since the calculated chi-square was less than the critical chi-square, we accept our null hypothesis that there was no statistical difference among ethnicities for protest and nonprotest reasons for not paying at least $1 for the mechanical fuel reduction program.

B. Willingness-to-Pay Analysis

A logit model including an ethnicity intercept and bid slope interaction terms was estimated to conduct a test of whether ethnicity had any effect on voting for the mechanical fuel reduction program. Responses classified as protest refusals to pay were removed from the logit analysis as is customary in CVM analysis (Mitchell and Carson 1989).

Table 3

Reasons Why Respondent Would Not Be Willing to Pay $1 for the Mechanical Fire Fuel Reduction Program

Reason	Caucasian	African American	Hispanic English	Hispanic Spanish	Total
Nonprotest Responses					
No value/benefits where I live	1	2	2	1	6
Cannot afford	1	2	0	0	3
Reduces food/vegetation for wildlife	4	7	4	1	16
Method is unnatural/leave nature alone	3	6	2	0	11
Dislike mechanical fuel program	3	4	3	1	11
Nonprotest Total	12	21	11	3	47
Protest Responses					
Prefer prescribed burning method	5	2	2	0	9
Others should pay for it	1	3	1	0	5
Does not need additional funding	2	2	0	0	4
Wants more information about program	1	1	0	0	2
Does not trust the state	0	1	0	0	1
Harms their business	0	1	0	0	1
Other	1	4	1	1	7
Protest Total	10	14	4	1	29
TOTAL	**22**	**35**	**15**	**4**	**76**

Table 4 displays the logit model with ethnicity intercepts and bid slope interactions. The Hispanic-Spanish intercept variable (HS) is significant and positive, indicating a relatively higher probability of voting for the mechanical fuel reduction program among this ethnic group. The Hispanic-English bid slope interaction variable (HE Bid) is also significant and positive. When combined with bid coefficient, this means that Hispanics given the survey in English have a higher probability of voting for the mechanical fuel reduction program at a given bid level; in other words, their demand is less price sensitive.

The bid amount, gender, income level, and income level squared are significant at the 10 percent level. In order to conduct the likelihood ratio test and calculate WTP for each ethnicity, we estimated four separate logit models. These four logit models are shown in Table 5.

Table 6 displays the likelihood ratio tests conducted using the logit models shown in Table 5. The pooled model of coefficient equality is rejected at the 5 percent level with a calculated chi-square of 34.36. It appears this is due to the Hispanic-Spanish logit WTP coefficients, which were statistically different from the other three ethnicities, including African Americans, at the 5 percent level.

These findings suggest that we should reject our null hypothesis of equality of WTP coefficients since the influence of education, gender, and income on the likelihood of paying for the program varies across ethnicities. Note that no statistical difference was found among groups receiving the survey in English. Thus the possibility exists that language, more than culture, influences the WTP responses.

As shown in Table 5, the individual ethnicity logit models were estimated, and the bid amount was statistically significant at the 5 percent level for Caucasians and Hispanics given the survey in Spanish. Therefore, we can evaluate whether there was any difference in the mean and median willingness to pay across these two ethnicities. Table 7 displays the mean and median willingness to pay for Caucasians and Hispanics given the survey in Spanish and the respective confidence intervals. We used the following formulae (Hanemann 1989) to calculate mean and median willingness to pay:

$$\text{Mean WTP} = (\ln(1+\exp(\alpha)))/\beta$$

$$\text{Median WTP} = \alpha/\beta.$$

Table 4

Logit Model with Ethnicity Intercepts and Bid Slope Interactions

Variable	Coefficient	z-Statistic	Probability
Constant	1.0525	1.2473	0.2123
Bid Amount	**−0.0024**	**−1.9357**	**0.0529**
Age	0.0099	1.3616	0.1733
Education	−0.0329	−0.7129	0.4759
Exp smoke	−0.0296	−0.1471	0.8831
Gender	**−0.3284**	**−1.6944**	**0.0902**
Income	−1.50E-05	−1.5719	0.1160
Income^2	**1.07E-10**	**1.9434**	**0.0520**
Own home	−0.1228	−0.5372	0.5912
People	0.0447	0.6879	0.4915
Resp prob	−0.2540	−1.0380	0.2992
African American (AA)	−0.0984	−0.2511	0.8017
AA Bid	0.0015	0.8086	0.4188
Hispanic-English (HE)	−0.1124	−0.2911	0.7710
HE Bid	**0.0027**	**1.6419**	**0.1006**
Hispanic-Spanish (HS)	**1.1234**	**2.2728**	**0.0230**
HS Bid	−0.0008	−0.4240	0.6716
Mean dependent variable	0.6287		
S.E. of regression	0.4727		
Log likelihood	−332.490		
Restricted log likelihood	−353.560		
LR statistic (16 df)	*42.141*		
Probability (LR stat)	0.0004		
McFadden R-squared	*0.0596*		
% correct prediction	*65.11%*		
Sample size	536		

Table 5

Individual Ethnicity Logit Models

	Caucasian			African American			Hispanic English			Hispanic Spanish		
Variable	Coef.	*z*-Stat.	Prob.	Coef.	*z*-Stat.	Prob.	Coef.	*z*-Stat.	Prob.	Coef.	*z*-Stat.	Prob.
Constant	−0.323	−0.276	0.782	1.698	1.303	0.193	1.974	1.833	0.067	8.719	3.741	0.000
Bid amt.	**−0.002**	−1.905	0.057	−0.002	−1.401	0.161	0.001	0.467	0.641	**−0.004**	−2.505	0.012
Education	0.109	1.386	0.166	−0.039	−0.417	0.677	−0.059	−0.659	0.510	−0.287	−2.405	0.016
Gender	−0.686	−1.982	0.048	0.579	1.339	0.181	−0.384	−1.130	0.259	−0.569	−1.160	0.246
Income	−1.6E-05	−1.165	0.244	−2.2E-05	−1.255	0.209	−2.5E-05	−1.264	0.206	−2.4E-04	−2.059	0.040
Income^2	1.1E-10	1.347	0.178	1.2E-10	1.115	0.265	1.6E-10	1.174	0.240	3.8E-09	1.935	0.053
Mean dep. var.			0.5472			0.5872			0.6178			0.7840
Log likelihood			−103.95			−71.70			−101.51			−54.79
Restr. log likelihood			−109.50			−73.89			−104.42			−65.22
LR statistic (5 df)			11.10			4.38			5.83			20.87
Probability. (LR stat.)			0.0494			0.4963			0.3231			0.0009
McFadden R^2			0.0507			0.0296			0.0279			0.1600
% correct prediction			61.01%			63.30%			62.42%			80.00%
No. of observations			159			109			157			125

Table 6

Likelihood Ratio Tests of Coefficient Equality across Ethnicities

Models	Log Likelihood	All 4 Models	HE vs. HS	HE vs. C	HE vs. AA	HS vs. C	HS vs. AA	C vs. AA
Caucasian (C)	−103.95							
African American (AA)	−71.70							
Hispanic English (HE)	−101.51							
Hispanic Spanish (HS)	−54.79							
Sum of Unrestricted	−331.95	−331.95	−156.30	−205.46	−173.21	−158.74	−126.49	−175.65
Pooled (Restricted)	−349.13	−349.13	−164.31	−209.40	−175.40	−170.56	−134.29	−179.14
Calc Chi-Square		**34.36**	**16.03**	7.89	4.38	**23.64**	**15.61**	6.98
Critical Chi-Square 0.05		28.87	12.59	12.59	12.59	12.59	12.59	12.59
Degrees of Freedom		18	6	6	6	6	6	6

Table 7

Caucasian and Hispanic Spanish Willingness-to-Pay Estimates

Mean Willingness to Pay		
	Mean	90% Confidence Interval
Caucasian	\$437	\$278–\$1,813
Hispanic Spanish	\$863	\$494–\$2,124
Median Willingness to Pay		
	Median	90% Confidence Interval
Caucasian	\$249	\$141–\$646
Hispanic Spanish	\$856	\$494–\$2,117

The product of the coefficient and mean values for all independent variables excluding the bid coefficient is denoted by α, and β is the absolute value of the bid coefficient (Park et al. 1991). As can be seen in Table 7, Hispanic households given the survey in Spanish were willing to pay at least twice as much for the mechanical fuel reduction program (\$863) compared to Caucasian households (\$437). One possibility that was suggested to us for the higher willingness to pay of Hispanic households for the mechanical fuel reduction program is that the mechanical fuel reduction program is very labor intensive and may provide a substantial and disproportionate number of jobs for Hispanics. Due to the inherent statistical inefficiency of the dichotomous choice method, the confidence intervals around willingness to pay are quite large, and overlap for Caucasians and Hispanics. Thus, there is no statistical difference between either the mean or median willingness to pay for each of these two groups. It is interesting to note that the mean and median WTP are nearly identical for Hispanics, indicating a more symmetric distribution of WTP. Caucasians have a more skewed distribution of WTP, with some very high values pulling the mean substantially above the median.

In order to determine if the mechanical fuel program would pass as a ballot issue, we multiplied the mean dependent variable for each ethnicity by the Census 2000 population. Since the survey was given to Hispanics in both languages, we averaged the mean dependent variable for Hispanics. The total weighted average is that 60 percent of the population would vote for the mechanical fire fuel reduction program given an average bid of $157.

VIII

Conclusion

In order to value the nonmarket benefits of a mechanical fire fuel reduction program, similar to President Bush's Healthy Forest Initiative, a referendum contingent valuation method was employed. Executive Order 12898, the National Environmental Policy Act, and the California Environmental Quality Act all require public agencies to evaluate the effects of their proposed policies with explicit attention to cultural diversity and minority populations. Our study evaluated differences and similarities in survey responses across ethnicities and language in their responses to questions of natural resource policy by focusing on potential differences in their survey response rate, protest responses, and willingness to pay. The response rate analysis included the initial interview and the in-depth interview responses. There was a statistically significant difference in response rates at the 1 percent level across all ethnicities for both initial and follow-up interviews. For the initial interview and the in-depth interview, there was a statistically significant difference in responses at the 5 percent level across Hispanics given the interview in English and in Spanish. There was a statistically significant difference (1 percent) in response rates to the in-depth interview between Caucasians and African Americans.

If respondents were not willing to pay $1 for the mechanical fire fuel reduction program, they were asked why. These responses were divided into protests (i.e., rejections) and nonprotest refusals to pay responses. Using the chi-square test, there was no statistical difference among ethnicities for protest and nonprotest responses.

The influence of ethnicity on voting for the program was examined by developing a logit model with ethnicity intercepts and bid slope interaction terms included and by estimating four separate ethnicity logit models. The Hispanic-Spanish intercept shifter variable was statistically significant and positive, showing that Hispanics given the survey in Spanish had a higher probability of voting for the program than anyone else. The Hispanic-English bid interaction term was also statistically significant and positive, meaning that Hispanics given the survey in English were less sensitive to the tax amount.

We also tested for differences in the influence of the independent variables on WTP by using the likelihood ratio test on four separate logit models. While the mean WTP of Hispanics given the survey in Spanish was twice that of Caucasians, the large variance on WTP resulted in us finding no statistically significant difference in mean and median willingness to pay across Caucasians and Hispanics given the survey in Spanish.

Overall, our results suggest the possibility exists that language, more than ethnicity, influences willingness to pay responses. This result suggests that surveys of the public regarding natural resource policy should be stratified by language, and potentially ethnicity, in order to properly gauge aggregate public opinion under conditions of ethnic and linguistic diversity. These results supports the importance of President Clinton's Executive Order 12898 to pay special attention to evaluate the effect of government programs on minority populations separate from majority populations. Therefore, further research should accommodate the increasingly diverse population of the United States when evaluating nonmarket public goods, beyond wildfire prevention programs, to see if the differences we found carry over to other public goods and to populations of nonnative English speakers in other states.

References

Arrow, Kenneth, Robert Solow, Paul R. Portney, Edward E. Leamer, Roy Radner, and Howard Schuman. (1993). "Report of the NOAA Panel on Contingent Valuation." U.S. Department of Commerce, *Federal Register* 58(10): 4602–4614.

Bush, George W. (2002, August). *Healthy Forests: An Initiative for Wildfire Prevention and Stronger Communities.* Available at http://www.whitehouse.gov/infocus/healthyforests.

Champp, Patricia A., Nicholas E. Flores, Thomas C. Brown, and James Chivers. (2002). "Contingent Valuation and Incentives." *Land Economics* 78(4): 591–604.

Clinton, William J. (1994). *Executive Order 12898*. Available at http://www.epa.gov/civilrights/eo12898.htm.

Cohn, D'vera. (2003). "Hispanics Overtake Blacks as Nation's Largest Minority." *Coloradoan* A1.

Cotter, Patrick R., Jeffrey Cohen, and Philip B. Coulter. (1982). "Race-of-Interviewer Effects in Telephone Interviews." *Public Opinion Quarterly* 46(2): 278–284.

Greene, William H. (1993). *Econometric Analysis*, 2nd ed. New York: Prentice Hall.

Gujarati, Damodar N. (2003). *Basic Econometrics*, 4th ed. New York: McGraw-Hill Company.

Hagen, Daniel A., James W. Vincent, and Patrick G. Welle. (1992). "Benefits of Preserving Old-Growth Forests and the Spotted Owl." *Contemporary Policy Issues* 10: 13–26.

Halstead, John M., A. E. Luloff, and Thomas H. Stevens. (1992). "Protest Bidders in Contingent Valuation." *Northeastern Journal of Agriculture and Resource Economics* 21(2): 160–169.

Hanemann, W. Michael. (1984). "Welfare Evaluations in Contingent Valuation Experiments with Discrete Responses." *American Journal of Agricultural Economics* August: 332–341.

——. (1989). "Welfare Evaluations in Contingent Valuation Experiments with Discrete Response Date: Reply." *American Journal of Agricultural Economics* June: 1057–1061.

Jorgensen, Bradley S., Geoffrey J. Syme, Brian J. Bishop, and Blair E. Nancarrow. (1999). "Protest Responses in Contingent Valuation." *Environmental and Resource Economics* 14(1): 131–150.

Lenski, Gerhard E., and John C. Leggett. (1960). "Caste, Class, and Deference in the Research Interview." *American Journal of Sociology* 65(5): 463–467.

Loomis, John B., Lucas S. Bair, and Armando Gonzalez-Caban. (2002). "Language-Related Differences in a Contingent Valuation Study: English Versus Spanish." *American Journal of Agricultural Economics* 84(4): 1091–1102.

Loomis, John B., and Armando Gonzalez-Caban. (1998). "A Willingness-to-Pay Function for Protecting Acres of Spotted Owl Habitat from Fire." *Ecological Economics* 25: 315–322.

Loomis, John B., and Richard G. Walsh. (1997). *Recreation Economic Decisions: Comparing Benefits and Costs,* 2nd ed. State College, PA: Venture Publishing.

Lotade-Manje, Justus. (2002). "Consumer Preferences for Fair Trade Labeled

Products." Unpublished master's thesis, Colorado State University, Fort Collins, CO.

Maddala, G. S. (1996). *Limited Dependent and Qualitative Variables in Econometrics.* New York: Cambridge University Press.

Mitchell, Robert Cameron, and Richard T. Carson. (1989). *Using Surveys to Value Public Goods: The Contingent Valuation Method.* Washington, DC: Resources for the Future.

Nash, J. Madeline. (2003). "Fireproofing the Forests." *Time* 162(7): 52–56.

Park, Timothy, John B. Loomis, and Michael Creel. (1991). "Confidence Intervals for Evaluating Benefits Estimates from Dichotomous Choice Contingent Valuation Studies." *Land Economics* 67(1): 64–73.

Reese, Stephen D., Wayne A. Danielson, Pamela J. Shoemaker, Tsan-Kuo Chang, and Huei-Ling Hsu. (1986). "Ethnicity-of-Interviewer Effects Among Mexican-Americans and Anglos." *Public Opinion Quarterly* 50(4): 563–572.

Schelhas, John. (2002). "Race, Ethnicity, and Natural Resource in the United States: A Review." *Natural Resources Journal* 42: 723–763.

Smedley, Audrey. (1993). *Race in North America: Origin and Evolution of a Worldview.* Boulder, CO: Westview.

Turner, Charles F., Susan M. Rogers, Tabitha P. Hendershot, Heather G. Miller, and Jutta P. Thornberry. (1996). "Improving Representation of Linguistic Minorities in Hearth Surveys." *Public Health Reports* 111(3): 276–279.

U.S. Census Bureau. (2000). *Ability to Speak English: 2000.* Available at http://factfinder.census.gov.

VanRensburg, Tom M., Greig A. Mill, Mick Common, and Jon Lovett. (2002). "Preferences and Multiple Use Forest Management." *Ecological Economics* 43(2–3): 231–244.

Weeks, Michael F., and R. Paul Moore. (1981). "Ethnicity-of-Interviewer Effects on Ethnic Respondents." *Public Opinion Quarterly* 45(2): 245–249.

Welch, Susan, John Comer, and Michael Steinman. (1973). "Interviewing in a Mexican-American Community: An Investigation of Some Potential Sources of Response Bias." *Public Opinion Quarterly* 37(1): 115–126.

White House. (2003). *Press Release.* Available at http://www.whitehouse.gov/news/releases/2003/05/print/20030520-1.html.

The Resource Economics of Grover Pease Osborne

Author of America's First Textbook on Resource Economics

By GERALD F. VAUGHN*

ABSTRACT. Grover Pease Osborne was the author of the first American textbook on resource economics. This field started during the Progressive Period in the United States (from 1890 to the 1920s). Osborne's pioneering work encourages us to ask what sorts of public policies towards the management of non-renewable resources are in the best interests of the nation.

Grover Pease Osborne's *Principles of Economics: The Satisfaction of Human Wants in so Far as Their Satisfaction Depends on Material Resources* (1893) was the first American textbook on resource economics available in the early years of our nation's initial conservation movement from 1890 to 1920. Osborne's remained the only relevant textbook for nearly a decade.

Osborne intended his *Principles of Economics*, I assume, as a general economics textbook. However, his book addressed more thoroughly, and with more acute insight, than other economics textbooks the problem of whether the American economic system will always make wise use of its natural resources. Noting the book's uniqueness, Bernhard E. Fernow, a renowned forester and leader in America's first conservation movement, remarked: "The economics of natural resources have received only incidental and scanty consideration by English writers. The only publication known to the writer which

*Professor Gerald F. Vaughn is Emeritus from the University of Delaware. He has reviewed books for this journal and contributed several pieces to the *Journal of Economic Issues* and *Land Economics* over the years. His specialties include economic and environmental history and biography.

American Journal of Economics and Sociology, Vol. 65, No. 3 (July, 2006).

discusses the subject in a broad manner is by G. P. Osborne, 'Principles of Economics'" (Fernow 1902: 415).

Fernow knew the relevant economics literature. Richard T. Ely recalled: "In the year 1896 Dr. B. E. Fernow gave a course of lectures on the economic aspects of forestry under the auspices of the Department of Political Economy in the University of Wisconsin, said to have been the first course of forestry lectures in the United States given within a department of political economy. In these lectures he advocated conservation measures and these lectures were published in 1902 under the title, 'Economics of Forestry'" (Ely, Hess, Leith, and Carver 1917: 17). When Fernow's lectures were published as the first American textbook on forest economics, Ely edited the book.

Ely knew the relevant economics literature even better than Fernow. He knew that economists such as Henry C. Carey, Henry George, Simon Newcomb, and himself, along with philosophical anarchists such as Joshua King Ingalls and Henry David Thoreau, had written on various aspects of natural resource use and conservation. Ely was in the position to have corrected Fernow's plaudit to Osborne's *Principles of Economics* if overstated. Though a member of the American Economic Association, Osborne (1847–1932) was an obscure economist. More accurately, I should indicate he was not a professional economist, but instead a Baptist clergyman and editor in Cincinnati. Obviously, Ely did not consider Fernow's plaudit to Osborne undeserved.

Had Osborne titled his book something like *Resource Economics and Policy*, his audience and market in 1893 would have been limited to Ely, Fernow, and very few others. America's first conservation movement began more than a century ago without an economic rationale. The conservation movement of 1890–1920 was led by natural scientists and engineers who were concerned about the physical waste of natural resources. They didn't look to economists for aid and would have gotten little or none if they had. Even the Michigan Department of Conservation's Land Economic Survey, started in 1922 after the conservation movement ended, was predominantly a physical inventory of land resources, although containing valuable economic elements. There were no economic land classification systems until Cornell University's system in the 1930s became the first that can be truly regarded as economic land classification.

Gerald Alonzo Smith indicates that prior to about 1910, the height of the first conservation movement: "Most economists did not (and do not) think that natural resources were unique factors of production but that they were similar to other factors and did not warrant special treatment in their analysis" (Smith 1979: 115). At Cornell, George F. Warren's research, writings, and intellectual development took him steadily further in the direction of agricultural economics. In 1918, a breakthrough came at the annual meeting of the American Farm Management Association held in Baltimore: Warren stated that he like many others previously had considered farm management and agricultural economics to be separate fields with little in common, but now he concluded the two fields were inextricably related and should not exist separately. This cleared the way for the merger in 1919 of the American Farm Management Association with the Association of Agricultural Economists, which created the American Farm Economic Association (today known as the American Agricultural Economics Association).

Harold J. Barnett and Chandler Morse observe: "In the vast Conservation literature of the period 1890–1920, there is no rigorous economic analysis of what is natural resource economic scarcity. Nor is there analysis of nineteenth-century economic history which, in a scholarly way, identifies and measures natural resource economic scarcity and its economic effects" (Barnett and Morse 1963: 75).

I know of a single important exception, Osborne's seminal work, which is both singular and important for many reasons, not least of which is that it anticipates by four decades Erich W. Zimmermann's technology-based concept of natural resource availability. Reviews of Osborne's book were scanty, mixed, and generally reflected each reviewer's lack of understanding of what must be acknowledged as unorthodox economics. The book seems to have been largely used as a textbook at Baptist colleges, where Osborne's reputation as a noted Baptist clergyman and editor were well known. Osborne's pioneering work was overlooked or ignored elsewhere.

Now, after more than a century has passed, new examination of Osborne's work reveals a hidden jewel in the economics literature.

Neither a defender of laissez-faire capitalism nor an advocate of socialism, Osborne rather took an institutionalist approach to analysis of America's land question, observing: "Usually he who has possession of the land may be trusted to make the best use of it, or at least some use which will be for the interests of Society. . . . But there are some exceptions. . . . If the use made of land by its holder is inconsistent with public policy, or if it is not used for the satisfaction of wants, the public have the right to interfere" (Osborne 1893: 236).

Let's look inside Osborne's book and examine the technological means that, he argued, would assure the continuing availability of productive natural resources. While Fernow and Ely could see the book's relevance during the conservation movement, I doubt that even they could foresee the eventual significance of this line of thought. Neither Fernow nor Ely knew there would be a theorist such as Erich W. Zimmermann.

I

The Content of Osborne's Book

OSBORNE'S ECONOMIC PRINCIPLES seem eclectic in origin: "I have sought to make an original investigation on what I believe to be the only true line of study; but truths once discovered, whether by English or German writers, must be used by all who come after them" (Osborne 1893: 3). He wrote his book at the time when classical economics was on its way out and neither neoclassical nor institutional economics were well established as replacements. Classical economics took Osborne only so far, and after that he was mostly on his own.

Osborne seems influenced by Cairnes's illustration of how the Malthusian tendency of population to increase at a geometric rate can be offset by opposing tendencies: "Mr. J. E. Cairnes has illustrated the truth by centrifugal and centripetal forces: the tendency of the attraction of gravitation is to draw a planet to the sun, while the tendency of its momentum is to fly off in a straight line; the result of the combined action is to cause the planet to move in an ellipse around the attracting body. We should make little progress in science if we did not understand and measure both these forces" (Osborne 1893: 120).

Beyond that influence, Osborne's book reflects far more his own perceptive thought about economic and social institutions. He presents a dynamic economics, premised on the forces of institutional change.

Following a preface and introduction, Osborne's book has six main divisions: (1) The Resources for the Satisfaction of Wants; (2) Population—The Number of People Whose Wants Are to Be Satisfied; (3) Ownership and Control of the Resources for the Satisfaction of Wants; (4) Economical Uses of the Resources; (5) Exchange; and (6) Distribution of Produced Wealth; plus an index.

The book's unique strengths are what Osborne thought and wrote about the institutional economics of natural resources in the first 268 of the book's 454 pages. Most significantly, Osborne's work appears to anticipate by four decades Erich W. Zimmermann's revolutionary *World Resources and Industries* (1933, rev. ed. 1951). While Osborne's principles were neither well developed nor cohesive in all respects, they nonetheless contain the key elements of Zimmermann's technology-based concept of natural resource availability.

In Division 1, "The Resources for the Satisfaction of Wants," Osborne defined resources thusly: "Utility is the quality of an object which makes it useful, or fits it to satisfy human wants. . . . All Resources for the Satisfaction of Wants possess utility. It is the possession of this quality that makes them resources. . . . Strictly speaking, the Utilities of the Resources of Nature can seldom be increased. They are as they were created, and nearly all additions to them are classed as Produced Wealth. . . . The utility of Produced Wealth, on the contrary, can be very greatly increased. Indeed, this is the main object of labor" (1893: 49, 51, 52).

By "Produced Wealth," Osborne meant technological progress. He says: "Produced Wealth is the Resources Produced by Human Industry" (1893: 45). And how does technological progress occur? Osborne says: "The knowledge of Nature's resources and laws, as discovered by physical and chemical research, and the practical spirit and power of invention which enable the men of the present day to make use of the idle forces of the past,—to use steam and electricity where past ages employed human muscle—these make up the greater part of the resource of Labor" (1893: 35). Osborne had some sense of the key role of energy in technological progress and natural resource availability.

In his book's second division, "Population—the Number of People Whose Wants Are to Be Satisfied," Osborne observed: "We realize now, as our fathers did not, something of the magnitude of the power stored up in nature which belongs to man to use. There is no reason to believe that discoveries on this line are at an end" (1893: 68). He realized: "When it is said, therefore, that the Resources of Nature can not be increased, it does not follow that provision may not be made for supplying the wants of more people. Natural Resources may be discovered, and we can never tell when we have completed our discoveries" (1893: 69). He understood that the "limits of diminishing returns may be reached in one country, but not in another; in one State, but not in the entire United States" (1893: 82).

Since education is essential to technological progress, Osborne envisioned a society in which "every boy and girl shall receive the best training and education that can be had from the most competent instructors" and "the enormous power of invention which would follow so general education, by means of which the Resources of Nature would be used to better advantage" (1893: 89–90). He saw the potential of technological innovations as instrumental means by which to produce greater and/or new utility: "With the progress of invention, vast wealth is profitable in the form of machinery, and the importance of Produced Wealth is greater than ever before" (1893: 99).

Osborne saw that the upper limit to population would be determined by the extent to which technology could in effect create natural resources: "It is evident that population can not increase beyond a certain ratio to the Produced Wealth of the land" (1893: 99). To Osborne, the economic returns to technology could stimulate technological progress, offset the threat of overpopulation, and actually expand the limits of population growth. Osborne could see, if not fully comprehend as Zimmermann later would, the possibilities of technology-based natural resource availability.

The book's third division focuses sharply on "Ownership and Control of the Resources for the Satisfaction of Wants," providing the institutional framework for his fourth division, "Economical Uses of the Resources." Osborne argued "that the use of land belongs to all the people alike, and that there can be no absolute ownership by any one" (1893: 169). He held that: "The man born a thousand years hence will have the same right to a share of the earth's surface as those now living" (1893: 169). He drew his ethical view of land use and ownership from the Land Laws of Moses in the Bible (see Leviticus 25, particularly verse 23 where God declares "the land is mine") (Osborne 1893: 168–169).

Combining his economic and ethical principles, Osborne reached highly normative conclusions. Because U.S. land belongs to all the people alike, he said: "If no private titles to land had been given, and we were at the beginning of a new nation, when all the land is 'government land,' the proper method [of land tenure] would be simple. The natural way of managing the land of the people is to rent it under a perpetual lease, with a revaluation of the rent every few years" (1893: 173). He felt that this leasing system, if immediately adopted for all unsold government land, would curb many of the abuses of previous disposition of public lands, thereby ending rampant land speculation and assuring more orderly settlement of the American West (1893: 176–177). Osborne, like Henry George before him, sought to capture economic rent for society.

In a cogent statement regarding disposition of the United States' remaining public domain, Osborne argued: "Not another acre should go into private ownership. No matter how poor or worthless at present, it belongs to the people; and there is no excuse for deeding away the people's heritage. . . . All public land should, hereafter, be leased on the condition that the occupant pays whatever it is worth each year" (1893: 179–180). In his book *The United States Oil Policy*, the great resource economist John Ise felt Osborne "showed an unusual grasp of the public land question, although he did not refer specifically to oil" (Ise 1926: 500). Osborne comes to discuss oil later.

In the fourth division, "Economical Uses of the Resources," Osborne says "modern methods of using the Resources" have contributed to better satisfaction of wants (Osborne 1893: 201). Moreover, "[n]ot only are wants better satisfied, but the world is able to provide for far larger numbers of people" (1893: 201–202). In using resources, he recognized not only economic efficiency but also, quite clearly, equity considerations: "Methods of using the Resources also determine to some extent whose wants shall be satisfied, irrespective of the question of the ownership of property" (1893: 202).

Osborne attached equal importance to the interests of the present population and those of people hundreds of years into the future. Thus, he was concerned about both present and future use of natural resources, and he was thinking about substitutes for resources that could be depleted. He knew that "we have discovered the reservoirs of petroleum beneath the surface of the earth, and the rock gas which, so long as it may last, takes the place of coal" (1893: 229). However, uncertain that oil and gas would prove to be enduring substitutes for coal, he wrote, "some substitute for coal may be discovered, but no one knows that it will be discovered. . . . All this is no reason for suffering the coal to remain unused—our interests are just as important as those of the people who shall come after us—but it is a reason for insisting that all such Resources of Nature should be used without waste, and the prevention of waste should be enforced by government authority" (1893: 231–232).

In achieving better satisfaction of wants, Osborne placed considerable emphasis upon governmental management of natural resources: "Neither ownership of land in fee, nor a perpetual lease to private persons at a variable rent, should prevent the general control of the Resources of Nature by Society" (1893: 235). Further, "the Resources of Nature are of right public property—the heritage of all the people, of each generation as it comes on the stage of action. Hence, it is impossible to treat of the use of Natural Wealth without more or less discussion regarding government management, since the government is the trustee of the people's property, and their agent for its use" (1893: 241). Ely and Fernow would agree.

To Osborne, wise use of natural resources meant producing not merely consumable wealth but, more importantly, accumulated permanent wealth. He predicted that: "[f]uture accumulations of Produced Wealth will be in the form of Permanent Wealth, both that which satisfies wants directly—such as dwellings, public buildings, parks, works of art, etc.—and that which satisfies wants indirectly—such as tools and machinery, roads and facilities for communication and transportation, improvements on land, etc." (1893: 267). Osborne's usage of "Permanent Wealth" corresponds generally to human-created capital in today's terminology, which Stephen

Viederman refers to as "products and technologies created by humans, including the built environment" (Viederman 1996: 47).

Osborne was saying that society's general control of natural resources, augmented by technological innovations, would assure the continuing availability of productive natural resources. Yet despite his book's strengths, readers didn't know what to make of it, judging from the reviews I have read. While offering a well-thought-out and engagingly written economic rationale that could have better informed the conservation movement, the book went largely unheeded. It came at a time when economics literature, as assessed by Arthur T. Hadley, "treated wealth as a matter of study by itself, apart from the history of institutions on the one hand, or the theory of motives on the other" (Hadley 1894: 251). Osborne dealt with all three, integrating institutional economics and social psychology, emerging approaches to inquiry for which his audience was unprepared.

Even as enlightened and progressive an economist as Ely was not fully prepared for Osborne's thinking. With foresight, Ely had devoted some pages to improving forestry in *An Introduction to Political Economy* (1889). However, regarding the overarching land question, he also had written "that important as this question is, the amount of land in proportion to our needs is still large, and it is a problem of tomorrow rather than of today" (Ely 1890: 112). The natural scientists and engineers who led the first conservation movement were of course much less prepared; among them, Fernow alone seemed to understand and appreciate what Osborne was groping for.

II

Richard T. Ely and Land Economics

EVENTUALLY, IN THE FIRST THIRD of the 20th century, Ely came to understand the land question better than anyone else. Ely's most widely known and enduring contributions to institutionalist thought were in the new field of land economics, which he fathered as a result of his work on the distribution of wealth.

Ely felt a debt a gratitude to Henry George, whose work was

probably the most instrumental in leading to scientific study of land economics in America. Ely wrote: "We need not speak about his devotion to the public good, about his integrity, about his sincerity of purpose; all of these have been abundantly recognized. I think Henry George is to be praised because he has brought forward the land problem as one of paramount importance. I agree with him that its solution is necessary for the salvation of society. It is the great economic problem of the twentieth century" (Ely 1917: 33).

In 1920, Ely founded the Institute for Research in Land Economics and Public Utilities at the University of Wisconsin, in which study of land and public utility problems was based on knowledge of the nature, significance, evolution, and operation of economic institutions and forces. Ely understood that land problems are basic to many economic activities and, therefore, chose as the institute's motto "Under all, the land." Land tenure was a major theme of Ely's early teaching and research, a theme aggressively carried forward by Henry C. Taylor and Benjamin H. Hibbard, who were aided by George S. Wehrwein (working with them first as a student and later as a research associate and faculty member). Wehrwein succeeded Ely as America's leading land economist.

Ely emphasized work on land tenure but was not unduly concerned about land as a monopoly: "Is land a monopoly? It is spoken of as a natural monopoly, but, as a matter of fact, it is far from being a monopoly. If there were unified control over the land, the land owners could starve the rest of the population to death and could absorb all of the wealth of the world, as John Stuart Mill has pointed out. On account of the limitation of the supply of better grades it has a value and this may sometimes be high. All values, however, imply limitation" (Ely 1920: 129). Land-use controls are an increasingly large part of this limitation.

In 1922 Ely, Mary L. Shine, and Wehrwein published a three-volume mimeographed work titled *Outlines of Land Economics.* Volume I dealt with *Characteristics and Classification of Land,* Volume II, *Costs and Income in Land Utilization,* and Volume III, *Land Policies. Outlines of Land Economics* represented the state of the art of the field at that time, and these volumes were a tremendous advance over previous literature in the field.

However, Ely tells about some criticism his land economics research institute received on these volumes: "We were criticized because in our early mimeographed volumes of the *Outlines of Land Economics* we stated certain conclusions. It was said that we should have waited until we had finished our researches and then given our conclusions. If this had been done, no conclusions or generalizations would ever have been reached, because researches must be continuous and never-ending . . . We stand ready to revise conclusions as we get new light. But first of all we must gather the essential facts; then we must look at these facts dispassionately and interpret them objectively. This is the spirit of the Institute's work" (Ely 1938: 239).

To Ely, land economics came to mean:

> the relation of the different kinds of natural resources to the distribution and movement of population, standards of living, industrial development, distribution of wealth, property rights, national policies, and international relations. The ultimate objective of this field of study and research is a group of integrated policies by national, state, and local governments and by business interests for the most effective utilization of all natural resources. (Ely 1938: 240)

Ely's and his associates' work tied in with Osborne's as foundational to the emerging thought of Erich W. Zimmermann regarding a technology-based and international concept of natural resource availability.

III

From Osborne and Ely to E. W. Zimmermann on Technology

Osborne's thinking about technology runs closely parallel to Zimmermann's. Osborne saw the possibilities of a technology-based concept of natural resource availability, by which technological innovations (including the development of substitutes) could overcome institutional resistances and indefinitely expand the wants-satisfying capacity of resources.

Osborne presented the concept's fundamentals in a tentative way. Zimmermann developed the concept into a well-rounded body of theory, with significant public policy implications nationally and internationally. Zimmermann exhaustively studied the availability of the world's agricultural and industrial materials that were functioning, or conceivably might function, as resources. Successive generations of economists, geographers, and conservationists have been appropriately indebted to Zimmermann.

Zimmermann defined "resourceship" (the quality of being a resource) as follows: "Neither the environment as such nor parts or features of the environment per se are resources; they become resources only if, when, and in so far as they are, or are considered to be, capable of serving man's needs" (Zimmermann 1933: 3). He later stated: "Resources are not, they become; they are not static but expand and contract in response to human wants and human actions" (Zimmermann 1951: 15). Zimmermann said a material is not a resource until people know how to use it: "Knowledge is truly the mother of all other resources" (Zimmermann 1951: 10). He argued that a material is not a resource unless it performs a function, or takes part in an operation, to satisfy a human want. To Zimmermann, energy (particularly inanimate energy) became the key to technological progress that creates resources.

Stephen L. McDonald, who studied under Zimmermann from 1947 to 1950 when the revision of *World Resources and Industries* was in preparation, emphasizes that to Zimmermann "natural resources are not fixed in either kind or quantity, but are themselves largely the 'creation' of man and his culture. Coal and oil cannot be understood as 'natural' resources except in the context of the industrial and transportation revolution" (McDonald 1998: 2). Osborne's fundamentals acquire greater clarity and meaning in the light of Zimmermann's application of the concept.

At a time when few professional economists regarded resource economics and policy as central to the study of economics, an Ohio clergyman and editor did. Grover Pease Osborne's *Principles of Economics* (1893) emerges as a seminal American treatise in resource economics and policy. His institutionalist perspective on wise use of natural resources, via society's general control and augmented by technological innovations, provided some of the first principles of this field of study. Osborne's basic treatment was unsurpassed until the abler and fuller treatments by Richard T. Ely and colleagues began to appear two decades later.

Osborne keenly understood the existing economic order and was prescient about its future tendencies, especially the potential of technology to increase natural resource availability. He inclined in the same direction as Zimmermann who, with a firmer grasp of the totality of economic science, eventually produced the more definitive, valuable, and enduring work. In the best exposition of Zimmermann, McDonald writes: "His explanation of the dynamics of resourceship (the quality of being a resource), with technological progress playing the active role, is today even more relevant to the problems of economic development and the management of depletable (or degradable) resources than in his time" (McDonald 1995: 152). This makes Osborne's much earlier inclination in the same direction all the more remarkable.

IV

Public Policy Implications

THE GENERAL PROBLEM of the economics of natural resources is as old as civilization. Ancient Greek literature contains a treatise that deals with how to best organize the mining of a steady flow of silver to increase the revenues of the City of Athens. But the modern theory of the economics of exhaustible resources began to take shape only in 1893, when Grover Pease Osborne authored his *Principles of Economics: The Satisfaction of Human Wants in so Far as Their Satisfaction Depends on Material Resources.*

I commend Osborne's work, especially for using an early form of general systems theory as his framework for analysis and solution of the problem of how to adjust to petroleum scarcity. The systems approach is an extraordinarily powerful tool, but difficult to apply. Osborne's usage has application even today.

Lack of economic freedom is the core of the problem that Osborne analyzes. The United States is not yet free to adopt the integrative technologies he advocates, partly due to economic reasons but ultimately due to geopolitical constraints. Let me briefly comment on a few of these institutional factors limiting our freedom to act and then suggest a scenario for institutional change.

A key premise underlying Osborne's public policy prescription is that oil is going to become more scarce. He didn't say how soon. Some supporters of the need for action insist that the imminence of the threat of oil shortages is irrelevant. They argue that since we know we'll run out of oil in the long run, we should start to shift to alternatives now. Unfortunately, redirection in major public policy seldom occurs until we are in a crisis. As long as oil shortages are perceived as only a distant threat, it is unrealistic to expect much of a shift away from petroleum.

Our nation's leaders aren't convinced that oil will be scarce anytime soon. They believe we can count on sizeable foreign supplies plus our own domestic production for many years and, thus, we have plenty of time to make any needed transition away from petroleum. After the repeated oil crises of the 1970s, some cautioned that the worldwide production of oil—and possibly of natural gas—would pass its peak by the year 2000. Others held that much of the world's oil and gas has not yet been found. If further exploration shows this assumption to be true, more time will, of course, be available for making the transition to other sources of energy.

During recent decades, significant advances in technology, such as three-dimensional and computer enhanced seismography, improved drill bits, horizontal drilling, and new offshore drilling platforms that permit drilling in up to 2,500 feet of water, have been made. Investment in development almost always increases the recoverable portion of oil and gas, creates additional capacity, and increases the feasible rate of extraction.

Unfortunately, the areas with high potential for exploration tend to be outside the United States. A decade ago, the major U.S. oil companies were decreasing their operations here and increasing operations in other parts of the world, planning to shift up to two-thirds of exploration and production to other countries.

Foreign oil has been underpriced, and this underpricing is a serious problem because it confuses the American people. As long as foreign oil was priced at $18–20/barrel delivered to the United States, gasoline prices stayed affordable and Americans perceived no threat to oil supplies. Our citizens don't understand that the underpricing of foreign oil destroys the incentive for U.S. oil companies to explore and extract within our boundaries. The result is a continuing rise in our dependence on oil from foreign sources. A sizeable portion of our oil supply will continue to be imported in the relatively near future, meaning heavy dependence on other countries to supply one of our most vital and strategic raw materials.

If as a nation we are concerned that such high dependence is far too risky, particularly in the event of war, the only way to reduce our dependence is to increase the price of imported oil to a price reflecting its full cost to us. During the period 1973–1985, when real oil prices rose 174 percent, our oil imports in relation to gross national product fell 40 percent.

As a long-term policy over many years, probably decades, a fixed import fee could stimulate greater domestic oil exploration and production, cover our defense expenditures in oil-producing regions, encourage energy conservation, increase energy efficiency, and induce some substitution of alternative energy sources (including sources more protective of the environment).

However, there is a formidable geopolitical constraint. Our imposing such a fixed import fee presumably would further upset the leaders of the oil-exporting countries of the Middle East, and we might lose more than we gain in terms of our overall national security interests in that volatile region. Also, the leaders of Mexico and Canada, two other major oil suppliers to the United States, likely would protest that the fixed import fee runs counter to the intent of the North American Free Trade Agreement. The United States has resisted imposing a fixed import fee because of geopolitical considerations.

Therefore, it seems we face not so much a question of technology policies as a question of geopolitics. I leave it to those of you who are more knowledgable than I am about geopolitics to help us find a way out of this dilemma.

I'll simply mention three options to a fixed import fee that might be beneficial to the United States while less antagonistic to our suppliers of imported oil: (1) a variable import fee, increasing when the world oil price falls and decreasing when the world oil price rises, (2) auctioned import quotas, or (3) a substantial tax on consumer purchases of gasoline. Of the three, the third option (a substantial consumer tax) probably would cause fewest geopolitical repercussions, but an uproar within our nation. Our nation's leaders face a difficult but inevitable choice.

The present balance of supply and demand for oil may remain relatively stable during the lifetimes of our elders. But young adults and their children and grandchildren may not be so fortunate. On behalf of future generations, we should aggressively pursue technological alternatives such as Osborne foresaw long ago. Substantially increasing the price of imported oil has to be the first step, probably via a higher consumer tax, and the sooner the better.

References

Barnett, Harold J., and Chandler, Morse. (1963). *Scarcity and Growth: The Economics of Natural Resource Availability.* Baltimore: Johns Hopkins University Press.

Ely, Richard T. (1890). *Problems of Today: A Discussion of Protective Tariffs, Taxation, and Monopolies,* 3rd ed. New York: Thomas Y. Crowell.

——. (1917). "Landed Property as an Economic Concept and as a Field of Research." *Papers and Proceedings of the American Economic Association, 4th Series* 7: 18–33.

——. (1920). "Land Speculation." *Journal of Farm Economics* 2(3): 121–135.

——. (1938). *Ground Under Our Feet.* New York: Macmillan.

Ely, Richard T., Ralph H. Hess, Charles K. Leith, and Thomas N. Carver. (1917). *The Foundations of National Prosperity: Studies in the Conservation of Permanent National Resources.* New York: Macmillan.

Ely, Richard T., Mary L. Shine, and George S. Wehrwein. (1922). *Outlines of Land Economics* (3 vols). Ann Arbor, MI: Edwards Brothers.

Fernow, Bernhard E. ([1902] 1972). *Economics of Forestry.* New York: Thomas Y. Crowell.

Hadley, Arthur T. (1894). "Recent Tendencies in Economic Literature." *Yale Review* 3: 251–260.

Ise, John. (1926). *The United States Oil Policy.* New Haven, CT: Yale University Press.

McDonald, Stephen L. (1995). "Erich W. Zimmermann, the Dynamics of Resourceship." In *Economic Mavericks: The Texas Institutionalists.* Ed. R. J. Phillips. Greenwich, CT: JAI Press.

——. (1998). Letter to the author, April 13.

Osborne, Grover Pease. (1893). *Principles of Economics: The Satisfaction of Human Wants, in so Far as Their Satisfaction Depends on Material Resources.* Cincinnati, OH: Robert Clarke and Company.

Smith, Gerald Alonzo. (1979). "Epilogue: Malthusian Concern from 1800 to 1962." In *The Morality of Scarcity: Limited Resources and Social Policy.* Ed. W. M. Finnin, Jr., and G. A. Smith. Baton Rouge: Louisiana State University Press.

Viederman, Stephen. (1996). "Sustainability's Five Capitals and Three Pillars." In *Building Sustainable Societies: A Blueprint for a Post-Industrial World.* Ed. D. C. Pirages. Armont, NY: M. E. Sharpe.

Zimmermann, Erich W. (1933). *World Resources and Industries: A Functional Appraisal of the Availability of Agricultural and Industrial Resources.* New York: Harper and Brothers.

——. (1951). *World Resources and Industries: A Functional Appraisal of the Availability of Agricultural and Industrial Materials,* rev. ed. New York: Harper and Brothers.

Escaping the Resource Curse and the Dutch Disease?

When and Why Norway Caught Up with and Forged Ahead of Its Neighbors

By ERLING RØED LARSEN*

ABSTRACT. In the 1960s, Norway lagged behind its Scandinavian neighbors in the aggregate value of economic production per capita, as it had for decades. By the 1990s, Norway had caught up with and forged ahead of Denmark and Sweden. When and why did Norway catch up? The discovery and extraction of oil in the early 1970s is usually suggested as the explanation. But oil alone cannot explain Norway's growth, since Sachs and Warner (2001) show that resource gifts often reverse growth, making oil a curse, not a blessing. Moreover, there is the possibility of contracting the Dutch Disease, which involves a rapid and substantial contraction of the traded goods sector. This article explains how deliberate macroeconomic policy, the arrangement of political and economic institutions, a strong judicial system, and social norms contributed to let Norway escape the Resource Curse and the Dutch Disease for more than two decades. Intriguingly, it appears that Norway in the late 1990s may show some symptoms: Norway has experienced reversed relative growth compared to Denmark and Sweden and a contraction of industrial activity. This article explores the political economy behind this recent slowdown.

*The author is Research Fellow at the Research Department, Statistics Norway, P. O. Box 8131 Dep., N-0033 Oslo, Norway; e-mail: erling.roed.larsen@ssb.no. The author grateful to insightful conversations with and comments on an earlier version of the paper from Ådne Cappelen, valuable suggestions from Hilde C. Bjørnland, Clair Brown, Dag Kolsrud, Knut Einar Rosendahl, Thor-Olav Thoresen, and Wei-Kang Wong, ideas from Ian McLean, and grants from the Norwegian Research Council, Project No. 149107/730. The author also thanks seminar participants at the Department of Economics, University of California, Berkeley; Department of Economics, the Norwegian Business School BI; and the Research Department, Statistics Norway. He shares merits with all of the above. Errors remain, of course, his sole responsibility.

American Journal of Economics and Sociology, Vol. 65, No. 3 (July, 2006).

I

Introduction

In the recent economic history of Scandinavia, one phenomenon is striking: the ascent of Norwegian gross domestic product (GDP) per capita compared to those of its neighbors Denmark and Sweden. Until the early 1970s, Norway had trailed its neighbors economically. For decades, or even centuries, Norway had been the poorest country of the three. However, by the turn of the millennium, Norway enjoyed the largest GDP per capita in Scandinavia, as is shown in Figure 1. This turn of events is conventionally and plausibly attributed to Norway's oil discovery in 1969 and subsequent extraction of oil starting in 1971. Today, Norway is one of the world's largest oil exporters. However, pointing to oil revenues is insufficient in explaining Norway's growth, since Sachs and Warner (1999, 2001) and others have documented the detrimental effects of newfound riches on economic growth. This counterintuitive effect is called the Resource Curse. It involves a surprising, negative relationship between resource

Figure 1

GDP per Capita, Scandinavia, 1960–2002, 1999 USD, PPP

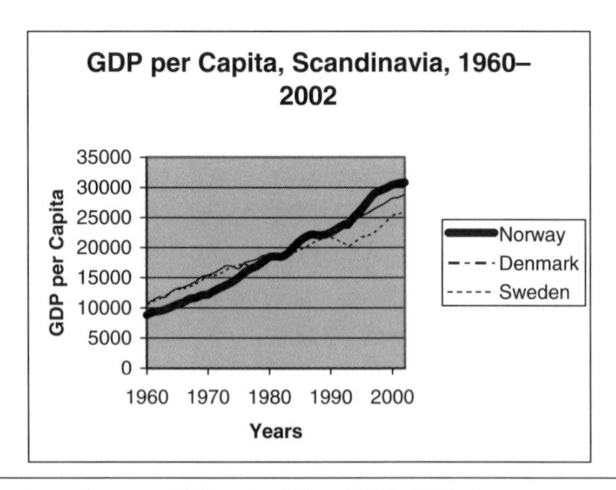

Note: Data from Bureau of Labor Statistics (2003), Table 1.

wealth and economic growth. Many resource-rich countries seem to suffer from it; only a few do not. Moreover, countries that are rich in resources may suffer from the Dutch Disease, an economic illness that involves factor movement, excess demand, and loss of positive externalities. The result of the disease is that the *traded goods sector*, which is exposed to foreign competition abroad or domestically, shrinks or disappears. In Norway, it did not. In other words, Norway's management of its oil wealth is no trivial feat. Based on the evidence of how other resource-rich countries have fared, Norway's growth is surprisingly successful, in fact. Stevens (2003) offers a survey of the literature and shows that there are only few exceptions to the curse and the disease, none of which are fully understood. This article focuses attention on how Norway did it and attempts to explain why Norway (for so long) avoided the curse and the disease. Then I go on to examine a twist in the tale: Norway's relative slowdown late in the 1990s. Norway possibly caught, presumably mild, strains of the illnesses. I examine why.

In order to explain how Norway succeeded in its oil management, this article first establishes the fact. I do this by comparing Norway to its two neighbors, Denmark and Sweden, and claim that these two neighbors illuminate the counterfactual rate of growth Norway would have followed had it not found oil. Simple first-order autoregressive regression models of GDP per capita time trends for each of the three countries show that Norway reached parity in the 1980s, some 10 years after oil extraction began. In addition, when we know, as is shown in Røed Larsen (2005), that the relative acceleration started *after* oil extraction began, oil may be thought of as an important element in Norwegian growth. I then proceed to discuss the policies Norway implemented.

I am, as the reader should be, aware of the lack of readily applicable frameworks in which to understand growth in general and the curse and disease in particular. My attempted answers, or sketches of answers, must by necessity involve some element of guesswork. Moreover, I will have to rely on several angles of evidence and substantiation, and sometimes even challenge parts of conventional economics.

What are the curse and the disease? While both presumably arise from resource riches, they take on different forms. The Resource Curse can be thought of as the phenomenon that resource-rich countries tend to grow more slowly—using aggregate output per capita as measure—than similar, not resource-rich countries do. The Dutch Disease, on the other hand, is a term most scholars use for the phenomenon that resource exports lead to a rapid contraction of the non-resource traded goods sector. Although research tends to concentrate effort on either the curse or the disease, I shall look at the two phenomena together. I do this because even though there is consensus that the two problems represent different aspects of resource wealth, they share the same origin: resource riches.

Many authors (e.g., Auty 2001a; Gylfason 2001a; Sachs and Warner 2001; Stevens 2003; Torvik 2002) point toward rent seeking and corruption as core elements of what causes the curse. The Dutch Disease, on the other hand, goes together with other mechanisms and is thought to be caused by something else. The literature has not reached a consensus on the nature of causes and symptoms, and Stevens shows the different aspects on Dutch Disease as they are laid out in Corden and Neary (1982), Corden (1984), Gylfason, Herbertson, and Zoega (1997), and Torvik (2001). Bjørnland (1998) and Brunstad and Dyrstad (1997) are recent Norwegian contributions on the Dutch Disease topic. There are other influential contributions; let me mention Chatterji and Price (1988), Hutchison (1994), and van Wijnbergen (1984). Instead of focusing attention on the differences, let me identify the essence of this large and growing literature: the Dutch Disease is intricately linked to a factor movement effect, a spending effect, and a spillover-loss effect. The factor movement effect is the reallocation of factors of production such as capital and labor from other activities to resource extraction. The spending effect arises from the increased aggregate demand created by resource receipts, which if converted to domestic currency may create periods of excess demand in the economy. The spillover-loss effect lies in the loss of positive externalities associated with the (crowded-out) non-oil traded goods sector.

Thus, an observer who wants to explain how Norway escaped the curse and the disease must argue how Norway minimized rent-seeking activities, limited the factor movement effect, curbed the spending effect, and dealt with the spillover-loss effect. However, the argument cannot be as clear-cut as we would like it to be. For example, Stevens (2003) points out that it is often not the question of what policies were followed, but why these policies were allowed and implemented at all. In every resource-rich country there exist coalitions that will oppose these policies. That allowance points toward complex political and economic interactions between institutions, special interest groups, and the electorate. In addition, it invokes social norms, the effect of which is notoriously hard to document.

I will make the case that the factor movement effect was dampened through income coordination. A highly centralized wage formation system made it possible to make the manufacturing sector the wage leader. This made it possible to limit wage increases to all sectors from an expanding resource sector. The spending effect was curbed because the government shielded the economy by fiscal discipline and investing abroad. The spillover-loss effect was small because losses were substituted for by gains in the highly technological offshore oil extraction sector, which requires more capital than on-land oil extraction. In addition, industrial policy sought to stimulate learning-by-doing and maintain diverse industrial activities. Social norms, transparent democracy, proper monitoring, an effective judicial system, and the wage negotiation system reduced the frequency of rent seeking. This article examines four factors of rent seeking and distribution conflicts: large-scale conflicts, small-scale illegal rent seeking, small-scale legal rent seeking, and political purchase of power through election promises. I discuss Norway's record on each.

Let me describe the structure of the article. The next section discusses the curse, the disease, and the interactions between the two. It also explains why I use Denmark and Sweden to sketch the counterfactual path of Norwegian growth. The third section presents empirical findings on intra-Scandinavian growth. It establishes when Norway reached parity and estimates the countercurse growth rates. The fourth section asks to what extent Norway suffered from the Dutch Disease, and the fifth section explains Norwegian policies to limit the Dutch Disease. The sixth section looks at the remedies to avoid the curse. The seventh section examines the political economy behind the recent slowdown. The final section concludes.

II

The Curse, the Disease, and Scandinavian Performance

ALTHOUGH BOTH THE CURSE AND THE DISEASE originate from sudden resource wealth, they involve different propagation mechanisms of problems. Auty (2001a), Gylfason (2001a), Mikesell (1997), Sachs and Warner (1999, 2001), and Torvik (2001) say that the curse mechanism may be generalized to conflicts over distribution, manifested in rent seeking. On the other hand, the disease may not be a disease at all, but part of a natural development path and actually an expected economic adjustment to new economic circumstances. Surely, if there occurs an exogenous shock to an economy, either in terms of a discovery of a valuable commodity or a discovery of a valuable technology, one would expect the economy to adjust to its new comparative advantage. It would be un-optimal not to utilize the new possibilities. Stevens (2003) discusses the controversy over the precise nature of the disease and the inconclusive evidence for its universal applicability. Van Wijnbergen (1984), for example, asked early on whether the term was a misnomer.

But we may not have to choose between classifying a contraction of the nonresource traded goods sector as either a disease or a natural adjustment. What it is, or turns out to be, depends. If the contraction is rapid and deep, and if there are irreversible losses of knowledge, then such a contraction bodes future problems. If the contraction is small, smooth, and slow, then it may simply signify a highly welcome new position given new parameters. In summary, the curse and the disease take on different shapes, and it is not surprising that many contributions look at them separately.

However, both the literature on the curse and the literature on the disease seem to share common traits. Neither the curse nor the disease is thought of as an inevitable outcome. Both seem attributed to some unwelcome arrangement of institutions or inappropriate policies. In both strains of literature it seems as if proper management can contain the problems. Stevens (2003) sums it up by saying that the problems have something to do with governance.

This article's view is that the governance of curse problems is related to the governance of disease problems. For example, I shall argue below that Norway's highly centralized wage negotiation system and income coordination scheme helped limit factor movement, a cause of the disease, and helped prevent rent seeking, a pathogen of the curse. Moreover, much of the literature, such as Torvik (2001, 2002) and Sachs and Warner (2001), argues that there are positive spillover effects from nonresource (export) manufacturing. The idea is that the activities of acquiring know-how and developing technology feed a virtuous circle that is beneficial to the whole economy. For example, van Wijnbergen (1984) says that it is a stylized fact that technological progress is faster in the traded nonsheltered sectors of the economy. But if that is the case, we would expect positive externalities to create an association between a sizeable nonresource traded goods sector and positive growth on the one hand and an association between a contraction of this sector and negative growth on the other. The latter entails a presence of *both* curse and disease. In other words, the stylized fact plays a role both for the Dutch Disease and the Resource Curse. Thus, they may be considered together because they share common roots. In Table 1, I tabulate the different combinations of the disease and the curse. Since I have argued that they may be linked, we would expect that resource countries are more likely to be positioned along the northwest-southeast diagonal than on the southwest-northeast diagonal. In the following, I shall explain why Norway ended up in the favorable northwest corner while many other resource-rich countries ended up in the highly unfavorable southeast corner.

I investigate whether or not Norway escaped the curse by comparing Norway to Denmark and Sweden. The rationale for this comparison should be clear: Norway's neighbors are intended to function as a control group. The idea is that the aggregate growth in Denmark and Sweden represents annual growth numbers that Norway would have followed had it not found oil. The strength of the idea of using Denmark's and Sweden's growth rates in this fashion depends on the

Table 1

Effects of a Resource Curse and a Dutch Disease

		Resource Curse	
		No	Yes
Dutch Disease	No	Overall growth and diverse export base	Stagnant growth, but diverse export base
	Yes	Overall growth, but strongly contracted manufacturing	Stagnant growth and strongly contracted manufacturing

soundness of premise that Norway would have grown at the same speed its neighbors did.

How sound is this premise? The Scandinavian countries are similar. They share a common history and an almost-identical political, linguistic, and cultural background. It seems as if the Scandinavian countries, historically, have progressed in tandem. Sweden blazed the trail, Denmark followed, and then Norway finally emulated its neighbors. The careful observer would perhaps expect this equality in factors to lead to economic parity. Røed Larsen (2005) examines the possibility that Norway's catch-up was in fact simply a laggard's leap, a convergence unrelated to oil. Similar circumstances should bring similar levels of GDP per capita. But evidence is inconsistent with a catch-up *not* related to oil since there is nothing to support the proposition that Norway started the catch-up in the 1960s. The catch-up started a few years *after* oil extraction started. Thus, it seems as if similar circumstances brought only similar growth rates—until oil happened.

This article claims, then, that it may be reasonable to believe that the relative economic growth in Scandinavia would have continued to follow the historical pattern had oil *not* been discovered off the Norwegian coast in 1969 and that only the enormous value of oil would change this order. In essence, the discovery of oil was one large natural experiment. Denmark and Sweden offer a glimpse into the counterfactual, the rate of growth that Norway would have followed had it not acquired oil receipts. It is Norway's acceleration, and the timing of it, that implies the escape from the curse.

III

Avoiding the Curse, Reaching Parity, and Moving Beyond

BEFORE EXPLAINING WHY Norway could catch up with and surpass Denmark and Sweden, I shall document that it did. To do this, I use a straightforward strategy. I estimate the underlying trend in the observed gross domestic product per capita numbers depicted in Figure 1 by using the standard regression technique. It is useful to think of such regressions as a way of simply compressing large data sets, that is, reducing the dimensionality of observations. More specifically, the idea is that the observed numbers of gross domestic product per capita contain a trend component and a stochastic component. We aim to estimate the trend, or deterministic, component by regressing gross domestic product per capita for each country onto a time variable while allowing for an appropriate stochastic process in the nondeterministic part. Here, I use a first-order autoregressive process. Norway's development is portioned into three periods, following Røed Larsen's (2005) finding that Norway accelerates relative to Denmark and Sweden in the 1970s and decelerates in the 1990s. Denmark's and Sweden's growth paths are modeled as following one period of smooth progression.

Table 2 tabulates the trend regression results. Details of estimation are included in the Appendix. We observe that the trend estimates for both Denmark and Sweden have higher starting points of GDP per capita. They start out in 1960 with trend levels of $10,743 and $10,847 in 1999 U.S. dollars (USD), respectively. Norway starts at $8,722. In the period 1960–1974, Norway grew approximately as fast as its neighbors, as this article proposed. Oil was not yet discovered. In fact, Norway's growth of $389 per capita per year is *between* the Swedish growth of $350 USD per capita per year and the Danish growth of $418 USD per capita per year. However, in the period 1975–1996, Norwegian growth accelerated to $597 (389 + 208) USD

Table 2

Estimation of Time Trends[a] for Norway, Denmark, and Sweden, 1960–2002

Parameter	Parameter Estimates, Norway (*t*-values)	Parameter Estimates, Denmark (*t*-values)	Parameter Estimates, Sweden (*t*-values)
Intercept, α	8,722 (15.9)	10,743 (29.68)	10,847 (17.59)
Time coefficient, β	388.97 (7.35)	417.82 (29.29)	349.81 (15.09)
Extra time coefficient, Norway, 1975–1996	207.91 (2.72)	—	—
Extra time coefficient, Norway, 1997–2002	−13.08 (−0.10)	—	—
Autoregressive factor, ϕ	−0.8382 (−7.70)	−0.7619 (−7.30)	−0.8759 (−11.79)
Number of observations, N	43	43	43
Regression R-squared	0.9614	0.9554	0.8506
Total R-squared	0.9981	0.9959	0.9930

[a]First-order autoregressive linear regression for Denmark and Sweden. First-order autoregressive linear three-step spline regression for Norway.

per capita per year. Using the estimates in Table 2 and inspecting Figure 2, we may compute that Norway was ahead of Sweden in 1981 and ahead of Denmark in 1988. After parity, Norway continued to expand GDP per capita at high speed until the mid-1990s. For the remaining period, 1997–2002, growth estimates are statistically insignificant, but economically important. The negative sign in the estimated −$13 USD per capita per year is evidence of the retardation found in Røed Larsen (2005).

The change in the speed of growth is consistent with an oil-induced acceleration. It made Norway a candidate for the curse and the disease. However, the continued fast growth over two decades is evidence of an escape from the Resource Curse. Norway did not

Figure 2

Trends in GDP per Capita, Norway, Denmark, Sweden, 1960–2002

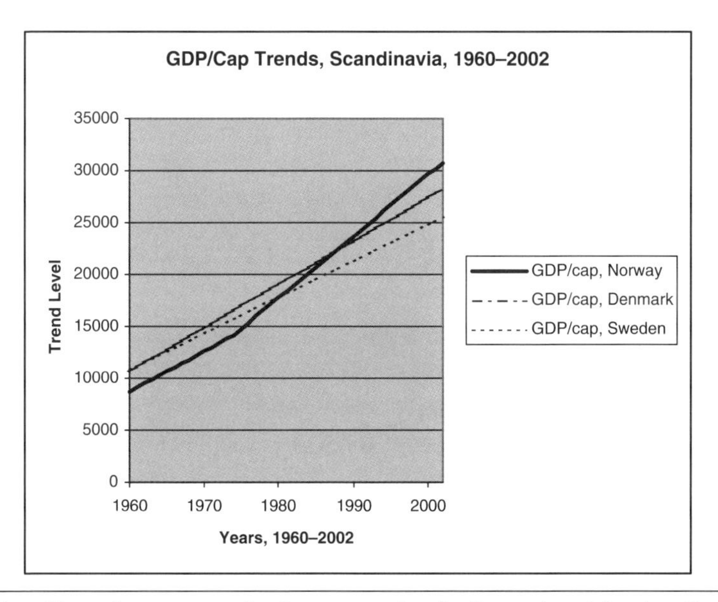

Note: First-order autoregressive linear regression for Denmark and Sweden. First-order autoregressive linear three-step spline regression for Norway. Data from Bureau of Labor Statistics (2003).

experience retardation. In the 25 years after oil extraction began, Norway grew more quickly than its neighbors Denmark and Sweden. Considering that Denmark and Sweden were ranked at number 4 and 6 in terms of GDP per capita in 2002, according to the U.S. Bureau of Labor Statistics (2003), the result is not a figment of poor performance in the control group. Thus, Norway's even faster growth implies that it did not suffer from the curse in the period of the mid-1970s to the mid-1990s.

IV

Did Norway Actually Avoid the Dutch Disease?

THERE IS A CONTRAST between the scrutiny of the Norwegian record on the curse and the examination of how it fared with the disease. While few commentators, except recently in contributions by Cappelen, Eika, and Holm (2000), Gylfason (2001a), and Stevens (2003), comment upon the lack of curse symptoms in Norway, many analysts have focused attention on how Norway dealt with the Dutch Disease. For example, Bjørnland (1998) finds that while there is weak evidence of Dutch Disease from North Sea oil in the United Kingdom, manufacturing output in Norway appeared to actually have benefited from the impact of higher oil revenues. Bye et al. (1994) find that Norway showed some symptoms in the 1970s and 1980s, when the Norwegian economy restructured to extract large quantities of oil, but that the Norwegian economy thereafter was poised to maintain a well-functioning nonoil traded goods sector. Cappelen, Eika, and Holm (2000) find that the traded goods sector contracted due to oil, but that manufacturing did not. Hutchison (1994) uses cointegration techniques and detects that the impulse response functions indicate an adverse effect on the manufacturing sector from the oil boom. Brunstad and Dyrstad (1997) note that there is evidence that the Norwegian petroleum sector has caused weak manufacturing performance, and that there are cost-of-living increases in areas close to the oil sector, the latter a symptom of real appreciation.

In other words, there is some dispute on Norway's performance in relation to the Dutch Disease. However, it may be possible to reconcile the different views into a synthesis. When oil extraction began and expanded rapidly, the oil sector demanded resources. Factors of production were moved to the North Sea. Thus, a buildup of real capital, labor, and expertise involved a reallocation of production capacities within the Norwegian economy. This resulted in a reduction of other production activities, a crowding-out. However, the essence of labeling the response a disease lies in whether the response signals future problems; for example, in a gradual erosion of the overall economic capacity, the production capacity in the nonresource traded goods sector, or both. However, even if the responding restructuring of the Norwegian economy followed the usual path of a disease by letting the oil sector expand quickly, the response appears not to become a disease. First, oil revenues' share of aggregate economic output became fairly constant quite early; see Figure 3. Thus, the *increasing* reliance on resource revenues, symptomatic of a disease, was not present. Second, labor movement into oil was modest, as shall be explained below.

Figure 3

The Importance of North Sea Oil and Gas, Fraction of GDP, Market Value, Norway, 1979–2002

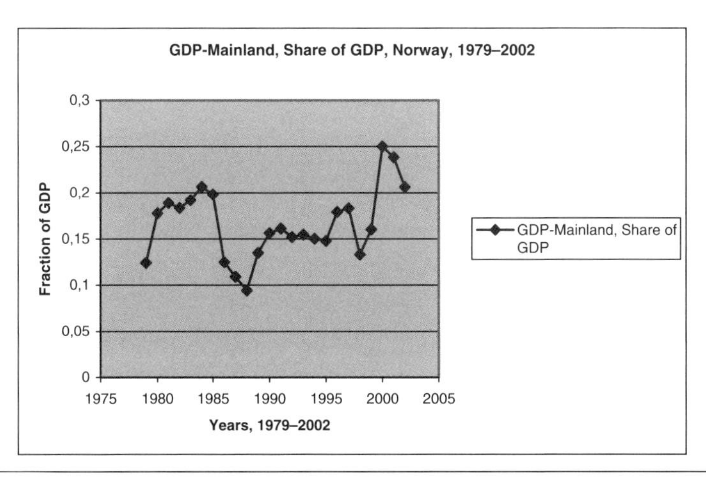

Source: National Accounts, Statistics Norway, Table 13.

The more recent fluctuations in oil's proportion of GDP are mainly the result of oil price fluctuations. The *volume* of Norwegian oil production per year is now rather stable. Whether or not Norway is suffering from the Dutch Disease depends on the definition of it. One definition for suffering from the Dutch Disease may simply involve the *level* of oil's share of economic activity; for example, the condition that oil revenues comprise a large share of gross domestic product. In Norway, that fraction is large. Another definition for suffering from the Dutch Disease includes scores on several facets of the economy such as both *levels* and *changes in levels* of output value shares, factor shares, and productivity increases in the oil sector, the nonoil traded goods sector, and the sheltered sector. Using all those indicators, the emerging picture of the last 20 years is much less conclusive in disease direction. In Norway, the oil sector commands a fairly small and constant share of total labor hours. The production volume is constant. The productivity increase in the manufacturing sector has been significant (see below). In fact, Bjørnland (1998) and Cappelen, Eika, and Holm (2000) claim that Norwegian manufacturing performed well in the 1980s and 1990s. Thus, while the oil sector obviously was and remains an important source for economic activity in the Norwegian economy, until recently Norway was not a typical Dutch Disease country.

V

How Did Norway Avoid a Full-Blown Dutch Disease?

NORWAY ESCAPED THE CURSE for two decades and avoided at least the most detrimental effects of the disease. How? I shall examine the disease first, then the curse. I do this because this way the presentation progresses in difficulty.

Norway managed to contain the factor movement effect, the spending effect, and the spillover-loss effect. This result may, of course, be part luck, part policy. However, we know from reports in Parliament in the 1970s and early documentation that Norwegian policymakers contemplated the dangers of the disease before they knew what label to put on it. Consequently, they attempted preemptive action. Explicit, deliberate policy schemes were implemented. What was the nature of these schemes? Let me present a brief summary of the policies.

1. *Factor Movement Policy.* Use the centralized wage formation system to limit general wage increases at the magnitude of productivity increases in the manufacturing sector. Use programs such as the "Solidarity Alternative" to coordinate income in order to moderate oil's effect on the nonoil traded goods sector.
2. *Spending Effect Policy.* Exercise fiscal discipline. Pay back foreign debts when possible. Thereafter, establish a Petroleum Fund abroad. Shield the economy from excessive demand and real appreciation when at full capacity, thus reducing loss of competitiveness. When not at full capacity, allow some increases in aggregate demand, but beware of the stickiness of public spending.
3. *Spillover-Loss Policy.* Encourage domestic accumulation of expertise in offshore oil extraction instead of using foreign specialists. Build up knowledge in technological centers. Educate oil experts. Invest in oil research. Attempt to maintain a diverse export base.
4. *Education, Research, and Development Policy.* Channel resources into education, research, and development. Stimulate scholarships for visits abroad. Increase labor hours in teaching and research. Establish centers of excellence.
5. *Active Countercyclical Policy.* Use enhanced political legitimacy of resource rents to counteract recessions. Exploit the feasibility of using returns to Petroleum Fund compared to other finance alternatives in active governmental capacity utilization.
6. *Labor Market Policies.* Maintain centralized wage negotiation system. Encourage the negotiating parties of employer and employee unions to keep in mind effects on the aggregate economy, not only special interests. Use a neutral agency to compute productivity increases in the manufacturing sector, and institutionalize these findings as ceilings of general wage increases. Stimulate female participation in the labor market. Enhance information coordination in the labor market through the establishment of vacancy and competence agencies.

7. *Industrial Policy.* Maintain and accumulate know-how in industrial activities. Keep variegated exports. Seek successful intertemporal paths for comparative advantage. Put emphasis on knowledge, technological progress, and human capital.

In assessing how these policies fared, we must relate them to the literature on the Dutch Disease. This literature proposes two main mechanisms of propagation, following the contributions from Corden and Neary (1982) onward. They are the *factor movement* effect and the *spending* effect. Later, various authors, such as Torvik (2001), categorized a subgroup of the former into a *spillover-loss* effect. The factor movement effect is caused by attractive favorable returns in resource extraction. Capital and labor move from other sectors to resource extraction, and in the process factor prices, such as wages, are bid up. Such higher prices make other sectors lose competitiveness, and they shrink. When they shrink, the fear is that several disadvantageous effects set in. First, know-how, physical plants, and technology may be lost irreversibly. Or they may be reversible only at substantial cost. Second, positive externalities from manufacturing are lost and not replaced by similar positive externalities from resource extraction. These positive externalities are learning-by-doing, technological progress, and innovative practices. This second effect may be categorized into the spillover-effect mentioned above. The spending effect results from converting resource revenue into domestic currency and using the currency domestically, thereby increasing aggregate, domestic demand. This procedure has two subeffects. First, the conversion entails increased demand for the domestic currency, which increases the nominal exchange rate. Second, the increased domestic demand for goods and services may create excess demand if the economy is at capacity. Both effects lead to real appreciation of the domestic currency, which leads to loss of competitiveness, which in turn leads to reducing the nonoil traded goods sector.

The following is the essence of whether adjustment to resource riches becomes a disease. Loss of competitiveness is not a problem in a highly flexible economy because factors will flow to the most efficient use and flow back when needed. But if there are path-dependencies, irreversible processes, lost spillover effects, and inflexibilities, an economy will suffer low capacity utilization in the aftermath of resource depletion before it returns to general equilibrium. Thus, a resource boom is more likely to lead to the disease in an inflexible economy. Most, if not all, real-world economies are less flexible than the textbook case.

These three mechanisms can be moderated through policy. Also, the technology content in the expanding sector may moderate the effects. In Norway, factor movement was atypical since offshore oil under the North Sea is more difficult to extract than on-land oil in the Arabic desert. More capital moved and less labor. In the 1970s, Norway invested heavily in real capital, technology, and human capital and built up large liabilities. To illustrate, *Statistical Yearbook of 2003* (Table 352) shows that Norwegian net assets were 41.1 percent of GDP in 2001. In 1980, nine years after extraction began, it was −32.6 percent, and −11.4 percent in 1970. Thus, Norway built up liabilities during the investments in the 1970s in the process of developing oil extracting capacity, and the labor movement effect was less acute in Norway. Moreover, wage increases were limited to the increases in manufacturing sector (see Cappelen, Eika, and Holm 2000). This was feasible because the parties in the labor market were large coalitions of employers and employees that were able to consider aggregate interests, not only special interests. Norway's centralized wage negotiation system may thus have contributed to preventing too fast and too much real appreciation. The extent of this moderating effect is linked to the degree the negotiating parties managed to use the productivity increases in the manufacturing sector as a ceiling for general wage increases. Notice that a prerequisite of a sustainable and controlled real appreciation in this manner is centralized wage formation. Wallerstein (1999) finds that Norway has one of the highest centralized wage formation systems in the world (see Table 3, Row 6). In effect, Norway succeeded in making manufacturing the wage leader, not the resource extraction sector, which may often be the case in resource-rich countries.

The spending effect was limited because the resource rent in its entirety accrued to the government, and because the government exercised fiscal discipline, paid back debts, and built up the

Table 3

Comparative Development, International Statistics, Selected Countries, 2001

	Norway	Denmark	Sweden	Germany	United States	Japan
Labor Force Participation Rate[1]	80.5	80.0	76.8	69.0	74.9	78.2
Corruption Index, Rank[2]	12	2	5	18	16	20
Human Development Index[3]	1	15	4	17	6	9
Wage Differences[4]	4.5	5.4	4.1	6.6	17.0	—
Index of Centralization in Wage Formation[5]	2.7	2.58	2.53	1	0.14	0.67

[1]OECD. Online: http://www.oecd.org/xls/M00037000/M00037562.xls. Euro-area is used for Germany.
[2]Transparency International. Online: http://www.transparency.org/cpi/202/cpi2002.en.html.
[3]Human Development Index, United Nations. Online: http://www.undp.org/hdr2001/.
[4]Barth and Moene (2000), Table 3.1; quoted statistics show standard deviation of relative wage differences between branches of the private sector, comparing employees with identical level of education, gender, and work experience.
[5]Wallerstein (1999). The index ranges from 0 to 3, in which 3 represents centralized negotiations on national level and 0 represents local wage formation within individual firms.

Petroleum Fund; see Bye et al. (1994), Cappelen, Eika, and Holm (2000), and Eika and Magnussen (2000). In comparison, if—hypothetically—individuals, coalitions, or an elite group had controlled the resource rent and brought it into the domestic economy, excess demand would likely have followed. In the 1990s, Norwegian authorities established the Petroleum Fund, specially designed to and with the express purpose of shielding the domestic economy from the spending effect; see Table 4. Mikesell (1997) reports that Chile also managed to create such a fund and that it helped stabilize revenue streams, and Stevens (2003) reports cases of successful funds. This fund protected the economy from excessive demand and the ensuing increase in domestic wages and prices. It also prevented nominal appreciation, since the fund is kept in foreign currencies.

The spillover-loss effect may have been less acute in Norway than in other cases. Sachs and Warner (2001) and Stevens (2003) say that it is a stylized fact that economic progress is intimately linked to a

Table 4

The Norwegian Petroleum Fund, 1996–2001

Year	Gross Domestic Product,[1] Billion NOK	Petroleum Fund,[2] Billion NOK
1996	1,027	47.6
1997	1,111	113.4
1998	1,132	171.8
1999	1,233	222.4
2000	1,469	386.4
2001	1,527	613.7
2002	1,521	609.0
2003	—	803.3[3]

[1]National Accounts, Table 1, Statistics Norway, market value. Online: http://www.ssb.no/emner/09/01/nr/.
[2]Central Bank of Norway; market value in NOK per December 31 each year. Online: http://www.norges-bank.no, use: petroleum fund.
[3]Market value per September 30, 2003.

diverse and entrepreneurial manufacturing sector. Van Wijnbergen (1984) also noted the stylized fact that technological progress is faster in the traded nonsheltered sectors than in the nontraded, and cites Balassa (1964) for an early reference. Torvik (2001) develops a model in which spillover effects play an important role. I suggest that it is plausible that the spillover-loss in Norway was much smaller than in other oil countries because of the immense real capital needed in the offshore oil industry and because of the inherent requirement of accumulation of know-how and expertise in offshore extraction. To illustrate, Cappelen, Eika, and Holm (2000) inform us that one labor hour in the Norwegian oil sector is combined with a capital stock 33 times that in manufacturing. This shows that Norwegian oil extraction is a high-technology sector, which we may assume has much the same positive spillover effects that manufacturing is assumed to have.

In my view, these three policies plausibly explain how Norway contained the factor movement effect, the spending effect, and the spillover-loss effect. They are also testable and documentable. More importantly, these policies are replicable. The other policies are increasingly more difficult to assess. In particular, I shall not present an assessment of the seventh policy type, industrial policy. As Stevens (2003) says, industry diversification is an obvious solution, but an extremely elusive one to achieve. Trying to identify winning sectors and activities is next to impossible. It may be as likely that Norway's growth has happened in spite of, not because of, its industrial policy. It is noteworthy, though, to observe that many authors, such as Mikesell (1997) and Sachs and Warner (2001), point toward the importance of a diverse nonoil traded goods sector. Rodrik (1995, 1997) demonstrates that some Asian countries may have succeeded because of governmental management and intervention.

As far as human capital policies go, Hægeland and Møen (2000) show that while Norway in the 1950s had 8 percent of the population with university degrees, which was the OECD mean, the proportion in Norway today is 20 percent. The OECD mean today is 15 percent. They further document that research and development at universities and research institutions in 1963 amounted to 4,500 man-labor years. In 1995, it was 14,500. This is, however, indicative of the effort to implement the policy, not its success. We can only speculate that the investments into human capital contribute to growth in a way alternative uses of investments could not.

How should we view the countercyclical and labor market policies? In Table 3, Row 2, we observe that Norway has succeeded in utilizing a large share of the labor force. This is indicative of an economically sound policy. We can presume that the current large share was, and the similar large historical shares were, results of countercyclical policies and labor market policies. Labor market design is of much interest because it is important in determining the flexibility of an economy and its capability of handling the restructuring implied by a resource gift and the disappearance of it. There is a large literature on the topic, and I shall not attempt a review here. Let me say that Bjørnland (1998: 558) discusses the labor market design and suggests that government consumption as a result of the oil-induced expansion of the public sector provided a stimulus for female participation in the labor market. That may prove to have been essential. Fifty percent of the talent in any economy is female. Engaging this talent into productive activity is likely to contribute to growth. The relevance to resource rents lies in the possibility of using, as Norway did, parts of the revenues to finance a growing public sector that increasingly employs female workers.

These policies, especially labor market policies, require no resource wealth, of course. However, resource wealth may make the policies easier to implement. In the daily practice of policy, prescribing and implementing countercyclical policies may be difficult because of the intricacies of political economy's nature. Changing tax receipts in excess of the built-in automatic cyclical changes in revenue is problematic. When the government commands revenues from resource rents, it is in a position to follow countercyclical policies more quickly, with more vigilance, and with more flexibility. Measuring the success of such policies is a challenge since we do not have the counterfactual growth path of how the economy would have fared without the policies. However, Cappelen, Eika, and Holm (2000) have examined the case in which the labor supply is not exogenous but endogenous, which is a variation of the countercyclical policy theme. In a multisectoral, large-scale macroeconomic model of the Norwegian economy (MODAG), they simulated how Norway would have performed on key macroeconomic variables without oil, using endogenous labor supply as a model feature. The results are tabulated in Table 5. We observe that while unemployment was of magnitude 3.3 percent in 1998, they compute that it would have been 6.3 percent in the counterfactual path without oil. One centerpiece of this model is that it allows wage formation to mimic the Norwegian institutional setup, in which the productivity increases in the manufacturing sector determine the ceiling of general wage increases. As a result of the wage limit and endogenous labor supply, they find that the real GDP growth for a quarter of a century might have been as low as 1.8 percent per year in the counterfactual without oil compared to the actual 3.3 percent with oil.

Table 5

Economic Development in Norway, Actual (with Oil) and Counterfactual (without Oil), 1974–1999, Compared to European Union

	European Union	Norway, Actual (w/Oil)	Norway, Counterfactual (w/o Oil)[1]
Total Growth, Real Hour Wage, Manufacturing, 1974–1998	7.4[2]	47.1	32.5
Unemployment, 1998	10.0	3.3	6.3
Governmental Net Financial Assets, Percentage of Nominal GDP, 1999	–55.6[3]	48.5	–65.8
Growth, Real GDP, 1974–1999	2.2	3.3	1.8
Growth, Real Private Consumption, 1974–1999	2.4	2.7	1.4

[1]Simulations in MODAG, a large-scale macroeconometric multisectoral input-output model of the Norwegian economy.
[2]EU15.
[3]Except Greece, Ireland, Luxembourg, and Portugal.
Source: Cappelen, Eika, and Holm (2000) and OECD, reproduced with permission from the authors.

VI

How Did Norway Escape the Resource Curse?

WHILE DEALING WITH the Dutch Disease involves mostly macroeconomic policy instruments, avoiding the Resource Curse may include more fundamental elements of society. The literature focuses on the presence of conflicts over distribution. Avoiding the curse, the literature says, reduces to *preventing rent seeking.* Most authors point toward political and economic institutions. The rationale for identifying rent seeking as the problem is that resource revenues constitute vast wealth, and when individuals or coalitions of individuals attempt to take control over it, they become less entrepreneurial. Thus, rent-seeking activity involves several detrimental aspects. First, the attempts themselves are time consuming and draw valuable labor hours away from productive, innovative activities. Talent is wasted in the pursuit of existing wealth instead of being employed at producing new growth. Second, when the activities are successful, the wealth may be disposed of in ways that are not conducive to growth. If the wealth is used for personal consumption abroad for the successful rent seeker and not invested in domestic technological progress and human capital, growth suffers. Probably, the wealth is acquired exactly for such purposes. Few agents acquire wealth to be able to act for the common good. The country's resource rent, then, is converted to luxury items, not research and development, so growth stagnates.

There exists a large literature on why conflicts over redistribution are so disruptive to economic performance. For example, Auty (2001a) uses predatory behavior and factional fights in some political states to explain stagnation, and Auty (2001b) argues that rent seeking degenerates into corruption, which discourages investment and limits growth. Moreover, Gylfason (2001b) emphasizes the quality of management and efficacious institutions in handling resource gifts. Baland and François (2000) argue that the opportunity cost of rent seeking is foregone entrepreneurship. Paldam (1997) demonstrates how rent seeking has affected Greenland negatively through such effects.

What political and economic institutions in Norway prevented rent seeking? What parts of the political economy helped reduce conflicts of distribution? I cannot hope to prove beyond doubt the presence of the factors I shall mention, nor can I identify the relative strengths of each factor. Instead, this article presents a story with some key elements. Some elements are part of what constitute a democracy and a developed country, and are therefore shared with other rich countries. Consequently, rich countries may be immune to the (worst cases) of the Resource Curse. Other elements are unique to Norway, and may not easily be generalized or replicated. More specifically, this article examines four factors: large-scale conflicts, small-scale illegal rent-seeking, small-scale legal rent-seeking, and political purchase of power through election promises. I argue that a social contract and social norms prevented the first. A legal system and norms reduced the frequency of the second. Politico-economic institutions reduced the impact of the third. Policymakers and politicians restrained themselves from the fourth—until erosion brought down some of the barrier in the late 1990s.

Conflicts, in the shape of rent seeking, involve coalition formation and result in coalitions seeking to prey on victimized weaker groups in a nontransparent way, affecting the country's production, labor effort, trust, and investment in the process. Such groups may be a ruling class or an elite of powerful allies, for example. But they could also be larger segments of society that come together in large coalitions, such as unions, and threaten major strikes, thereby initiating a rush to relatively higher compensation and conflicts of relative position. Norway was able to maintain low frequencies and small amplitudes of labor conflicts, plausibly because of what Eichengreen (1996) would call a *social contract*, similar to Rodrik's (1995, 1997) concerted collective actions. Both employers and employees were satisfied with contract conditions and the eventual outcome. Table 3 shows why: Norway is now ranked number one on the U.N. Human Development Index. With such rewards, the incentives to defect were smaller, given that norms helped agents feel obliged to participate in efforts toward the common good. Moreover, Norway has had no ruling elite that could funnel revenues into small, private ends. Instead, it is a highly egalitarian society that prides itself on being that. This article's view is that Eichengreen's notion of a social contract, therefore, is particularly relevant to understanding Norway's success in avoiding conflicts over distribution. When the public sentiment is one of satisfaction with and acceptance of the way society is organized, each individual feels less inclined to participate in conflicts such as strikes, sitdowns, or walk-slows. In Norway, laborers appeared content with the visible economic growth, knowing that profits would be plowed back into growth. The perception was that resource revenues were used to the benefit of all, in investments, technological advance, and education. Laborers found support for this perception in evidence: real capital accumulated, economic growth was reported, and levels of education grew.

Even if norms are hard to document, their impact on growth may be substantial. Norms may prevent individuals from seeking effortless reward and may reduce the frequency of conflicts. The early sociologist Max Weber pointed out the importance of the Protestant work ethic, which is often thought to be very much in effect in Norway. Norms for effort and equal reward receive popular support and function as a behavior guideline in Norway. Claims of such norms are presented, examined, and discussed in Barth and Moene (2000). When the judicial system cannot capture individual defections from the social contract, norms may deter them. But how can we substantiate the presence of norms? This article does not attempt to, but let me instead indicate how norms may serve as building blocks in Norway. Norms ensure wide support for a public school system, which every pupil attends. They lead to a public health care system, in which every citizen is automatically entitled to state-of-the-art medical treatment with only very small co-payments. They allow a system in which tax payments and taxable income actually are public information, published by newspapers and accessible on the Internet. Consider that for a moment. Citizens may easily find out others' contribution to finance collective goals in media channels. This creates not only a sense of monitoring but also a sense of common destiny and team spirit, which few seem to desire to challenge or are dissatisfied with. In summary, norms constitute institutions that affect actions like laws do, even if the sanctions are different in nature and differently imposed.

In general, then, open large-scale conflicts of distribution and disruptions of economic performance were few in Norway. But what about small-scale individual attempts to confiscate parts of resource revenues, the typical rent-seeking activity? Individual rent seeking may take two forms: illegal and legal. When norms fail, a strong and swift judicial system appears to detect and thus deter individual, fractional, and unlawful enrichment. The latter may be seen as one of the reasons why developed countries possibly are less likely to be affected by the curse. Illegal confiscation of collective wealth through corruption, theft, and misreporting is plausibly relatively infrequent in Norway due to the transparency of a small country, a well-functioning legal system, intense media attention, and surveillance and monitoring by public agencies. Table 3 shows that studies find relatively low levels of corruption in Norway.

Transparency, media scrutiny, rule of law, and politico-economic institutions prevent easy access to the public funds of resource rents for small coalitions. The following two factors may help to understand why. First, oil revenues go through the government. Thus, any attempt at securing for oneself a larger share of the pie must include negotiations with the government. Second, the avenues that ensure illegitimate access to oil revenues for individuals from the government are in fact limited, quite transparent to all, and always under media scrutiny. To see how oil revenues are channeled through the government to special interests, recall the following structural relations: the government taxes profits of private oil companies; the government owned its own oil company and owns part of an oil company; and the government owns the ground from which oil is extracted. The avenues to revenue access may take another form then, legal rent seeking. This may also be called negotiation or lobbying.

Legal acquisition of funds is possible for any Norwegian citizen through negotiation with the government. The negotiation may take the form of direct wage negotiation; lobbying in Parliament for subsidies and support to specific sectors; tariffs; and tax relief. There are some examples of powerful unions and coalitions securing favorable wage increases, and some evidence that pressure groups have been quite successful in lobbying for transfers, subsidies, and other benefits; see, for example, Brunstad (2003). Overall, however, this activity does not seem widespread. Moreover, since the public sector employs a large share of the Norwegian labor force, and since the government may finance this service stream by converting oil assets denominated in dollars into Norwegian kroner, it is at any time possible for individuals or groups to seek an appropriation of larger parts of the oil wealth, either by persuading the public sector to employ them favorably or by persuading the public sector to accept large wage raises. This would be especially acute in situations in which individual employees negotiated with an individual public servant, where the latter would not face the costs of yielding to pressure, thus creating moral hazard possibilities. This principal-agent problem was and still is largely avoided, however, since most wage negotiations go through a collective and transparent forum biannually, in large-scale negotiations between employer unions and employee unions, results of which are reported in the media. Thus, the centralized wage formation system may not only have helped avoid the Dutch Disease, but also helped avoid the curse.

The fourth factor in avoiding rent seeking is especially elusive to scrutiny and tests. It is related to the interplay between economics and politics and how the election system works. Explaining election results is an emerging branch of economic inquiry; see, for example, Persson and Tabellini (2000). Politicians seeking power may purchase this power and pay with promises. Thus, when resource rents accrue and the government controls a massive chest of resource rents, it is possible for politicians to promise special interests that they will be rewarded if they elect the politicians. This may create a rush of promises, a massive race, which in turn may erode resistance to shielding the economy through a resource fund. Tapping into this fund to pay back election promises would first create excess demand and real appreciation and, second, reward special interests. The latter would make it tempting to position oneself as recipient, and thus encourage rent seeking. It appears that through fiscal discipline, debt repayments, and the establishment of a Petroleum Fund, Norway managed to avoid this phenomenon in the 1980s and early 1990s. However, the success has bred its own complications. The accumulation of reserves abroad is now highly visible and is daily reported in the media. This has created a not unfounded perception of immense wealth among citizens. Even if accurate, this wealth cannot without difficult repercussions immediately be brought into an economy at full capacity, and explaining this has proven difficult. Instead, recently, politicians—seeking popularity votes—have been seen to demand utilization of this fund, ignoring advice from economists and prudent finance ministers. In short, the containability of resource wealth may be inversely related to the magnitude of accumulated reserves.

VII

But All Is Not Well: When Political Resistance Erodes

As late as the mid-1990s, Norway seemed to have avoided both the curse and the disease. But Røed Larsen (2005) detects a negative structural break around 1997. In Table 2, we observe that the period-specific additional growth coefficient for Norway from 1997–2002 is negative. Norway's faster growth relative to Denmark and Sweden weakens. Close scrutiny of Figure 1 shows that Norway reached its maximum lead over Denmark in 1998, when the difference in GDP per capita was of magnitude USD$2,506 (in 1999 dollars), and its maximum lead over Sweden in 1997, in which the difference was USD$6,431 (in 1999 dollars). After that, the lead diminishes and is reduced to USD$2,007 and $4,837 per head, respectively, in 2002. Table 6 summarizes the growth numbers of the relative slowdown. It is still too early to conclude, but it seems as if Norway may be experiencing some curse symptoms.

Table 6

Recent Growth in 1993–1998 and 1999–2002, Scandinavian Countries, Gross Domestic Product per Capita, 1999 USD at PPP

	Norway	Denmark	Sweden
GDP per Capita, 1993–1998	21.5%	15.8%	14.8%
GDP per Capita, 1999–2002	3.1%	4.6%	6.7%

Source: BLS (2003). "U.S. 1999 Dollars, PPP." Online: http://www.bls.gov.

Table 7

Changes in Total Labor Hours in Key Sectors, 1993–1998 and 1999–2002

Period	Oil and Natural Gas[2]	Industry	Public Sector
1993–1998	11.1%	11.4%	4.8%
1999–2002	–5.4%	–7.3%	–1.9%

Sources and Notes: Statistics Norway, National Accounts. Online: http://www.ssb.no/emner/09/01/nr. Table 21.

Moreover, Table 7 uncovers that, as Norway grew relative to its neighbors into the 1990s, it initially avoided the labor hours displacement from industry that the Dutch Disease involves. In fact, during the period 1993–1998, the labor effort exerted in the industrial sector increased *more* than it did in the oil sector, by 11.4 percent compared to 11.1 percent. The second row, however, in Table 7 reveals two other phenomena in the period 1999–2002. First, there is a contraction of exerted labor hours in all three key sectors. This accompanies the structural break retardation (see Røed Larsen 2005) and may again be indicative of a mild curse. Second, the relative magnitude of contraction is a symptom of the disease. Industry shed the most labor hours, and the public sector shed the least. This may still only be indicative of a natural progression of development, but nevertheless breaks the earlier trend from 1993–1998.

Observers of Norwegian economy and society have noted that the pressure from oil revenues intensified during the 1990s. By the turn of the millennium, the pressure from the public on policymakers was so intense that the government felt forced to institutionalize the so-called action rule, an oil management strategy that specified that only *returns* to the Petroleum Fund could be used domestically, not the fund itself. This was a political attempt to bind oneself to the mast, after the advice of prominent economists, in order to insulate oneself against popular calls from sirens. It was institutionalized to buttress the effort of shielding the economy from the excess demand in the spending effect so that real appreciation and loss of competitiveness could be avoided. However, the rope was not bound tight. Norway has spent much more than the action rule prescribes each year, and in 2003, public spending was about 50 percent larger than the prescribed dosage. We may interpret this as an erosion of the forward-looking policies discussed above.

The erosion of proper management is driven by popular demand and challenges analysts to understand the sociology and political economy of resource wealth. The perception is that Norway is extremely wealthy. The attitude is that it is inadmissible that certain tasks are left unattended to when the nation is fabulously rich. The riches are visible for all through the accumulation of the Petroleum Fund, now exceeding 50 percent of GDP; see Table 4. The political mechanism amounts to this: politicians promise to use oil receipts as remedies. There is thus a pulverization of responsibility due to moral hazard. Politicians who are elected on a platform of spending receive much support, but face a small risk. The disadvantages of real appreciation and Dutch Disease lie in the future and are shared by many, and the links to promises are obscure. Thus, the political mechanisms offer little incentive for restraint.

In return for promises, generous politicians are elected to Parliament. Sociologically, this creates an atmosphere of spending. The resulting economic mechanism is that the government channels oil fund money to public finance. This allows public bids in competition with private enterprise. Labor flows to the higher bidder, which is publicly controlled, prices and wages increase, the interest rate rises, and the nominal exchange rate appreciates. We observe real appreciation and reallocation of labor into domestic production of non-tradable goods and services. Private, industrial activities are crowded out. Possibly then, Norway's successful track record may largely be understood by saying that oil revenues are more manageable when they go directly to debt repayment than when they go directly into a fund. Since the Petroleum Fund is quite recent, it was too early in the 1980s and 1990s to say definitively that Norway escaped the temptation of resource riches. However, it is also still too early to assess the magnitudes of the political spiral of resource rent utilization. Norway may still escape from the curse and disease with only moderate effects, depending on the efficacy of the institutions discussed above.

VIII

Concluding Remarks and Policy Implications

MANY COUNTRIES EXPERIENCE slow or negative growth after discovery of a valuable resource. This phenomenon has been termed the Resource Curse. Factors of production move to extraction activities, and this factor movement effect is part of the mechanism in the Dutch Disease. The disease also includes a spending effect, since resource revenues allow increases in aggregate demand, creating excess demand domestically, which puts additional pressure on manufacturing through real appreciation and loss of competitiveness. Moreover, if there are positive externalities connected to having a large nonresource traded goods sector, resource countries also face a spillover-loss effect since resource extraction may imply slower technological progress. Thus, moving factors from the traded goods sector to the resource-extraction sector leads to loss of positive externalities. Both the curse and the disease are resource-extraction phenomena. Some economists believe that only poor countries are afflicted by the curse, and that rich countries are immune to it. Even if true, this begs the question of why. Some economists believe that the disease is not a disease, but merely structural adjustments that may easily and quickly be reversed. This article has examined both positions, and suggests that they may be mistaken. I do this by using Norway's management of its oil wealth as a case study.

First, I show that Norway caught up with and passed its highly similar neighbors in the 1980s. This involved an escape of the Resource Curse. Second, during this growth, Norway did not seem to reallocate factors from industry to resource extraction and public service at an alarming rate or lose competitiveness at an exceedingly large rate due to real appreciation from the spending effect. It does not seem that Norway lost important spillover effects. Instead, it appears as if Norway avoided both the Resource Curse and the Dutch Disease. Moreover, since growth accelerated only after oil discovery, it seems clear that oil was the engine of growth. When so many resource-rich countries have experienced growth problems after resource discovery, Norway's management of its oil riches deserves scrutiny. The question is what Norway did right.

This article argues that rent seeking is the pathogen of the curse and agrees with other authors that labor displacement, spending, and spillover-loss effects lead to the disease. Social norms, a social contract, transparency, and rule of law may contribute to limiting rent seeking. Thus, explanations involve sociology. This article examines four such political and sociological factors of rent seeking: large-scale conflicts, small-scale illegal rent seeking, small-scale legal rent seeking, and political purchase of power through election promises.

I sort Norway's macroeconomic policies implemented to handle the Dutch Disease into seven categories, not all of them immediately consistent with the conventional economic formulae. Most notably, since spending of oil money may involve deterioration of future comparative advantage, there may exist a negative externality in such spending. Concerted public and governmental effort may shield the economy from this effect in a way that individual behavior in a laissez-faire economy cannot. For example, the government may invest the revenues outside the economy. This shields the economy from the spending effect and limits real appreciation and labor displacement. Moreover, the erosion of the traded goods sector or manufacturing may be prevented through income coordination. Income coordination may improve upon laissez-faire since wage increases in the public sector can be limited to productivity increases in the internationally competing industrial sector. In essence, income coordination internalizes the externality resulting from intertemporal adjustment problems. Since Norway has one of the world's most centralized wage negotiating systems, it managed to make manufacturing the wage leader, in contrast to other resource economies, where resource extraction is the wage leader. In addition, this article discusses to what extent spillover losses from a contraction in manufacturing may be compensated for by spillover gains in the establishment of a capital-intense, technologically advanced offshore oil sector. This would make the Norwegian oil sector different from the oil sectors in countries that build up on-land oil-extracting expertise.

Intriguingly, it appears as if Norway did not continue its relative growth compared to Denmark and Sweden. A diagnostic test shows that it went through a structural break in the 1990s. Growth slowed down. Domestically, industry rapidly decreased in 1999–2002 compared to 1993–1998. This stagnation may be the result of a late onset of a curse and a disease. I inspect the political economy of the political and popular pressure that resulted from building up vast financial wealth abroad. Even if rent seeking is avoided, another form of un-optimal rent utilization may arise. Politicians may purchase political power in elections by extending to special interests generous promises of using oil revenues. When they keep such promises, excess demand may arise, which creates real appreciation, loss of competitiveness, de-industrialization, and both the curse and the disease.

References

Auty, R. M. (2001a). "The Political Economy of Resource-Driven Growth." *European Economic Review* 45(4–6): 839–846.

——. (2001b). "Transition Reform in the Mineral-Rich Caspian Region Countries." *Resources Policy* 27: 25–32.

Baland, J.-M., and P. François. (2000). "Rent-Seeking and Resource Booms." *Journal of Development Economics* 61: 527–542.

Balassa, B. (1964). "The Purchasing Power Doctrine: A Reappraisal." *Journal of Political Economy* 72: 584–596.

Barth, E., and K. O. Moene. (2000). "Er lønnsforskjelle for små?" ["Are Wage Differences Too Small?"]. In *En strategi for sysselsetting og verdiskaping*, NOU 21. Oslo: Norges Offentlige Utredninger.

Bjørnland, H. C. (1998). "The Economic Effects of North Sea Oil on the Manufacturing Sector." *Scottish Journal of Political Economy* 45(5): 553–585.

Brunstad, R. J. (2003). "Hvorfor følger ikke politikerne økonomenes råd i jordbrukspolitikken?" ["Why Don't Politicians Follow Advice from Economists in Agricultural Policy?"]. *Økonomisk Forum* 57(1): 30–34.

Brunstad, R. J., and J. M. Dyrstad. (1997). "Booming Sector and Wage Effects: An Empirical Analysis on Norwegian Data." *Oxford Economic Papers* 49(1): 89–103.

Bureau of Labor Statistics, U.S. Department of Labor. (2003). *Comparative Real Gross Domestic Product per Capita and per Employed Person: Fourteen Countries, 1960–2002.* http://www.bls.gov/fls.

Bye, T., Å. Cappelen, T. Eika, E. Gjelsvik, and Ø. Olsen. (1994). *Noen konsekvenser av petroleumsvirksomheten for norsk økonomi* [*Some Consequences of the Petroleum Activities on the Norwegian Economy*]. Report 94/1. Oslo: Statistics Norway.

Cappelen, Å., T. Eika, and I. Holm. (2000). "Resource Booms: Curse or Blessing?" Paper presented at the Annual Meeting of the American Economic Association. Oslo: Statistics Norway.

Chatterji, M., and S. Price. (1988). "Unions, Dutch Disease and Unemployment." *Oxford Economic Papers* 40: 302–321.

Corden, W. M. (1984). "Booming Sector and Dutch Disease Economics: Survey and Consolidation." *Oxford Economic Papers* 36: 359–380.

Corden, W. M., and J. P. Neary. (1982). "Booming Sector and De-Industrialization in a Small Open Economy." *Economic Journal* 92: 825–848.

Eichengreen, B. (1996). "Institutions and Economic Growth: Europe After World War II." In *Economic Growth in Europe Since 1945*. Ed. N. Crafts and G. Toniolo. Cambridge: Cambridge University Press.

Eika, T., and K. A. Magnussen. (2000). "Did Norway Gain From the 1979–1985 Oil Price Shock?" *Economic Modelling* 17: 107–137.

Gylfason, T. (2001a). "Natural Resources, Education, and Economic Development." *European Economic Review* 45: 847–859.

——. (2001b). "Nature, Power, and Growth." *Scottish Journal of Political Economy* 48(5): 558–588.

Gylfason, T., T. T. Herbertson, and G. Zoega. (1997). *A Mixed Blessing: Natural Resource and Economic Growth*. Discussion Paper No. 1668. London: CEPR.

Hægeland, T., and J. Møen. (2000). "Kunnskapsinvesteringer og økonomisk vekst" ["Investments in Knowledge and Economic Growth"]. In *Frihet med ansvar*, NOU 14. Oslo: Norges Offentlige Utredninger.

Hutchison, M. M. (1994). "Manufacturing Sector Resiliency to Energy Booms: Empirical Evidence from Norway, the Netherlands, and the United Kingdom." *Oxford Economic Papers* 46(2): 311–329.

Mikesell, R. F. (1997). "Explaining the Resource Curse, with Special Reference to Mineral-Exporting Countries." *Resources Policy* 23(4): 191–199.

Paldam, M. (1997). "Dutch Disease and Rent Seeking: The Greenland Model." *European Journal of Political Economy* 13: 591–614.

Persson, T., and G. Tabellini. (2000). *Political Economics. Explaining Economic Policy*. Cambridge: MIT Press.

Rodrik, D. (1995). "Getting Interventions Right: How South Korea and Taiwan Grew Rich." *Economic Policy* 20: 53–97.

——. (1997). "The 'Paradoxes' of the Successful State." *European Economic Review* 41: 411–442.

Røed Larsen, E. (2005). "Are Rich Countries Immune to the Resource Curse? Evidence from Norway's Management of Its Oil Riches". Discussion Paper 362. Oslo: Statistics Norway. *Resources Policy* 30(2): 75–86.

Sachs, J. D., and A. M. Warner. (1999). "The Big Push, Natural Resource Booms and Growth." *Journal of Development Economics* 59: 43–76.

——. (2001). "The Curse of Natural Resources." *European Economic Review* 45: 827–838.

Statistical Yearbook, 2003. (2003). Oslo: Statistics Norway. Available at http://www.ssb.no.

Stevens, P. (2003). "Resource Impact: Curse or Blessing? A Literature Survey." *Journal of Energy Literature* 9(1): 3–42.

Torvik, R. (2001). "Learning by Doing and the Dutch Disease." *European Economic Review* 45: 285–306.

——. (2002). "Natural Resources, Rent Seeking, and Welfare." *Journal of Development Economics* 67: 455–470.
Usui, N. (1997). "Dutch Disease and Policy Adjustments to the Oil Boom: A Comparative Study of Indonesia and Mexico." *Resources Policy* 23(4): 151–162.
van Wijnbergen, S. (1984). "The 'Dutch Disease': A Disease After All?" *Economic Journal* 94: 41–55.
Wallerstein, M. (1999). "Wage-Setting Institutions and Pay Inequality in Advanced Industrial Societies." *American Journal of Political Science* 43(3): 649–680.

Appendix

A. Data and Estimation Method

Real GDP per capita converted to U.S. 1999 dollars using a technique involving purchasing power parity can be found in Table 1 at page 9 in "Comparative Real Gross Domestic Product Per Capita and Per Employed Person: Fourteen Countries, 1960–2002," U.S. Department of Labor, Bureau of Labor Statistics, Office of Productivity and Technology, July 2003. Online access is possible using data at http://www.bls.gov/fls/flsgdpdf. However, the BLS publishes the most recent data, currently covering the period 1960–2004, using 2002 U.S. dollars as base. This article's analysis is based on the period 1960–2002, using 1999 U.S. dollars as base. This data set may be obtained through communication with the author or by correspondence with the BLS.

Gross domestic products measured in national currencies are converted to comparable entities using purchasing power parity (PPP). For a given country, a ratio is computed that consists in the numerator of the monetary units needed to purchase a common basket of goods and in the denominator of the monetary units needed to purchase the basket in the United States. This ratio is then used to compute an international equivalent of a country's gross domestic product.

B. Details of Trend Estimation

Equations (1)–(3) describe a first-order autoregressive process, estimated through feasible general least squares:

$$GDP_i^t / cap_i^t = \alpha_i + \beta_i t + e_i^t, \; i \in \{Denmark, Sweden\}, \quad t \in \{0, 1, 2, \ldots, 42\}, \tag{1}$$

$$e_i^t = \o e_i^{t-1} + \varepsilon_i^t, \tag{2}$$

$$\varepsilon_i^t = IN(0, \sigma_i^2), \tag{3}$$

in which the variable t is a time counter, the stochastic term e follows a first-order autoregressive process governed by the autoregressive factor ϕ, and the deterministic trend component is governed by the intercept α and the slope β. The stochastic term ε represents white noise and is identically, independently, and normally distributed with mean zero and constant variance. Subscripts i and t refer to country and time. I use maximum likelihood estimation in the *SAS* statistical analysis and programming package.

In this package, the reported total R-squared in Table 2 is unity minus the ratio of the sum of squared differences between model predictions and original observations to the sum of squared differences between the original observations and their mean. The regression R-squared is unity minus the ratio of the sum of squared differences between model predictions and the AR-1 transformed response variables to the sum of squared differences between the transformed response variables and their mean. The large difference between the two R-squared measures indicates presence of an AR-1 process that has been accounted for.

Since Røed Larsen (2005) documents an acceleration and deceleration in Norway's relative development, Norway's growth is estimated through a slightly different model, a three-step linear spline. It is similar to the one presented in Equations (1)–(3) for Denmark and Sweden, except that I have modeled an additional coefficient for the growth speed in the period 1975–2002, and a third growth speed component in the period 1997–2002. Thus, Equation (1) is changed to (1′) for Norway:

$$GDP_n^t / cap_n^t = \alpha_n + \beta_{1,n} t + \beta_{2,n} t_2 + \beta_{3,n} t_3 + e_n^t t_3, \begin{cases} t_2 = t - 14 & if \quad t > 14 \\ t_3 = t - 36 & if \quad t > 36 \\ t_2, t_3 = 0 & otherwise. \end{cases} \tag{1'}$$

Heavy Constraints on a "Weightless World"?

Resources and the New Economy

By JONATHAN PERRATON*

ABSTRACT. Late 1990s claims of a shift toward a new economy in the United States and other developed economies were said to accelerate earlier trends reducing the material content of production. The shift toward a postindustrial services economy is said to have been accentuated by application of new information and communications technologies, dramatically reducing the material content of production. This offers the possibility for continuous economic expansion unconstrained by resources supply. This paper provides a critical analysis of these trends in relation to resource use by developed economies. It shows that trends toward a lower material content of production are occurring, and this has led to poor demand conditions for primary producers. Nevertheless, these trends fall well short of eliminating Western economies' dependence on key resources. This paper shows the changing role of resources in economic activity among developed economies.

> It is simply wrong to believe that nature sets physical limits to economic growth—that is, to prosperity and the production and consumption of goods and services on which it is based. . . . Although raw materials will always be necessary, knowledge becomes the essential factor in the production of goods and services. (Sagoff 1997: 83, 90)

*The author is at the Department of Economics and Political Economy Research Centre, University of Sheffield, UK; e-mail: j.perraton@sheffield.ac.uk. An earlier version of this paper was presented at the Resource Politics and Security in a Global Age International Conference, University of Sheffield, July 2003. I thank the participants for useful comments but absolve them of responsibility for any remaining errors.

American Journal of Economics and Sociology, Vol. 65, No. 3 (July, 2006).

I

A Little Context

THE RELATIONSHIP BETWEEN natural resources and economic performance has periodically held center stage in political economy. The early 19th-century Ricardo-Malthus doctrine of diminishing returns predicted that economic growth would ultimately grind to a halt: even if manufacturing exhibited increasing returns to scale and technical progress, diminishing returns to agriculture (as expanding farming moved onto successively less fertile land) would limit the growth of the food supply and thus economic expansion, leading to a long-run steady state.[1] In 1865, Jevons made similar predictions for the exhaustion of coal supply as the key energy resource. Subsequent history demonstrated much greater potential for technical progress in agriculture, and the discovery of new energy deposits and resources in oil enabled further expansion. In the early 1970s, however, an influential Club of Rome report prophesied limits to economic expansion from finite supplies of fossil fuels and key raw materials, as well as limits to the environmental capacity of the earth to absorb rising pollution (Meadows et al. 1972). The 1970s oil crises, heralding the end of the postwar Golden Age, appeared to some to confirm this vision. Subsequent analysis indicated that oil price rises were only one of a range of factors explaining the end of the postwar boom, but still a key one (e.g., Bruno and Sachs 1985; Glyn et al. 1990). The persistence of the effects was such that growth rates among developed countries did not return to their 1950–1973 rates even during the 1980s, when oil prices fell back to levels comparable (in real terms) to those before the 1970s price hikes.

Although economists have often been skeptical that resources are likely to constrain expansion over the longer term, in the late 1970s many would have taken the ecologists' side over the economist Julian Simon's side in a famous bet over the future course of raw materials' prices—that if economic growth were to lead to scarcities of finite natural resources, one would expect to see sharp price rises. In the event, Simon clearly won: over the 1980s in real terms, the prices of primary commodities dropped to levels not seen since the Great Depression. In this some saw a vindication of the 1950s Prebisch-

Singer hypothesis that in the long run relative prices of primary commodities would fall. In part this was attributed to dematerialization among developed economies, associated with the shift from heavy to light manufacturing and from manufacturing to services. These trends were discussed in the 1980s, but shifts among key developed economies in the 1990s appear to have heightened them. Several economists—not to mention any number of "futurologists"—have characterized the impact of new information and communications technologies (ICTs) as creating a "weightless world" economy with limited or even minimal material content of production and very strong economies of scale and technical progress (Chichilnisky 1998; Coyle 1999; Quah 1997, 2001a, 2001b). The result of this is that the standard limitations on economic expansion are relaxed, if not eliminated. In part this is expressed in terms of standard growth models, with potentially a more rapid rate of technical progress and possible increasing returns to scale at the macroeconomic level. But this is also claimed to enhance the possibilities for reducing the material content of production and substitution of nonmaterial inputs. A classic example here is software, which made Microsoft *the* company of the 1990s. The fixed costs of developing software may be high but are largely human, not material; once developed, the material content and production costs of software are negligible.

With the end of the dotcom bubble, rising demand for and prices of resources, and renewed concern over the stability of oil supplies, the "weightless world" claims may now appear over optimistic. Whereas in the 1990s many analysts were predicting sustained low oil prices, in the first decade of the 21st century, predictions of sustained higher oil prices are commonplace. For all the emphasis on the new economy, U.S. production of "light" autos rose by almost 50 percent over the 1990s and was key to sustaining economic expansion after the collapse of the dotcom bubble (Rutledge 2005: ch. 9). The key question here is whether the new economy effectively decouples economic growth from natural resources as production becomes dematerialized and nonmaterial inputs can be substituted for material ones. This paper proceeds as follows. Section II briefly surveys orthodox economic theory on natural resources and economic growth. Section III presents evidence on structural change and the new economy that might be expected to reduce demand for resources and selects economies for examination based on the contribution of ICT investment to their growth. Section IV considers the general impact of economic expansion on resource use. Section V considers trends in energy use among these economies and Section VI considers trends in use of the main metals. Section VII briefly considers the global impact of these trends. Section VIII concludes.

II

The Economics of Finite Resources

DESPITE THE MALTHUS-RICARDO results, orthodox postwar growth theory, either of Solow or endogenous type, almost invariably abstracts from natural resource inputs. Until the 1980s, most short-run macro adjustment models also assumed away resource inputs and any changes in their price. Various interconnected lines of defense are offered for this approach (Neumayer 2000; Nordhaus 1974; Solow 1974). Given futures markets and backstop technology, the price profile of a scarce resource ensures substitution at exhaustion. Optimal depletion theory would lead one to expect a predictable rate of depletion rather than the sudden emergence of shortages. Of course, this is subject to key "second-best" caveats—notably over whether the private and social discount rates coincide, imperfect property rights, and the market structure of extractive industries—and it cannot simply be assumed that observed rates of depletion are in any sense optimal. As the price of the depleted resource rises, one would expect to see standard economic responses: other, more abundant resource inputs will substituted for it, and consumption of goods intensive in the scarce resource will be discouraged by rising prices. Rising prices will tend to increase incentives for more efficient use of the resource and recycling. It will also increase incentives to improve search and extraction technology and open up previously economically unviable deposits. Changes in technology and patterns of demand make previously essential resources obsolete, whereas others assume center stage: the Stone Age didn't end from a shortage of stones; before the 20th century there were few uses for oil.

Two mechanisms within orthodox growth theory could in principle act to sustain expansion even with finite resources. First, technical progress may be resource augmenting (possibly stimulated by rising resource prices) so that continuously higher output is possible for given resource inputs. Alternatively, if it is possible to substitute reproducible capital for natural resources, then indefinite expansion may be possible. Under Hartwick's rule, if rents from exhaustible resources are invested in creating reproducible capital, then an economy's total capital stock is maintained (Hartwick 1977); "weak sustainability" effectively is maintained, as the rate of change of net wealth over time is not negative. Indefinite growth of consumption is possible if the elasticity of substitution between reproducible capital and finite resources is unity and the reproducible capital share exceeds the resource share. Thus, economic expansion could in principle be sustained with diminishing material inputs. Ironically, perhaps, the science of choice under ubiquitous scarcity denies that finite resources limit expansion. This approach has fed directly into green national accounting, correcting GDP estimates for resource depletion (see Section VII below).

Several objections have been advanced to this line of argument by ecological economists, notably Daly (1977), who argues that the natural resources are complements, not substitutes, to other inputs (as well as seriously underestimating the resource demands of growing economies); and Georgescu-Roegen (1971), who argued that it neglects the role of entropy processes.[2] Daly rejects the substitutability argument, arguing that creating the "substitute" entails using more of the resource that it is supposed to be substituting for. Just as economists criticized the Club of Rome and others for naively extrapolating from past trends, even if technological progress has been resource augmenting and/or the elasticity conditions are currently satisfied, there is no guarantee that will continue to be the case in the future, which is inherently uncertain, particularly as key resources are depleted (Neumayer 2000). As noted below in relation to energy inputs, there are unresolved difficulties in estimating elasticities of substitution between energy inputs and reproducible capital.

Arguably, these models are best viewed as "parables," rather than definitive claims that "the world can, in effect, get along without natural resources, so exhaustion is just an event not a catastrophe . . . at some finite cost, production can be freed of dependence on exhaustible resources altogether" (Solow 1974: 11). Although by definition there must be some physical limits to economic expansion, economists' grounds for believing that these would be unlikely to be binding in the foreseeable future are partly empirical. The absence of clear upward trends in resource prices—rather the reverse—and evidence that resources remain abundant relative to likely future demands are typically taken as grounds for believing resource constraints are unlikely to be binding.

Short-run macro adjustment models, at least since Bruno and Sachs (1985), have often included the effects of raw materials and/or energy prices on output. Nevertheless, these effects are assumed not to affect the long-run growth rate by imposing the assumption of unit elasticity between material inputs and capital, but sometimes evidence is produced to indicate that these effects do not persist. Thus, even a sustained rise in the price of materials inputs is predicted to have a level effect on income but not affect its growth rate.

III

Structural Change and the "New Economy"

THE MAIN LONGER-TERM TRENDS among developed economies are too well known to need documenting in detail. The decline of agriculture followed by deindustrialization as production (and, more clearly, employment) shifts from manufacturing to services (and within manufacturing from heavy to lighter industries) have been analyzed extensively. These trends would be expected to lead a diminishing material content of output with development. In cross-section, there is a hump-shaped relationship between the intensity of resource use and economic development (e.g., Rowthorn and Wells 1987: ch. 2). At low levels of development, resource use tends to be low. Industrialization raises demand for resources but, once deindustrialization sets in, the resource content of production tends to decline with it, but not necessarily absolute resource use. Similar trends are observed with energy use in production (Snil 2003: ch. 2). This is the material input version of the "environmental Kuznets curve" (EKC); similar trends have been claimed for greenhouse gases emissions and other pollutants, both from these production trends and from increased demand for environmental goods at higher income levels.

Taken at face value, the "new economy" has elements of both continuity and radical novelty in relation to these trends. Authors differ in their use of the term, but a consensus interpretation would be that it entails significant productivity growth from enhanced use of new ICTs (Cohen, Garibaldi, and Scarpetta 2004; Temple 2002; Stiroh 2002b), but also that information has become increasingly important to economic processes. Information has unusual characteristics as an economic good: it has strong public good characteristics and is not intrinsically scarce (e.g., Hodgson 1999: pt. 3; Quah 2001a). On the contrary, agents often face an abundance of information relative to their cognitive capacities. Although not necessarily disembodied from physical capital accumulation—indeed, depreciation of ICT capital equipment is relatively rapid due to high rates of technological obsolescence—the resource demands for ICT equipment are claimed to be relatively low, and the driving force of growth is the development of new ideas. In growth theory terms, this would be associated with raising the productivity growth rate and possibly increasing returns to scale at the aggregate level. All this is associated with higher productivity growth, lower inflation, falling unemployment, and an apparent loosening on the constraints of expanding output (Jorgenson and Stiroh 2000).

Several developed economies saw heavy investment in ICTs in the 1990s and to a greater or lesser degree saw higher productivity growth. In part, these trends would be expected to increase the shift toward services; although ICTs would be expected to enhance productivity in both manufacturing and services, they are particularly important in providing potential productivity growth in services, are typically assumed to have much less (but nonzero) potential for productivity growth than manufacturing, and are important in increasing the variety of services available and thereby enhancing the shift in demand patterns toward services. Quah (2001a) in particular argues that focusing exclusively on supply-side productivity effects of ICTs misses their key role in raising product innovation in the services sector and thus increasing demand. More generally, although investment in ICTs entails investment in physical assets, the key point here is that much of the increases in productivity comes from application of human capital, information, and technology, so that the expansion of the economy is driven by nonmaterial factors. As one analyst of the 1990s new U.S. economy boom concluded: "Investment in software is more important than investment in hardware" (Jorgenson 2002: 47).

Both the trends toward services in general and the new economy—and with it the implications for resource use—are subject to major caveats. Because the prices of services tend to rise more rapidly than manufacturing due to their lower productivity growth, much of the apparent rise in the services share is a price effect. Evidence on sector shares in constant price terms indicates the shares of manufacturing and services in output have remained fairly stable among the industrialized countries (Summers and Heston 1991; Rowthorn and Ramaswamay 1997). Moreover, many services are intrinsically tied to production of material goods, and part of the apparent rise in services production may reflect the contracting out of services by manufacturers that were previously done in-house (Blades 1987). Following this, Table 1 provides rough indicators for changes since 1990 in the shares of national value added (excluding government services) accounted for by goods, services whose production is directly linked to goods production, and free-standing goods production. Among the developed economies, although there has been some relative decline in goods production, it is the share of services directly related to goods production that has risen, with the share of free-standing serv-

Table 1

Shares in Total Value Added of Market Activities
OECD Countries, 1990–2000

	1990	2000
Goods	35.9	33.8
Services linked to goods production	39.5	41.9
Free-standing services	25.0	24.9

Source: Calculated from OECD, *National Accounts of OECD Countries: 1970–2000, Volume 2: Detailed Tables*, using classifications from Blades (1987). Shares are in constant price terms.

ices barely changing. This is consistent with other evidence that the growth in services connected to material production has been relatively rapid (Oulton 2001).

Thus, while goods output only accounts for a minority of economic activity and employment in developed countries, this cannot simply be taken as evidence that their economies are becoming increasingly dematerialized. Other misconceptions about the services sector can be dispelled (Gallouj 2002). Conventional approaches typically underestimate innovation in the services, while more detailed studies indicate significant innovation in this sector for developed economies over the 1990s, consistent with new economy arguments. In the United States at least, productivity grew strongly over the 1990s in those services that used ICTs intensively (Ark, Inklaar, and McGuckin 2003; Triplett and Bosworth 2003). Moreover, far from being universally low-capital intensive, services now account for around half of capital accumulation in developed economies (Gallouj 2002). This of course implies material input to services production. These trends have key potential implications for the sustainability of growth processes with increasing demand for services: under Baumol's "cost disease," if equalization of wage growth rates (but not necessarily wage levels) is maintained between technologically dynamic and stagnant sectors of the economy, then relative costs of the latter will grow without limit and growth will decline as resources shift toward that sector. Oulton (2001) shows that the decline in the growth rate is dependent on the assumption that the stagnant sector only supplies final goods. If the "stagnant" sector supplies intermediate goods and experiences some productivity growth, albeit at a lower rate than the dynamic sector, then a shift of resources to that sector will not lead to a decline in the growth rate. The reason for this apparently paradoxical result is that productivity grows in the dynamic sector both directly and indirectly through the contribution of productivity improvements in the inputs produced in the other sector. ICT investment in the services economy may therefore play a key role in sustaining economic expansion. Nevertheless, this is also consistent with goods production continuing to rise in absolute, if not relative, terms.

Arguably there are even stronger caveats with respect to the "new economy" than with the rise of the services economy. The collapse of the dotcom stock market bubble and the end of the 1990s U.S. boom potentially helps the task of disentangling structural changes from purely cyclical factors. Much of the productivity gains experienced by the U.S. economy in the 1990s were in the computer and allied industries, with limited spillover effects on the rest of the economy (Gordon 2000; Stiroh 2002a); although the vast majority of studies of the new economy focus on the United States, similar results were found for Finland (Daveri and Silva 2004). While phenomenal growth in the computer industry did play a key role in the 1990s U.S. expansion, other factors—notably favorable demand conditions and rising profits from sluggish or nonexistent wage growth and low real interest rates—were also crucial (Cornwall and Cornwall 2002; Thompson 2004). Gordon (2000) not only attributes much of the 1990s U.S. growth to cyclical factors but also queries whether the ICTs will lead to radical product innovations and increased demand on the scale of the major 20th-century innovations.

Beyond the United States, the impact of the new economy is even more ambiguous. As Table 2 shows, the major economies varied in their levels of ICT investment, with some approaching U.S. shares. However, there was no clear relationship across countries between ICT investment levels and growth rates (Cohen et al. 2004; Colecchia and Schreyer 2002; Daveri 2002). Ark et al. (2003) found comparable levels of productivity growth in ICT industries themselves among the major EU economies, Australia, and Canada; however, the United States saw significantly faster productivity growth in ICT-using industries. It may be that with investment lags and learning effects these technologies will have a clearer impact in years to come. Among developed economies' booms beyond the United States, only six EU economies—Denmark, Finland, Greece, Ireland, Sweden, and the United Kingdom—show clear evidence of the contribution of ICT investment to growth risen in the second half of the 1990s, as also do Australia and Canada.[3] Greece is excluded from the analysis here since ICT investments have grown rapidly but from very low levels. The largest continental European economies—France, Germany, Italy, and Spain—saw stagnant or declining contributions of ICT investment to growth over this period. In the following analysis, we focus upon

Table 2

Percent Share of ICT Investment in Nonresidential Investment, 1990–2000

		Australia	Canada	Finland	France	Germany	Italy	Japan	U.K.	U.S.
IT Equipment	1990	5.5	4.5	3.6	3.5	5.5	4.2	3.8	6.0	7.0
	1995	8.4	5.7	4.0	3.9	4.6	3.5	4.6	8.6	8.7
	2000	7.2	7.9	2.9	4.4	6.1	4.2	5.2	8.4	8.3
Communication	1990	3.8	3.8	3.9	3.2	4.8	5.7	4.0	2.0	7.5
Equipment	1995	4.7	4.0	9.3	3.5	4.2	6.7	5.3	3.6	7.3
	2000	5.6	4.2	15.3	3.9	4.3	7.2	6.9	3.6	8.0
Software	1990	4.6	4.9	5.2	2.6	3.7	3.8	3.1	2.1	8.0
	1995	6.4	7.1	9.2	3.5	4.5	4.3	4.0	3.5	10.1
	2000	9.7	9.4	9.8	6.1	5.7	4.9	3.8	3.0	13.6

Source: Cohen et al. (2004: 22).

these eight countries where the impact of new economy technologies appears to have been greatest.

IV

Economic Expansion and Intensity of Resource Use

Standard analyses proceed by estimating an economywide intensity of use index (apparent consumption of the resource relative to GDP). Trends can then be divided into "weak" dematerialization, a fall in the intensity index, and "strong" dematerialization, an absolute decline in materials use with GDP growth. Although common in the literature, this form of analysis has been subject to several criticisms (Cleveland and Ruth 1999; Labson 1995). As an identity, it may provide useful summary information but does not explicitly identify economic relations driving these changes; in particular, it does not directly account for the roles of relative prices, elasticities of substitution between inputs, and technical progress. It also fails to take into account the derived nature of demand for resources—they are typically demanded as inputs to final goods. This may be crucial to assessing price effects: technical progress leading to greater efficiency effectively lowers the price of inputs to users and therefore could operate through a "rebound" effect to raise consumption, which was noted as far back as Jevons. Further, many studies fail to account adequately for the time series properties of the data; this is crucial to assessing whether there is a long-term trend toward greater efficiency or largely a one-off shift. In particular, recent studies have found that although both material use and GDP are clearly trended for developed economies and have unit roots, they are not cointegrated, so that stochastic shocks tend to introduce permanent drift in the relationship between material use and GDP (de Bruyn 2000; Labson 1995); this would also hamper attempts to forecast future demand for resources. Earlier results indicating an EKC were based on inappropriate time series analysis (cf. de Bruyn 2000: ch. 6). Once reestimated, taking into account the degree of integration of the data series, the expected relationship between pollution and income no longer holds and there is no presumption that growth, by raising income levels, will lead to lower harmful emissions. Finally, earlier analyses do not account for the extent to which increased trade specialization leads to developed countries importing resource-intensive products, so that domestic resource consumption increasingly understates the resource content of final consumption in developing countries.

While some studies have acknowledged these problems, data limitations hamper attempts to address them. Nevertheless, some national statistical authorities recently have begun to collect materials use statistics systematically. Unfortunately, many of these are only available for recent "snapshots," and limiting meaningful timeseries analysis. For the United States, 20th-century figures up to 1995 indicate rising total consumption of nonrenewable materials until the 1960s but a roughly constant level since then, thus implying declining intensity of use (Matos and Wagner 1998). This study does not estimate the material content of imports into the United States, although it does note more rapid increases in material consumption in the rest of the world. The aggregate picture for the European Union over 1980–2000 was roughly unchanged for total materials consumption, implying a trend decline in intensity in use. Eurostat (2002) found broadly similar trends for the United States and Japan, although with the United States, intensity in use was around 30 percent higher than the European Union and fell at a slower rate over this period. Within these EU-wide trends there was considerable variance between countries, with "strong" dematerialization only clearly observed in France, Germany, the Netherlands, Sweden, and the United Kingdom. There was no clear association between ICT-based growth among EU countries and dematerialization. Moreover, although there is evidence of an EKC relationship in terms of material consumption for Denmark, the Netherlands, and the United Kingdom, such a relationship was not found for the initially poorer EU members, with their material consumption per capita continuing to rise with income. This suggests that, for material consumption at least, if an EKC relationship does exist, then the turning point is at a relatively high income level, although the high level of dispersion across countries precludes clear conclusions. Other studies of developed economies point to weak dematerialization trends but continued absolute growth of material use and considerable levels of waste (World Resources Institute 2000).

Of particular relevance here is the material content embodied in ICT equipment. As noted above, rapid technological obsolescence leads to relatively rapid depreciation and replacement of ICT equipment, although the new economy view has tended to assume that this equipment has relatively low material content compared to capital equipment in more traditional industries. Information is limited, but piecing together available data, Williams, Ayres, and Heller (2002) found that the total material use in semiconductor manufacturing was over 600 times the mass of the final product, a figure orders of magnitude larger than for traditional manufacturing. The complexity of the final product and the levels of refinement and purification of material inputs required leads to high material requirements relative to the final product. Moreover, these products typically require (albeit small) inputs of relatively scarce specialist materials. This suggests more generally that while the final output of new economy industries may have relatively low material content, the impact of increasingly complex products and organization of production on secondary material inputs cannot simply be read off from this. Analysis at this level of aggregation can provide only broad indicators, and we now move to analyzing more specific materials.

V

Energy Demand and Economic Expansion

THE RISE OF MODERN ECONOMIES was built upon the shift from biomass to fossil fuels as chief energy providers (Snil 2003: ch. 1; Roberts 2004: ch. 1). While generally, cheap fuel masked the importance of energy for much of the 19th and 20th centuries' economic expansion, since the 1970s oil crises, the economic and strategic significance of energy hardly needs to be stressed.

Over the 20th century as a whole energy consumption rose at roughly the same rate as GDP, so that energy intensity of real income was similar at the end of the century as at the start; but this picture conceals rises in the intensity up to 1973 and falls thereafter (Hannesson 2002; Snil 2003: 65–69). In the aftermath of the 1970s oil price rises, the energy intensity of GDP was reduced in the developed economies, as higher oil prices led to energy conservation. Structural changes in these economies might also be expected to continue this trend; the evidence is less clear from the 1980s when oil prices fell back in real terms. Nevertheless, improvements in energy efficiency tend to induce "rebound" effects: consumers heat their homes warmer in the winter, use more air conditioning in the summer, and drive larger cars. Thus, post-1973 trends toward more fuel efficient autos in the United States were reversed in the 1990s, with falling real fuel prices and the rise of SUVs (Roberts 2004: chs. 6 & 9; Rutledge 2005: ch. 9).

In the 1990s, there was widespread optimism that for the developed countries "additional energy use is likely to be minimal, and there appears to be a consensus that little or no growth in overall demand is likely. . . . Most of the countries that record high per capita use of energy have reached the point at which energy demands are unrelated to the level of economic growth" (Churchill 1993: 442–443). We pursue this issue further here for our ICT-intensive growth countries. Table 3 shows the evolution of the ratio of total primary energy supply in relation to real GDP (calculated on a PPP basis) for these countries. Trends are clear: although the energy required to produce a unit of GDP has continued to decline, energy consumption has largely continued to rise with GDP growth, although there is some evidence of stabilization with Denmark, Sweden, and the United Kingdom. Economic expansion has not therefore become decoupled from energy use. There remains considerable dispersion in the energy-GDP intensities, so that although both Canada and the United States in particular have seen improvements in energy efficiency, they remain relatively high energy-intensity economies in comparison with other developed countries.

The question remains, though, whether the 1990s experience is anything more than a continuation of earlier trends toward lower energy intensity of GDP. Further testing for a structural break in 1990 is reported in Table 4. For all sample countries except Denmark, this indicates a break in this series (although the Australian and Finnish cases are marginal). For four of the countries, the rate of growth of energy efficiency increased in the 1990s, consistent with a new economy thesis. Ireland is a particularly striking case of combining rapid economic growth with gains in energy efficiency, although this

Table 3

Energy Consumption-GDP Ratios, 1960–2003

	1960	1973	1980	1990	1995	2003
Total energy consumption: (Millions of tons oil equivalent)						
Australia	31.55	57.62	70.37	87.54	94.38	112.65
Canada	76.27	159.84	193.00	209.09	231.74	260.64
Denmark	9.03	19.83	19.78	17.58	20.05	20.76
Finland	9.78	21.35	25.41	29.17	29.63	37.55
Ireland	3.78	7.19	8.49	10.57	11.36	15.09
Sweden	20.54	39.32	39.91	46.66	50.00	51.53
U.K.	160.46	220.72	201.28	212.18	223.18	231.95
U.S.	1,021.4	1,736.5	1,811.7	1,927.6	2,088.5	2,280.8
Energy-GDP ratios:						
Australia	30.27	28.78	29.32	26.60	24.98	22.44
Canada	46.14	42.63	40.29	33.23	33.82	28.98
Denmark	19.15	23.20	21.24	16.15	16.71	14.56
Finland	29.54	34.40	33.94	28.86	30.66	29.55
Ireland	24.45	27.18	23.32	20.34	17.39	12.51
Sweden	27.61	31.22	27.81	26.31	27.45	23.06
U.K.	32.01	30.32	25.79	20.95	20.29	17.07
U.S.	42.97	42.93	37.68	29.29	28.22	23.93

Source: See Appendix.

wasn't sufficient to prevent rises in Irish total energy consumption. For Finland, the United States, and the United Kingdom, though, the rate of growth of energy efficiency declined in the 1990s relative to earlier in the post-1973 period. While the Finnish case may be affected by its severe recession in the 1990s, the key U.K. and U.S. cases indicate that new economy trends do not necessarily lead to faster declines in energy intensities. Only limited reliance can be put on these results with limited observations, but they do indicate trending in the data rather than simply a structural break around 1973.

It was noted above that past studies have been criticized for being

Table 4

Rates of Change in Energy-GDP Ratios, 1973–2003

	1973–2003	1990 Break?	1973–1990	1990–2003
Australia	–0.39	4.11**	–0.38	–0.60
Canada	–0.54	4.74**	–0.64	–0.68
Denmark	–0.67	1.82		
Finland	–0.21	2.63*	–0.40	–0.32
Ireland	–0.97	33.11***	–0.59	–1.68
Sweden	–0.37	5.23**	–0.25	–0.70
U.K.	–0.77	14.48***	–0.95	–0.85
U.S.	–0.84	20.94**	–1.09	–0.72

1990 Break—Chow test for structural break in trend series. ***, **, *—significance tests at 1%, 5%, 10% levels.
Source: See Appendix.

atheoretic and/or not adequately accounting for the time series properties of the data. Advances in time series analysis means we cannot simply view an observed time trend as evidence of a deterministic trend. In particular, if energy consumption and GDP are not cointegrated, although both trended, there may be no clear trend in energy intensities. In an earlier study, de Bruyn (2000) found evidence of this for energy (and steel) use among four developed economies. An intuitive explanation for this is that shocks to material consumption and to GDP may be largely unrelated. Thus, a material-saving innovation could lead to a reduction in the material intensity of production, but material intensity subsequently may rise again for "rebound" effect reasons; technological innovations may raise income levels or even growth rates without having an impact on energy intensities. Nevertheless, it is not possible to pursue detailed time series econometric analysis meaningfully with limited observations, particularly with evidence of a structural break around 1973. Instead, following Ormerod (1994: ch. 7) and de Bruyn (2000: ch. 8), we can use the concept of "attractor points." By plotting annual observations for (logged) energy intensity relative to the previous year's value, it is possible to discern several points. It provides an analysis of whether intensities show clear trends or are largely cyclical; if they are cyclical, it can indicate the attractor point around which intensities fluctuate and the magnitude of these cycles. Moreover, it can indicate whether there are periodic shifts before cycles settle down around a new attractor point. This approach is particularly useful here, where there are pronounced cyclical variations and we are interested in determining whether there is a trend fall in resource intensity or periodic shifts in response to regime changes; in other words, in distinguishing between cases of continuous technological change from shifts associated with a particular cluster of innovations (cf. de Bruyn, 2000: ch. 8). Thus, it can help illuminate whether there has been a clear shift toward lower intensities of resource use with new economy trends.

The results in Figure 1 display a diversity of experience, to some extent in contrast with the smaller sample in de Bruyn (2000: ch. 8). For Australia, Canada, Denmark, and Sweden, energy-GDP ratios are consistent with attractor point patterns for much of the post-1973 period, but some evidence of trends toward lower intensity are observed in the most recent years. However, as this only applies for a small number of recent observations, it cannot be presumed to indicate a continuous trend and it may resume attractor point behavior in the future. The Finnish case indicates attractor point behavior for energy intensity throughout the post-1973 period, although this may partly reflect the movements in GDP with its 1990s recession. Among Ireland, the United Kingdom, and the United States, though, there is clearer evidence of trend decline in intensity over this period; thus, although econometric estimates above pointed to a slowing down in the rate of decline of energy intensity for the United Kingdom and the United States, this analysis points to greater evidence of a continuous decline in energy intensity than for other sample countries.

We can investigate this further by decomposing changes in energy intensity into the effects of technological change and structural change in the economy (Howarth et al. 1991; de Bruyn 2000: ch. 9). This utilizes Laspeyres indices of change due to intensities holding sectoral shares constant, and change due to structural change holding intensities constant; over discrete intervals, this necessarily will produce a

Figure 1

Connected Scatterplots of Energy Intensities, 1973–2003

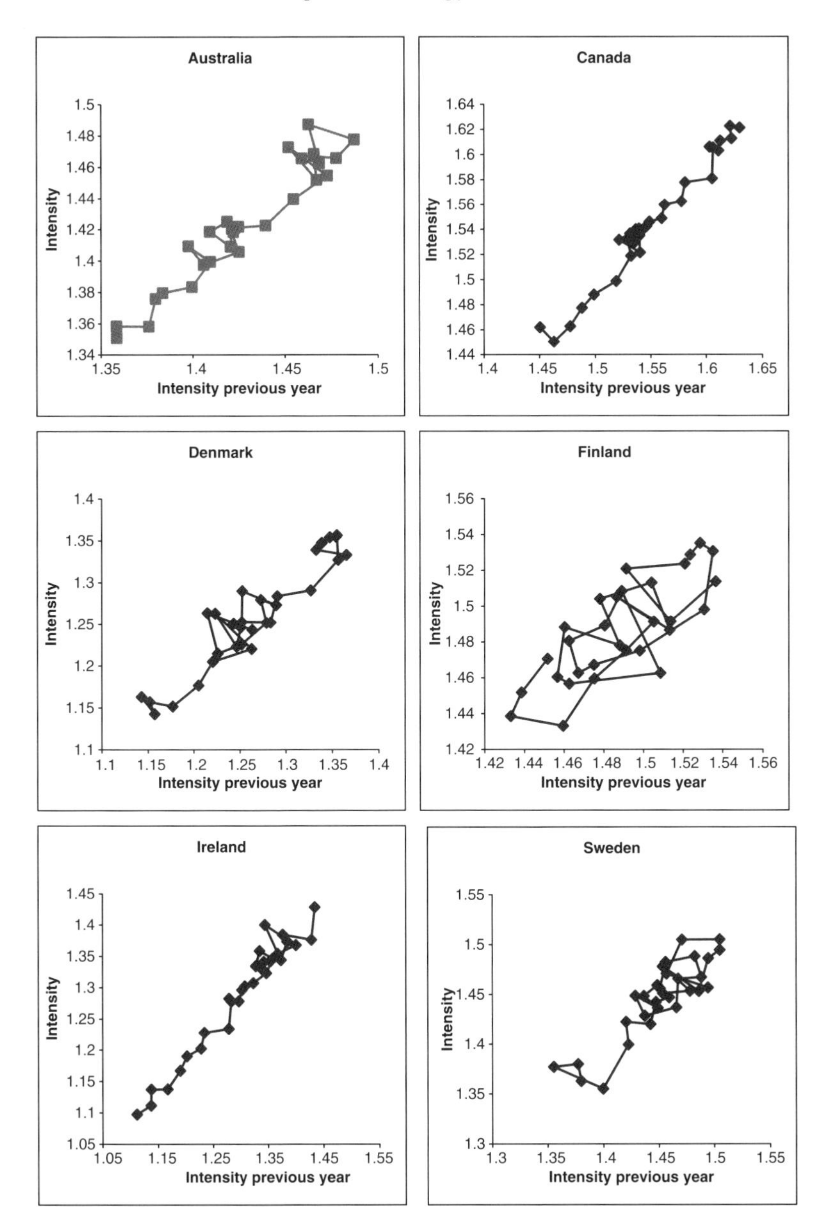

Figure 1 *Continued*

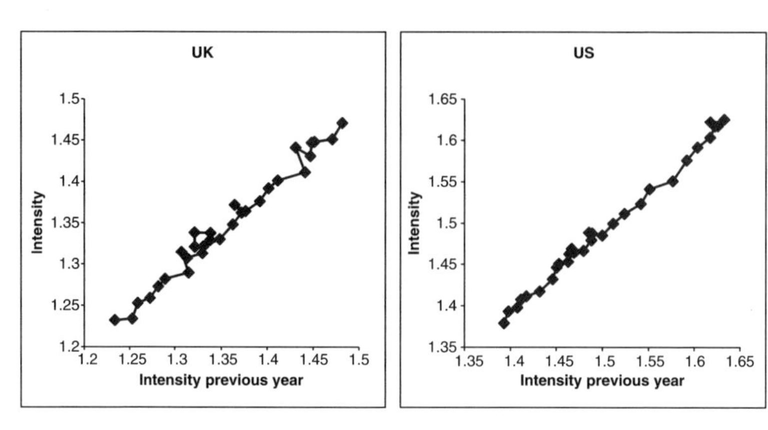

residual, but as Howarth et al. (1991) point out, this can be interpreted as an interaction term. While in principle ICT use could have a positive impact on energy intensity via both channels, the expected direct impact of ICT use in reducing energy consumption is limited; rather, the implication of the hypotheses discussed above is that the main impact of ICT use would be in promoting the expansion of sectors with low energy use. Unfortunately, for most of our sample countries suitably disaggregated data are not available; nevertheless, typically the transportation and residential sectors are the largest single consumers of energy and are unlikely to be significantly affected by ICT use. Table 5 reports the contribution of increased energy consumption by these sectors to total rise in energy consumption over 1990–2000; for the Anglo-Saxon countries, these sectors continue to account for the majority of increased energy consumption to the extent that in the Canadian case transportation accounts for more than the rise in all energy consumption, implying strong dematerialization in other production sectors. Patterns among the Scandinavian economies are more varied, but transportation is a major contributor in all cases except Finland, with the variation being largely down to differences in residential energy consumption. In this sense, arguments that continued output growth has become decoupled from growth in energy demand are borne out, at least for the

Table 5

Sectoral Contributions to Energy Consumption Growth, 1990–2000

	Transportation	Residential
Australia	43.0	13.5
Canada	115.3	15.2
Denmark	34.3	8.6
Finland	7.6	−2.2
Ireland	61.4	12.5
Sweden[1]	86.5	−76.4
U.K.	52.0	43.1
U.S.	62.3	17.5

[1] 1993–2000.

Anglo-Saxon countries; growth in production of goods and services makes a relatively small contribution to the growth in energy demand.

Detailed sectoral evidence on energy consumption is available over this period for U.S. manufacturing output, Canadian nonagricultural production, and U.K. output in respect of consumption of energy from fossil fuels. Table 6 reports decomposition results for these data. For Canada and the United Kingdom, the results clearly indicate that the effects of structural change are outweighed by the effects of falling intensities within sectors. The U.S. results are harder to interpret; the overall effects are distorted by the petroleum refining and coal sector, the most energy-intensive sector by an order of magnitude. Excluding this sector reverses the relative importance of sectoral change and changing intensities. Isolating the combined effects of those sectors that either produce ICT goods or use them intensively (following the classification of Ark et al. 2003) indicates that this only accounts for a minority of the decline in energy intensities,[4] which appears largely to be due to structural shifts and technological progress unrelated to new economy trends. Note that this is the whole contribution of these sectors to changes in energy intensities: not all of the industries within these categories will be ICT-intensive, and not all of their relative

Table 6

Decomposition of Change in Energy Intensity of Value Added, 1990–2000

	Canada	U.K.	U.S.[1]	U.S. (excl. petrol)
Change in Energy Intensity	−14.6	−17.3	−5.0	−5.4
Change due to:				
Structural change	−3.1	−4.8	16.5	−6.5
Sectoral intensities	−8.2	−11.6	−12.3	1.1
Interaction effect	−3.3	−0.9	9.2	0.0
Percentage due to:				
ICT sectors	31.5	7.0		4.0

[1] 1991–1998.

expansion or change in energy intensities can be attributed to ICTs. The relatively high Canadian figure partly reflects the inclusion of a catch-all "offices" sector in the ICT-using group. Thus, these decompositions are likely to overestimate the contribution of ICTs to reducing aggregate energy intensities.

Thus, although the energy *intensity* of GDP has fallen in developed economies, total *use* of energy continues to rise. With rising absolute levels of energy use, economic expansion won't necessarily be freed from material constraints, even if new technologies offer the potential for continuous productivity growth.

As noted above, orthodox accounts have typically imposed the assumption that changes in material supplies and their prices do not affect the growth rate because of substitution effects. This turns on whether capital and energy are substitutes—and, if so, at what elasticity—or complements in production. Standard analysis since Bruno and Sachs (1985) views energy as an input to the production process with limited substitution possibilities, at least in the short run.[5] In the short run, therefore, oil price rises have a significant negative impact on output and employment. The predicted absence of a longer-term impact on growth is dependent on substitutability of capital for energy. However, econometric studies have produced a wide range of estimates, some finding that capital and energy are complements (Frondel and Schmidt 2002; Neumayer 2000: 322–323), although recent studies find that over the medium term, capital and energy are substitutes at approximately unit elasticity, consistent with standard analysis (e.g., Thompson and Taylor 1995). Thus, energy price levels would not affect the long-run growth rate. Nevertheless, only limited reliance can be placed on this: "Overall, it has to be said that a satisfactory explanation for the variation in econometric studies has not been found yet and that we do not have a reliable answer on the question whether energy and manmade capital are substitutes or complements" (Neumayer 2000: 323–324). In particular, there are well-known conceptual problems with estimating aggregate production functions using cost data, and Frondel and Schmidt (2002) find that these results can be explained as an artifact of using cost data: with the standard static translog approach, the larger the cost shares of capital and energy the more likely the estimated cross-price elasticity is to be positive and thus indicate that the two are complements. Reported elasticity estimates in the literature vary in accordance with capital and energy cost shares as predicted.

Whatever view is taken of the long-term effects of energy prices, in the short term higher energy prices, particularly oil prices, significantly affect economic activity. Carruth et al. (1998) found that the real price of oil was strongly significant in explaining the U.S. unemployment rate throughout the postwar period, with better predictive powers than standard macroeconomic forecasts, even after accounting for the effects of any anti-inflationary policy response. Murchison and Siklos (1999) found that the price of oil and unemployment were cointegrated for developed economies. Some studies indicate that the effect of oil price changes is asymmetric, with price rises having a significantly larger effect on output than falls (Hamilton 2003). Recent simulations indicate that a substantial rise in energy prices would have a smaller negative impact than in earlier decades, in line with falling energy intensities, but would still be significant (Hunt et al. 2002; IMF 2005: ch. 4; OECD 2004: ch. 4). However, unlike earlier analyses, these simulations focus on the impact on inflation (and sometimes terms of trade effects), with limited modeling of the effects of energy prices on production. It thus remains unclear what the effects on output of sustained higher oil prices would be.

Forecasting energy supply and demand is notorious for providing either inaccurate forecasts or ones whose accuracy derives more from error canceling than accurate estimation of the underlying mechanisms (Snil 2003: ch. 3). The price of oil is scarcely determined in free market conditions, with partial cartelization of producer countries through OPEC, concentration among oil companies, and domination of demand by large consumer countries. Among producer countries, there is wide variation in extraction costs, with Middle East producers having by some distance the lowest extraction costs. The price of oil thus does not simply reflect its scarcity or extraction cost, but is the result of a complex interaction between partially cartelized producer countries, oligopolistic companies, and major purchaser countries. Real oil prices declined over the first half of the 1980s and remained low by post-1973 standards from 1986 (excepting the first Gulf War), until by 1998 they had fallen to levels comparable (in real terms) to pre-1973 prices; since then, oil prices have sharply risen in real and nominal terms from 1999 (IMF 2005: ch. 4; OECD 2004: ch. 4; Rutledge 2003). Real prices remain below 1979–80 peaks, which would be around $80 a barrel in current prices (IMF, 2005: ch. 4). Whereas analysts in the late 1990s could be found predicting continued low oil prices—even the 2005 U.S. Energy Department report expected prices in the $27–35 a barrel range (EIA 2005: 3)—one can now readily find prophesies of oil prices rising to $100 a barrel and persisting at such levels. Recent rises reflect not emerging scarcity, but in part attempts by OPEC to raise prices by restricting supply, limited investment in new extraction, and continued disruption to Iraqi supplies against growing global demand from expansion in China, the United States, and elsewhere (e.g., IMF 2005: ch. 4; Stevens 2005). Analyses of oil price developments point to at least some convergence of interests and tacit cooperation between the United States (as the world's largest oil consumer and second largest producer) and OPEC to maintain prices above some putative free market level (Goldstein et al. 1997; Rutledge 2003). While OPEC clearly wishes to maintain high prices, above a certain level these would trigger a global economic downturn in the short run and greater conservation measures and development of alternative energy sources over the longer term. For the United States, particularly after the fall of oil prices in the late 1990s, lower prices reduce the profitability of, and hence capacity investment in, oil production outside the Middle East (including in the United States itself). It is thus not just a defense of domestic oil producers but a more general security concern that lower prices would discourage development in the relatively costly Central Asian, Latin American, and West African oil fields and thus undermine U.S. efforts to diversify its supply away from Middle Eastern sources.

Despite earlier forecasts of flattening U.S. energy demand, it accounted for a fifth of the increase in global oil demand over 1995–2004 (OECD 2004: 129). While higher energy prices would be expected to have some effect on restraining demand growth, rebound effects tend to undermine gains from greater efficiency. The U.S. Department of Energy forecasts continued energy demand growth, with oil demand predicted to rise by 37 percent in 2020 and demand for transportation fuels to rise by 40 percent over the same period; by then, transportation alone is predicted to account for the same level of petroleum demand as total U.S. demand in 1999 (Rutledge 2005: ch. 9). Particularly in the United States, continued growth in oil demand is likely because of transportation, reflecting not only consumer decisions but also path-dependency effects of 20th-century urban planning and energy policy decisions to construct urban and suburban America around private car use and downgrade public transportion (Rutledge 2005: chs. 2 & 9; Roberts 2004).

Despite the dominance of the developed countries in energy demand—by 2000, the United States accounted for over a quarter of global primary energy demand and the G7 countries 45 percent, whereas the poorest quarter of the world's population only account for 2.5 percent (Snil 2003: 50)—the industrialization of major developing economies is projected to account for much of the future rise in energy demand. China alone accounted for a quarter of the rise in global oil demand over 1995–2004 (OECD 2004: 129). Nevertheless, over the 1980s, the energy intensity of China's GDP fell by 40 percent, and energy efficiency continued to rise at similar rates during the

1990s (Snil 2003: 139; Fisher-Vanden et al. 2004). Rising relative energy prices and greater economic efficiency helped achieve this, but its sustainability is less clear as China's industrial structure develops and demand for private autos continues to grow rapidly. In general, there is considerable uncertainty over the growth of energy demand in developing economies. Over the past 20 years, total commercial energy intensity of GDP has fallen among the lowest-income countries but has grown slightly among middle-income countries.[6] Optimistic scenarios expect increasing energy efficiency with industrialization among these countries but, as already noted, the presumed EKC relationship is not robust, and these scenarios are critically dependent upon these countries having access to relevant technology and pursuing appropriate energy policy and pricing choices (cf. Churchill 1993).

On the supply side, in the medium term, attention has focused on potential disruption to Middle Eastern oil supplies in particular (e.g., Rutledge 2005: chs. 10–11). One result of rising U.S. oil demand has been to increase U.S. dependence on imported oil, leading to concerted efforts to diversify its supplies away from Middle Eastern producers given obvious security concerns there. Attention has shifted to what has been dubbed "the new great game," with strategic rivalry over the oil fields of the former Soviet central Asian republics (Kleveman 2003; Rutledge 2005: ch. 8). Key players here are the United States, China through its growing energy demand, and Iran as a major regional oil producer. China in particular has been actively pursuing agreements with key producer countries to secure supplies and has been willing to pay a premium price to ensure this. Although the Caspian Sea region reserves may be large, they are landlocked; since the mid-1990s, the United States has been committed to a 1,000-mile pipeline project linking oil fields in Azerbaijan, and potentially Kazakhstan, through the southern Caucasus to refineries in Turkey in an effort to secure oil supplies. Nevertheless, early projections that Caspian region oil had the potential to rival major Middle Eastern producers appear over optimistic, with estimated reserves now comparable to smaller OPEC producers (Roberts 2004: ch. 2; Rutledge 2005: ch. 8); moreover, Central Asian oil production costs are three to four times those of Middle Eastern producers. For all the strategic significance attached by U.S. administrations since the mid-1990s to Central Asian oil, it is unlikely to be able to provide a major alternative source of supply.

Relatively high levels of medium-term investment levels would be required to meet projected increased demand (Birol 2005). Conventional forecasts, though, assume that ample global reserves of oil remain (EIA 2005; OECD 2004: ch. 4). As with coal previously, a century of analysts have prophesied the end of oil supplies only to be confounded: new reserves are discovered, search and extraction technologies improve, fuel efficiency rises, and alternative energy sources are being developed and falling in price (e.g. Snil 2003: ch. 4).

Matters may not be so simple over coming decades. Despite the slightly conspiratorial air of some of their writings, there are grounds for taking contemporary oil pessimists seriously. The key issue is not the end of oil supplies as such, but of consumption exceeding production levels as rising demand hits oil production, passing its supply peak with concomitant effects on prices. In the "Hubbert's peak" interpretation, supplies of oil follow a bell-shaped distribution over time. For these analysts, estimates of initial global oil reserves have been relatively steady since the 1970s, indicating that around 1.1 trillion barrels remain (of an estimated 2 trillion originally), about 40 years' consumption at current rates (Aleklett and Campbell 2003; Holmes and Jones 2003; Heinberg 2003: ch. 3; Roberts 2004: ch. 2).[7] New oil discoveries peaked in the 1960s and have been in decline since; there have been no major new oil field discoveries since the 1970s; oil extracted per foot drilled has also declined. Alternative energy sources are being developed, but their supply remain limited and their substitutability for oil is limited for key uses, particularly transportion (Heinberg 2003: ch. 4; Roberts 2004: ch. 8; Snil 2003: ch. 5).[8] World oil production is predicted to peak within the next 30 years on these projections, as early as the next 5 to 15 years on the most pessimistic end of the scale.

Critics charge that this confuses known with potential reserves and that production:reserves ratios have been fairly steady for decades (Lynch 2002, 2003: Snil 2003: ch. 4). The question is less whether oil supplies can be analyzed as following a "Hubbert's peak" than interpretation of relevant data. Essentially, critics accuse the pessimists of using very limited data to place a geological interpretation on an economic phenomenon: the rate of extraction, and exploration for new reserves, depends on price incentives, so that lower rates of output and/or of discovery of new reserves cannot simply be taken as indicating depletion. New technologies of discovery and extraction emerge, and their development is likely to be stimulated by higher prices. Since the 1980s, several major OPEC countries have significantly revised upward their estimated reserves. The U.S. Geological Service (USGS) estimates that proven oil reserves stand at around 1.7 trillion barrels (most of which is in the Middle East), with an estimated 900 billion barrels further lying undiscovered, although with continued growth in world oil consumption, this would lead to hitting peak production around 2040 (Holmes and Jones 2003; Roberts 2004: ch. 2). However, the USGS regards these estimates for undiscovered oil as at the lower end of their predictions, with the likely level of undiscovered oil to be another 1–1.5 trillion barrels.

Nevertheless, there are grounds for skepticism about these (revised) estimates. Enough is known about the geographical distribution of oil reserves to be reasonably sure that the "easy" oil discoveries are likely to be over. While OPEC countries had probably underestimated their reserves before the 1980s, subsequent revaluations appear to have gone beyond levels that might be justified by new discoveries or improved technologies and reflect instead OPEC policy of relating country sales quotas to estimated reserves levels (Roberts 2004: ch. 2). In 2005, the International Energy Agency, the International Monetary Fund, and the G7 all pressured OPEC to open its oil field estimates to independent audit. In 1985, Kuwait announced that it possessed 50 percent more oil than it had previously declared, and other OPEC states followed suit. In 2005, Kuwait still claims the same reserve levels as in 1985.[9] Matthew Simmons, an energy investor who advises the Bush administration, has recently produced detailed evidence indicating significant overestimation of Saudi oil reserves that may soon pass their peak (Simmons 2005). More evidence of the questionable nature of estimated reserves was provided in 2004 when Shell Oil Company reclassified a fifth of its "proven" reserves as "unproven" (Roberts 2004: ch. 7). The official U.S. estimates largely rely on estimates from producer countries and curiously described their upward revision in the late 1990s in these terms: "these adjustments to the USGS and MMS estimates are based on non-technical considerations that support domestic supply growth to the levels necessary to meet projected demand levels" (Energy Information Administration 1998: 217). Moreover, the dynamics of peak production would be likely to increase consuming nations' reliance on Middle Eastern oil supplies, as peak production is likely to occur earlier among other producer countries. The U.S. Department of Energy was sufficiently concerned to commission a report in February 2005 on the "Peaking of World Oil Production: Impacts, Mitigation, & Risk Management," although it has not officially been made publicly available.[10] The report concludes that "the bottom line is that no one knows with certainty when world oil production will reach a peak, but geologists have no doubt that it will happen," and that this has the potential for unprecedented disruption to developed economies. A nongeologist cannot be expected to adjudicate on these disputes, and the safest conclusion is surely that oil reserves remain fundamentally uncertain and that we lack the information necessary to estimate them reliably; but there are grounds for expecting limits to the rate of expansion of supply relative to demand over the short term and potentially beyond.

Rising demand for oil relative to limited supplies has led some analysts, notably Klare (2002), to predict intensifying geopolitical conflict over these resources. It hardly needs pointing out that oil is the key strategic resource and much of it is based in unstable countries with ongoing conflict in the Middle East and strategic rivalries over Central Asian oil fields. Nevertheless, there are a range of possible strategic responses, as indicated by the differences between the United States and European countries' policies toward Iran and Iraq (among other countries). These analysts tend to assume that dependence on oil determines foreign policy, particularly in the United States, whereas it may be more useful to examine how the underlying strategic assumptions of countries' foreign policies determine their policy toward oil producers (cf. Bromley 2005). Intensifying conflict over oil is clearly a distinct possibility, but it is not inevitable.

The optimists may well yet trump the pessimists with new oil discoveries, but even some relatively optimistic predictions predict a peak in oil extraction within decades against continued growth in demand. Despite dismissing pessimistic forecasts, one analyst concludes: "Whatever the future gains may be, the historical evidence is clear: higher efficiency of energy conversions leads eventually to higher, rather than lower, energy use, and eventually we will have to accept some limits on the global consumption of fuels and electricity" (Snil 2003: 317). We now turn to the issue of metals.

VI

Economic Expansion and Demand for Metals

DECIDING WHICH METALS to focus on requires some judgment. In 1946, the U.S. Congress passed the Strategic and Critical Materials Stockpiling Act, resulting in the U.S. government maintaining stocks of strategic materials for which the United States has high import dependence and for which overseas sources of supply face potential disruption. With the end of the Cold War, some of these threats have receded; many of the metals are only used in small quantities, but in strategic military industries. Some key metals are used in small quantities but with limited substitution possibilities in ICT equipment production, but our focus here is on the most heavily used metals, in descending order: steel, aluminum, copper, zinc, lead, and tin.

The difficulties encountered when attempting to model energy demand are, if anything, greater still with metals. Again, demand is derived rather indirectly for the product: growth in demand for metals can be decomposed into output growth, the material intensity of production of each commodity, and the changing product composition of output. However, sectoral data on use of metals is not generally available. Earlier studies found clear evidence of a structural break around 1973 in demand for the major metals, with demand for them falling significantly below their postwar trend growth rates (Cleveland and Ruth 1999; Roberts 1996; Tilton 1990: appendix). In part, this reflects the slowdown in GDP growth rates, but it also marks a switch in the nature of growth among the developed economies, from rising intensity of metal use in production to falling intensity (both from shifts in product mix and through resource-saving technology in particular industries). Both declining intensity of use and falling growth rates therefore contributed to lower growth or stagnation in demand for metals. Examining earlier versions of the argument that structural changes in the developed economies had decoupled economic growth from metals demand, Tilton (1989) concluded that this was premature and that declines in the metal intensity of output still left demand for metals dependent on economic growth, and argued that industrialization elsewhere in the world might act to increase demand for metals. Capital equipment, construction, and consumer durables production all tend to use metals intensively, although high-technology production and its associated capital equipment are relatively light in metals use. Recently, the industrialization of China and other major developing economies has sharply increased demand for metals. Although economic development may lead to a declining material content of production over time, in the short run, expansions often raise demand for raw materials as capital goods and construction tend to be intensive in these resources. Thus, in the short run, income elasticities of demand for metals may be relatively high (Tilton 1989).

However, it is unclear whether the material intensity of production declined continuously over the postwar period or whether there was a one-off shift in the intensity of metals use. Labson (1995) found that there was a clear structural break around 1973 in metals intensity by the major economies; however, he found that there was no clear trend in the intensity of metals use once this break was allowed for. Long series of data are only available for the world and the United States, but this does allow extension of post-1973 data points. Figures 2 and 3 show (logged) intensities of metal use relative to industrial production indices for the world and the United States.

At the global level, there are downward trends for lead, tin, and zinc, although this may be due more to substitution of other metals than a general dematerialization trend (Tilton 1990). Steel intensity also displays some decline since the 1970s, although this has risen again recently. Aluminum intensity has risen over the postwar period, while copper intensity fell to low levels in the 1970s and 1980s but has since recovered. Data for countries besides the United States are only available for a limited time period, and ratios may be

Figure 2

World Intensity of Use of Main Metals

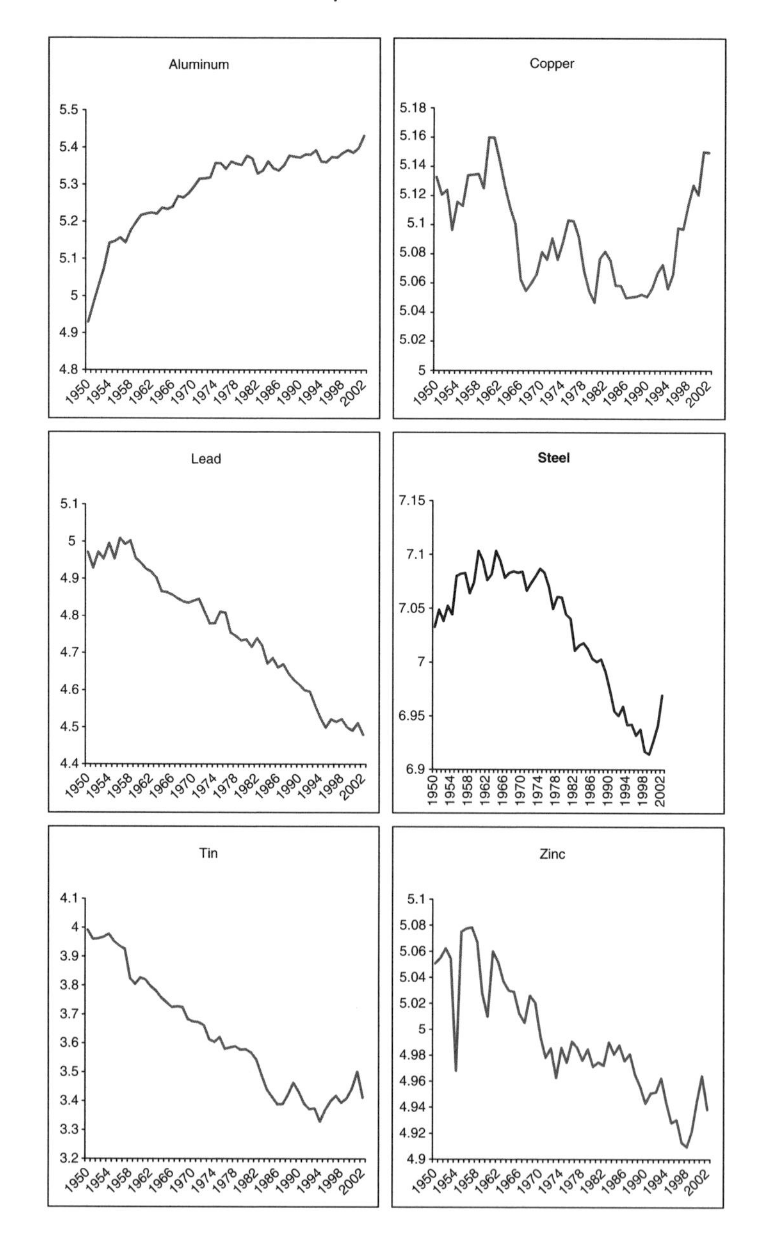

Figure 3

U.S. Intensity of Use of Major Metals

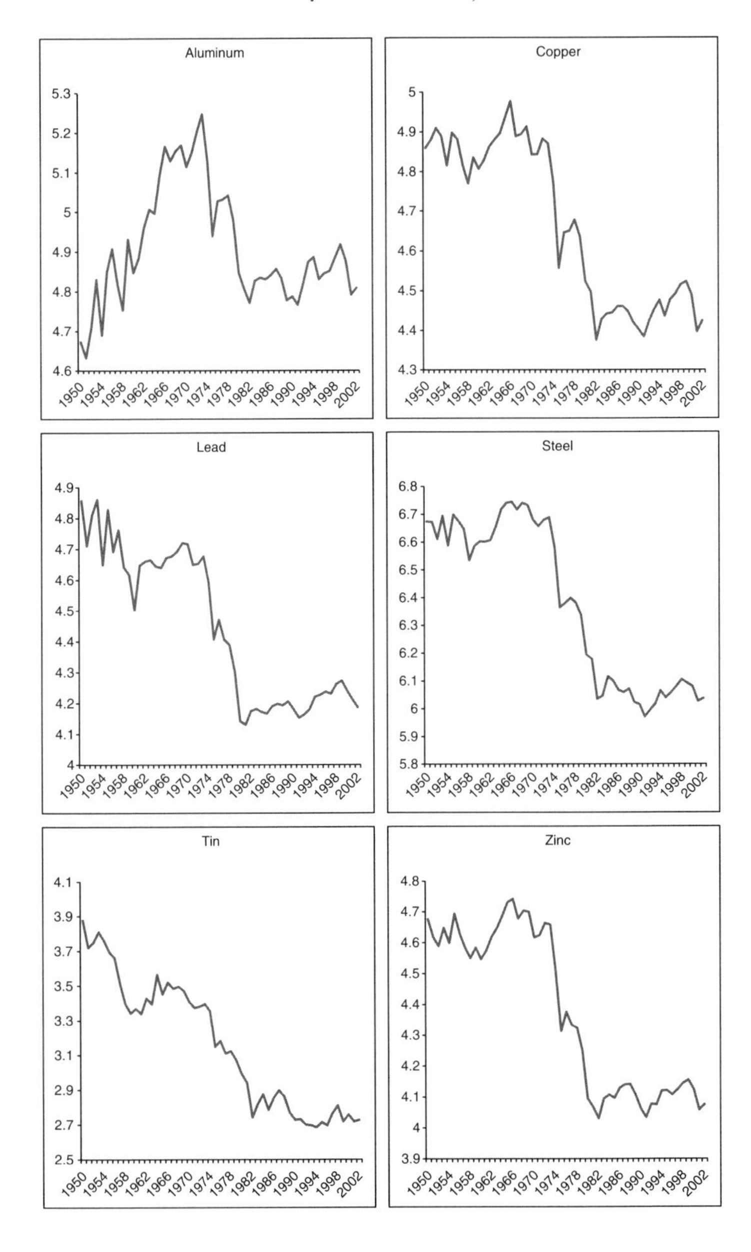

misleading with relatively small metal consumption figures. Therefore, further analysis is confined to the United States, although available figures do not indicate any clear trends in metals intensities over the 1990s for the other new economy countries.

For the United States, as shown in Figure 3, there are similar trends for all six metals: sharp falls in intensities over the 1970s and early 1980s and few clear trends thereafter. Not surprisingly, from a visual inspection of metals intensity trends, there is varied evidence of cointegration between metals consumption, industrial production, and relative prices either for the world or the United States. For four of the metals, a Johansen test indicated a cointegrating relationship for world demand; Table 7 reports estimates of an error correction equation of the following form:

$$m_t = \beta_1 \Delta q_t - \beta_2 \Delta p_t + \beta_3 (m_{t-1} - \alpha_0 - \alpha_1 q_{t-1} - \alpha_2 p_{t-1} - \alpha_3 t) + \mu_t, \quad (1)$$

where m is demand for the metal, q is the index of industrial output, p is the price of the metal relative to the producer price index, t is a time trend designed to capture technical progress (broadly defined to include the effects of structural change), and μ is a random error term (all variables except t are in logs). Structural stability tests were performed to test for breaks in 1973, 1980 (consistent with metals price data discussed below), and 1990 (to test the new economy thesis).

For these metals, predictably, the long-run results are largely driven by the close correlation between industrial production and metal demand, but for lead and steel there is also a significant—if small—trend decline in demand consistent with resource-saving technical progress. For aluminum, there is no evidence of a structural break in 1990 but strong evidence of earlier ones. For both lead and steel, there is evidence of a significant structural break around 1990, although with lead this does appear to be consistent with a new economy thesis, with actual values largely below fitted ones over the 1990s. This is less clear with steel. There is no evidence of a structural break for zinc.

For those metals for which the Johansen test did not indicate a cointegrating relationship, and in light of results indicating that industrial production and (less clearly) metals demand, was I(1), the following difference relationship was estimated:

Table 7

Error Correction Estimates of World Demand for Metals, 1957–2002

		Aluminum	Lead	Steel	Zinc
Short-run	β_1	0.99	0.48	1.11	0.97
adjustment		(6.30)***	(2.76)***	(8.58)***	(5.47)***
parameters	β_2	0.04	−0.02	0.04	−0.06
		(1.17)	(−0.53)	(0.33)	(−1.85)*
	β_3	−0.12	−0.44	−0.10	−0.44
		(−0.96)	(−3.47)***	(−1.10)	(−3.40)***
Long-run	α_0	4.64	4.88	6.76	5.18
equilibrium		(47.53)***	(43.93)***	(60.57)***	(48.52)***
parameters	α_1	1.41	1.10	1.26	0.92
		(20.18)***	(13.29)***	(14.30)***	(12.42)***
	α_2	0.05	−0.03	0.02	−0.08
		(1.31)	(−1.03)	(0.20)	(−2.83)***
	α_3	−0.001	−0.01	−0.01	−0.002
		(−1.12)	(−10.40)***	(7.06)***	(−2.02)**
R^2		0.99	0.89	0.97	0.98
DW		1.83	1.28	1.44	1.61
LM1		1.78	3.03*	0.78	1.19
ARCH		3.46*	0.27	0.49	0.17
JB		2.78	4.84*	7.52*	1.68
CHOW(73)		9.64***	1.66	5.57**	1.50
CHOW(80)		6.23***	2.67**	2.50**	0.64
CHOW(90)		0.73	7.28***	2.86**	0.96

DW is the Durbin-Watson statistic; *LM* is the F-Form of the Breusch-Godfrey test for a second-order serial correlation process in the residuals; *ARCH* is the F-Form of the LM test for a first-order ARCH process in the residuals; *NORM* is the Jarque-Bera test for normality of the residuals distributed as $\chi^2(2)$; *CHOW* is the F-Form of the Chow forecast test for parameter stability with a breakpoint in 1973, 1980, and 1990. Significance levels: ***, **, *—1%, 5%, 10% levels. *t*-statistics in parentheses.

$$\Delta m_t = \gamma_0 + \gamma_1 \Delta q_t - \gamma_2 m_{t-1} + \gamma_3 q_{t-1} - \gamma_4 p_{t-1} + \mu_t. \quad (2)$$

The results are reported in Table 8. For the two metals estimated using Equation (2), the fit is generally poor; although the estimates imply a long-run income elasticity of demand above unity for copper, it is not significantly above unity on a Wald test; the results for tin indicate a long-run income elasticity of demand significantly below unity. Although Chow tests indicate a structural break in 1990 for both metals, it is not clear that this indicates systematically lower-than-predicted demand over the 1990s, and in both cases there is also evidence of significant structural breaks earlier. In general, little reliance

Table 8

Estimates of World Demand for Metals, 1957–2002

	Copper	Tin
γ_0	0.13	0.72
	(0.27)	(1.48)
γ_1	0.64	1.03
	(4.41)***	(3.13)***
γ_2	–0.03	–0.16
	(–0.32)	(–1.68)
γ_3	0.04	0.07
	(0.48)	(2.11)**
γ_4	0.01	0.01
	(0.35)	(0.82)
R^2	0.35	0.29
DW	1.71	1.52
LM1	0.53	1.21
ARCH	0.91	8.78***
JB	7.33**	0.67
CHOW(73)	2.95**	2.07*
CHOW(80)	1.69	4.57***
CHOW(90)	3.04**	5.18***

Key: See Table 8.

can be put on results for these two metals given the low power of the estimation.

Estimating Equation (1) for U.S. data where a Johansen test indicates a cointegrating relationship and Equation (2) where it does not gives the results reported in Tables 9 and 10. Not surprisingly, from a visual inspection of the plots in Figure 3 for most metals, testing did not indicate a cointegrating relationship. The fit for aluminum and steel is generally good, with significant negative price coefficients as well as significant positive income elasticity estimates. For aluminum, the trend is segmented for 1950–1973 and 1974–2002, indicating that technological change was resource-saving over the latter period.[11] For steel, there is evidence of resource-saving trends throughout the postwar period, but only weak evidence of a structural break after 1990 (and roughly as strong evidence for a structural break in 1980).

For the other four metals, estimates of U.S. demand produced mixed results. Only for zinc was the price coefficient negative and significant. Except for tin, the lagged industrial output coefficient was positive and significant; in all cases the implied long-run income elasticity of demand was below unity (significantly so on Wald tests in each case). There is some evidence, except for tin, of earlier structural breaks but no clear evidence for the new economy thesis. Overall, the picture for these four metals is one of sluggish postwar demand growth in the United States, a trend that the new economy has not significantly accelerated. (For estimates of Equation (2), including a trend term was not significant for either world or U.S. demand.)

Demand trends can also be examined in terms of price movements. Ocampo and Parra (2003) model the behavior of primary commodity prices (both mineral and agricultural) since 1900. In general, the relative decline of primary commodity prices over the 20th century can be accounted for by a steep fall around the time of the Great Depression with no trend thereafter and another structural break around 1980 followed by a negative price trend. These findings hold for metals in general, too; modeling specific metals prices produces more variable results, but the general picture is of falling metals prices since 1980. This is consistent with the new economy thesis to the extent that it is consistent with dematerialization trends among

Table 9

Error Correction Estimates of U.S. Demand for Metals, 1950–2002

		Aluminum	Steel
Short-run adjustment parameters	β_1	0.99	2.33
		(6.30)***	(13.11)***
	β_2	0.04	−0.07
		(1.17)	(−0.37)
	β_3	−0.12	−0.22
		(−0.96)	(−3.22)***
Long-run equilibrium parameters	α_0	4.47	6.30
		(21.65)***	(23.97)***
	α_1	1.59	1.56
		(20.24)***	(6.97)***
	α_2	−0.33	−0.37
		(−2.57)***	(−2.30)**
	α_3		−0.02
			(−6.78)***
	α_4	0.001	
		(0.62)	
	α_5	−0.01	
		(−5.70)***	
R^2		0.95	0.51
DW		1.89	1.64
LM1		1.01	1.62
ARCH		1.59	0.56
JB		1.95	3.73
CHOW(73)		—	1.91
CHOW(80)		—	3.11**
CHOW(90)		—	2.36*

Key: See Table 8.

Table 10

Estimates of U.S. Demand for Metals, 1950–2002

	Copper	Lead	Tin	Zinc
γ_0	2.90	2.61	1.19	1.74
	(6.24)***	(4.09)***	(2.39)**	(5.06)***
γ_1	1.94	1.47	1.50	1.74
	(11.33)***	(3.79)***	(3.60)***	(9.62)***
γ_2	–0.53	–0.49	–0.26	–0.29
	(–6.42)***	(–4.31)***	(–2.87)***	(–4.89)***
γ_3	0.24	0.17	–0.04	0.07
	(6.16)***	(3.25)***	(–0.71)	(3.61)***
γ_4	0.01	0.07	0.03	–0.11
	(0.20)	(0.89)	(0.63)	(2.55)***
R^2	0.81	0.44	0.36	0.73
DW	2.43	2.15	2.48	1.87
LM1	2.15	0.51	2.59*	0.25
ARCH	0.00	5.66**	1.68	1.72
JB	2.32	0.31	0.23	0.80
CHOW(73)	0.71	3.25**	1.03	2.98**
CHOW(80)	2.15*	2.36**	1.01	1.75
CHOW(90)	0.92	0.76	0.65	0.29

Key: See Table 8.

developed countries, although it predates the ICT investment boom. Nevertheless, short-term volatility tends to predominate any long-term trends in commodities prices. As Table 10 indicates, in the first half of this decade the composite metals price index has risen sharply relative to post-1980 values. Part of this reflects cyclical upturns in demand relative to inelastic short-run supply; it also reflects sharply increased demand from developing country industrialization, especially in China. Markets for leading commodities remain cartelized (LeClair 2000), even after the collapse of price stabilization schemes in the 1980s, but the effects of this on prices and depletion rates remains unclear. Recent price rises are unlikely to reflect significantly increased scarcity, with estimated reserves ample relative to demand (Tilton 2003). The longer-term trends of demand for resources have largely continued, with sluggish growth in demand (with primary energy sources) or little trend growth (with metals). In this sense, the demand problems that have faced resource exporters in the 1980s and 1990s are likely to continue. Cyclical factors continue to be particularly important, both on the demand and the supply sides, and the potential for short-run price rises is evident for some metals. The safest prediction is that cyclical factors will continue to dominate.

Earlier evidence indicated that energy prices played a significant causal role in postwar economic activity among developed economies, and this appears to be the case for commodity prices in general (Labys and Maizels 1993). The early 1980s collapse in commodity prices played a key role in reducing developed country inflation and laying the basis for subsequent expansion (Beckerman and Jenkinson 1986). The ability of the United States and other economies to sustain expansion over the 1990s without igniting inflation was underpinned in part by continued low commodity prices; thus, low commodity prices played a role in sustaining the 1990s new economy boom. As noted above, recent macroeconomic forecasting models frequently don't examine the direct effects of higher energy (or com-

Table 11

Indices of Metals Prices, 1974–2005 (2000 = 100)

	All Metals	Aluminum	Copper	Lead	Tin	Zinc
1974	78.7	49.3	113.3	130.1	150.6	109.7
1980	119.0	114.4	120.4	199.3	308.6	67.5
1990	120.0	105.7	146.7	178.2	111.9	134.6
2001	90.2	93.2	87.1	104.9	82.6	78.6
2002	87.7	87.1	86.0	99.6	74.7	69.1
2003	98.4	92.4	98.1	113.2	90.0	73.4
2004	134.0	110.8	157.8	194.2	156.0	92.9
2005*	164.4	118.6	191.1	209.3	141.6	114.8

*First three quarters.

modity) prices on output and focus instead on their inflationary impacts. Although the resource intensity of GDP has fallen, it remains unclear precisely how commodity prices contributed to sustaining noninflationary growth and what impact sustained higher energy and commodity prices would have on output and inflation among developed economies, even if the source of such price rises is as likely to come from increased demand by industrializing developing economies as from developed economies.

Overall, the evidence for the leading metals does not support notions that a new economy based on ICT technologies is leading to reduced material intensity of production. In keeping with earlier patterns, economic upturns tend to increase demand for materials in the short run, at least since equipment and construction are relatively intensive in the use of materials. After sharp declines in metals prices in the 1980s and early 1990s, they showed signs of recovery in this decade, with rising demand for aluminum and copper in particular. One can thus concur with Tilton's conclusion before the rise of the new economy that, although the metals intensity of income has declined, growth has not been decoupled from metal resources (Tilton 1989).

VII

Global Dimensions

ONE OBVIOUS QUESTION HERE is whether trade specialization means that developed economies have reduced their domestic resource use but increasingly source resources from abroad instead. If this is the case, then derived demand for natural resources would be satisfied by overseas production, but the effective resource intensity of GDP (and its growth) would have fallen by less than these figures would indicate.

This can partly be addressed by updating the analysis of Atkinson and Hamilton (2002). They examined the difference between resource depletion in various countries and found that, once adjusted for an "ecological balance of payments," the depletion was accounted for by domestic absorption. Assuming that production of traded goods depletes resources at the same rate as national income, exports are netted out of this depletion, but the derived impact of imports on resource depletion in the source country is added in. Such an analysis is subject to two key limitations, both of which are likely to bias estimates of the impact of domestic absorption on depletion downward: first, in assuming that a country's exports deplete resources at the average rate for national income, whereas for resource exporters in particular the export sector is likely account for more than an average share of depletion; second, this approach cannot account for the derived demand effect, in which imports from one country are produced using depleted resources imported from a third country. Nevertheless, in the absence of detailed input-output data for many countries, this approach can provide an approximate indicator of the global impact on resource depletion of domestic economic activity. Recent advances in green national accounting mean that estimates of domestic resource depletion are readily available; the average (national income-weighted) global rate of depletion fell slightly on these figures, from 4.1 percent to 3.8 percent of global income over this period; nevertheless, this is still above the 1 percent figures often suggested. These are presented for sample countries in Table 12; they show no clear pattern over the 1990s.

These adjustments do not make a dramatic difference to depletion rates in these economies, although they do reduce the dispersion among this group. Australia shifted from being a net resource-depletion exporter to importer over this period; reforms to the economy over this period saw a shift away from resource-based industries. Canada has a relatively high rate of resource depletion by developed country standards, and this rose during the 1990s; it was also the only country of this sample that was consistently a resource-depletion exporter over this period. The relatively high depletion import figures for Finland and Sweden largely reflect their high level of trade with Russia. While acknowledging the caveats with this analysis, the analysis does not indicate that low rate of depletion among developed economies simply reflects displacing this onto other economies through trade. It is not surprising that this adjustment makes relatively little difference to the impact of these economies on resource depletion, since most of developed countries' trade is with other developed countries that typically have low rates of resource depletion.

Table 12

Ecological Balance of Payments, 1990–2000

	Energy Depletion	Mineral Depletion	Total	Depletion Exported $ Millions	Depletion Imported $ Millions	Adjusted Total Depletion
Australia						
1990	1.38	2.05	3.42	1,332.6	1,094.6	3.33
2000	1.81	1.53	3.34	2,108.5	2,166.7	3.35
Canada						
1990	2.60	0.67	3.27	4,134.8	2,703.0	2.98
2000	4.93	0.18	5.11	14,061.9	5,139.6	4.02
Denmark						
1990	0.30	0.02	0.32	108.9	496.6	0.70
2000	0.87	0.00	0.87	441.6	727.3	1.06
Finland						
1990	0.00	0.05	0.05	13.3	890.4	1.04
2000	0.00	0.02	0.02	9.2	1,637.9	1.27
Ireland						
1990	0.17	0.28	0.45	107.0	204.0	0.69
2000	0.05	0.10	0.15	114.5	558.7	0.59
Sweden						
1990	0.00	0.26	0.26	138.0	877.5	0.74
2000	0.00	0.07	0.07	61.5	1,382.5	0.65
U.K.						
1990	0.72	0.00	0.72	1,332.7	3,889.2	0.97
2000	1.15	0.00	1.15	3,252.3	6,330.2	1.34
U.S.						
1990	1.49	0.07	1.56	6,132.5	27,143.4	1.92
2000	1.15	0.02	1.17	9,032.3	63,754.3	1.73

Resource exporters may still be depleting their deposits at unsustainable rates in order to earn foreign exchange in the face of prices that do not fully reflect social costs.

VIII

Conclusions

THIS PAPER HAS EXAMINED whether the new economy based on ICT economies has led to a "weightless world" in the sense of significantly reducing the material content of production. The conclusion is largely negative. The expansion of the services sector is not clearly leading to a growth of a dematerialized weightless sector, as much of the services sector remains associated with goods production. The energy intensity of GDP in developed economies declined in the 1990s, but this is largely a continuation of earlier trends. The relative growth of sectors that produce ICT goods or use them intensively do not appear to explain much of the decline in energy-GDP intensities among the largest new economy countries. Nevertheless, much of the growth in demand for energy is driven by the transportation and residential sectors, rather than by growth in production of goods and services. It remains unclear, though, whether this is driven by continuous technical progress (and structural change) more or less associated with economic growth processes, or whether it is driven by discrete changes in energy efficiency. If the latter is the case, then forecasting future demand is rendered particularly problematic. It seems likely, though, that developed country demand will continue to grow as demand rises from industrialization in key developing economies. With short-term limitations of capacity and possible longer-term supply constraints, this is likely to increase the importance of strategic interaction between major oil producer nations and consumer nations, although the dynamics of this relationship are not predetermined.

With demand for the major metals, there was again considerable ambiguity as to whether there were clear declines in demand over time or discrete changes at key intervals. In general, it does not appear either that growth of income in developed economies has a simple link with resource demand or that the technological processes driving growth in these economies is clearly associated with greater resource efficiency and declining material content. Technology does offer possibilities for enhanced efficiency in resource use—and, associated with this, moderating the emission of greenhouse gases—but this is not automatic and requires supporting institutions and public policy.

As noted at the outset, predictions about the future demand and supply of resources have been fraught with dangers. Claims that structural changes will substantially reduce the material content of production have often turned out to be overstated; authors have often overlooked that declining content is still consistent with rising absolute consumption levels. The evidence does not point to resource demands simply being displaced onto importing countries, although this needs further investigation. Nevertheless, the low commodity prices that have helped limit inflation during upturns over the past two decades show signs of coming to an end. Resources may have a more significant impact on output and inflation in the 21st century than they had at the end of the 20th.

Notes

1. This pessimistic prediction lay behind Thomas Carlyle's oft-cited characterization of economics as "the dismal science."
2. See also the special issue of *Ecological Economics* 2(3) (Sept. 1997).
3. Note that the success of Scandinavian economies in ICTs indicates that the new economy is not limited to deregulated market economies but is also compatible with market economies characterized by centralized wage bargaining systems and extensive welfare states financed by relatively high taxation levels.
4. Although some have claimed that direct use of ICT equipment gives rise to significantly increased electricity demand, this appears to be based on faulty estimates (Laitner 2003).
5. Some authors have pointed out that all work is applied effort and therefore requires energy. While this must be true from the laws of physics, it still leaves open precise elasticities in relation to particular energy inputs and product outputs. More fundamentally, there are formidable problems in deriving an economic value theory from energy concepts, as Mirowski (1988, 1989) has explored in detail.
6. From: "Energy Indicators," *OPEC Review* 28(4) (Dec. 2004).
7. These assessments are based in part on an analysis from former oil

company geologist Colin Campbell, available at http://www.mbendi.co.za/indy/oilg/p0070.htm.

8. Supplies of natural gas are estimated to be greater than for petroleum, and gas has considerable potential as a "bridge fuel" to reduce reliance on oil; but its location tends to be closely related to petroleum, and there are greater logistic difficulties in its transportation and use. The controversial nature of nuclear power hardly needs stating. Potential energy from renewable sources is unlikely to provide more than a fraction of current energy demand in the near future.

9. See Adam Porter (July 15, 2005). "How much oil do we really have?" Available at http://news.bbc.co.uk/1/hi/business/4681935.stm.

10. A copy has been posted online. See Robert L. Hirsch, Roger Bezdek, and Robert Wendling. (February 2005). "Peaking of World Oil Production: Impacts, Mitigation, & Risk Management." U.S. Department of Energy. Available at http://www.hubbertpeak.com/us/NETL/OilPeaking.pdf.

11. This was estimated for all metals where a cointegrating relationship was indicated, but in all other cases did not significantly improve the fit of the estimation.

References

Aleklett, K., and C. Campbell. (2003). "The Peak and Decline of World Oil and Gas Production." *Minerals and Energy* 18(1): 5–20.

Ark, B., R. van Inklaar, and R. McGuckin. (2003). "'Changing Gear': Productivity, ICT and Services Industries in Europe and the United States." In *The Industrial Dynamics of the New Digital Economy*, eds. J. Christensen and P. Maskell. Cheltenham, UK: Edward Elgar.

Atkinson, G., and K. Hamilton. (2002). "International Trade and the 'Ecological Balance of Payments'." *Resources Policy* 28: 27–37.

Beckerman, W., and T. Jenkinson. (1986). "What Stopped the Inflation: Unemployment or Commodity Prices?" *Economic Journal* 96: 39–54.

Birol, F. (2005). "The Investment Implications of Global Energy Trends." *Oxford Review of Economic Policy* 21(1): 145–153.

Blades, D. (1987). "Goods and Services in OECD Economies." *OECD Economic Studies* 8: 159–184.

Bromley, S. (2005). "The United States and the Control of World Oil." *Government and Opposition* 40(2): 225–255.

Bruno, M., and J. Sachs. (1985). *Economics of Worldwide Stagflation*. Oxford: Basil Blackwell.

Bruyn, S. de. (2000). *Economic Growth and the Environment*. Dordrecht: Kluwer.

Carruth, A., M. Hooker, and A. Oswald. (1998). "Unemployment Equilibria and Input Prices: Theory and Evidence from the United States." *Review of Economics and Statistics* 80: 621–628.

Chichilnisky, G. (1998). "The Knowledge Revolution." *Journal of International Trade and Economic Development* 7: 39–54.

Churchill, A. (1993). "Energy Demand and Supply in the Developing World, 1990–2020." In *Proceedings of the World Bank Annual Conference of Development Economics 1993*. World Bank.

Cleveland, C., and M. Ruth. (1999). "Indicators of Dematerialization and the Materials Intensity of Use." *Journal of Industrial Ecology* 2(3): 15–50.

Cohen, D., P. Garibaldi, and S. Scarpetta. (2004). *The ICT Revolution: Productivity Differences and the Digital Divide*. Oxford: Oxford University Press.

Colecchia, A., and P. Schreyer. (2002). "ICT Investment and Economic Growth in the 1990s." *Review of Economic Dynamics* 5(2): 408–442.

Cornwall, J., and W. Cornwall. (2002). "A Demand and Supply Analysis of Productivity Growth." *Structural Change and Economic Dynamics* 13(2): 203–229.

Coyle, D. (1999). *The Weightless World: Strategies for Managing the Digital Economy*. Cambridge: MIT Press.

Daly, H. (1977). *Steady-State Economics*. San Francisco: W. H. Freeman.

Daveri, F. (2002). "The New Economy in Europe, 1992–2001." *Oxford Review of Economic Policy* 18(3): 345–362.

Daveri, F., and O. Silva. (2004). "Not Only Nokia: What Finland Tells Us about New Economy Growth." *Economic Policy* 38: 117–163.

Energy Information Administration. (Various years). *Annual Energy Outlook*. Washington, DC: Energy Information Administration.

Eurostat. (2002). *Material Use in the European Union 1980–2000*. Luxembourg: Office for Official Publications of the European Communities.

Fisher-Vanden, K., G. Jefferson, H. Liu, and Q. Tao. (2004). "What is Driving China's Decline in Energy Intensity?" *Resource and Energy Economics* 26: 77–97.

Frondel, M., and C. Schmidt. (2002). "The Capital-Energy Controversy: An Artifact of Cost Shares." *Energy Journal* 23(3): 53–79.

Gallouj, F. (2002). "Innovation in Services and the Attendant Old and New Myths." *Journal of Socio-Economics* 31(2): 137–154.

Georgescu-Roegen, N. (1971). *The Entropy Law and the Economic Process*. Cambridge, MA: Harvard University Press.

Glyn, A., A. Hughes, A. Lipietz, and A. Singh. (1990). "The Rise and Fall of the Golden Age." In *The Golden Age of Capitalism*, eds. S. Marglin and J. Schor. Oxford: Oxford University Press.

Goldstein, J., X. Huang, and B. Akan. (1997). "Energy in the World Economy, 1950–1992." *International Studies Quarterly* 41(2): 241–266.

Gordon, R. (2000). "Does the 'New Economy' Measure Up to the Great Inventions of the Past?" *Journal of Economic Perspectives* 14(4): 49–74.

Hamilton, J. (2003). "What Is an Oil Shock?" *Journal of Econometrics* 113(2): 363–398.

Hannesson, R. (2002). "Energy Use and GDP Growth, 1950–1997." *OPEC Review* 26(3): 215–233.

Hartwick, J. (1977). "Intergenerational Equity and the Investing of Rents from Exhaustible Resources." *American Economic Review* 67(5): 972–974.

Heinberg, R. (2003). *The Party's Over: Oil, War and the Fate of Industrial Societies.* Gabriola Island, Canada: New Society Publishers.

Hodgson, G. (1999). *Economics and Utopia: Why the Learning Economy is Not the End of History.* London: Routledge.

Holmes, B., and N. Jones. (2003). "Brace Yourself for the End of Cheap Oil." *New Scientist* 2(August): 9–10.

Howarth, R., et al. (1991). "Manufacturing Energy Use in Eight OECD Countries." *Energy Economics* 13(2): 135–142.

Hunt, B., P. Isard, and D. Laxton. (2002). "The Macroeconomic Effects of Higher Oil Prices." *National Institute Economic Review* 179: 87–103.

International Monetary Fund. (2005). *World Economic Outlook.* April. Washington, DC: International Monetary Fund.

Jorgenson, D. (2002). "Information Technology and the U.S. Economy." In *Economic Policy Issues of the New Economy*, ed. H. Siebert. Berlin: Springer.

Jorgenson, D., and K. Stiroh. (2000). "Raising the Speed Limit: U.S. Economic Growth in the Information Age." *Brookings Papers on Economic Activity* 1: 125–211.

Klare, M. (2002). *Resource Wars: The New Landscape of Global Conflict.* New York: Henry Holt.

Kleveman, L. (2003). *The New Great Game: Blood and Oil in Central Asia.* New York: Atlantic Monthly Press.

Labson, B. S. (1995). "Stochastic Trends and Structural Breaks in the Intensity of Metals Use." *Journal of Environmental Economics and Management* 29: S34–S42.

Labys, W., and A. Maizels. (1993). "Commodity Price Fluctuations and Macroeconomic Adjustments in the Developed Economies." *Journal of Policy Modeling* 15(3): 335–352.

Laitner, J. (2003). "Information Technology and U.S. Energy Consumption." *Journal of Industrial Ecology* 6(2): 13–24.

LeClair, M. (2000). *International Commodity Markets and the Role of Cartels.* Armonk, NY: Sharpe.

Lynch, M. (2002). "Forecasting Oil Supply: Theory and Practice." *Quarterly Review of Economics and Finance* 42(2): 373–389.

——. (2003). "The New Pessimism about Petroleum Resources." *Minerals and Energy* 18(1): 21–32.

Matos, G., and L. Wagner. (1998). "Consumption of Materials in the United States, 1900–1995." *Annual Review of Energy and the Environment* 23: 107–122.

Meadows, D., et al. (1972). *The Limits to Growth: A Report for the Club of Rome's Project on the Predicament of Mankind.* New York: Universe Books.

Mirowski, P. (1988). "Energy and Energetics in Economic Theory." *Journal of Economic Issues* 22(3): 811–830.

——. (1989). *More Heat than Light.* Cambridge: Cambridge University Press.

Murchison, S., and P. Siklos. (1999). "A Suggestion for a Simple Cross-Country Empirical Proxy for Trend Unemployment." *Applied Economics Letters* 6(7): 447–451.

Neumayer, E. (2000). "Scarce or Abundant? The Economics of Natural Resource Availability." *Journal of Economic Surveys* 14(3): 307–335.

Nordhaus, W. (1974). "Resources as a Constraint on Growth." *American Economic Review* 64(2): 22–26.

Ocampo, J. A., and M. A. Parra. (2003). "The Terms of Trade for Commodities in the Twentieth Century." *CEPAL Review* 79: 7–35.

Organisation for Economic Co-Operation and Development. (2004). *Economic Outlook No. 76.* Paris: Organisation for Economic Co-Operation and Development.

Ormerod, P. (1994). *The Death of Economics.* London: Faber and Faber.

Oulton, N. (2001). "Must the Growth Rate Decline? Baumol's Unbalanced Growth Revisited." *Oxford Economic Papers* 53(4): 605–627.

Quah, D. (1997). "Increasingly Weightless Economies." *Bank of England Quarterly Bulletin* 37(1): 49–56.

——. (2001a). *Technology Dissemination and Economic Growth: Some Lessons for the New Economy.* London School of Economics, Centre for Economic Performance Discussion Paper No. 522.

——. (2001b). "The Weightless Economy in Economic Development." In *Information Technology, Productivity, and Economic Growth,* ed. M. Pohjola. Oxford: Oxford University Press.

Roberts, M. (1996). "Metal Use and the World Economy." *Resources Policy* 22: 183–196.

Roberts, P. (2004). *The End of Oil: The Decline of the Petroleum Economy and the Rise of a New Energy Order.* London: Bloomsbury.

Rowthorn, R., and Ramaswamay. (1997). "Deindustrialization: Causes and Consequences." *Staff Studies for the World Economic Outlook.* Washington, DC: International Monetary Fund.

Rowthorn, R., and J. Wells. (1987). *Deindustrialization and Foreign Trade.* Cambridge: Cambridge University Press.

Rutledge, I. (2003). "Profitability and Supply Price in the U.S. Domestic Oil Industry: Implications for the Political Economy of Oil in the Twenty-First Century." *Cambridge Journal of Economics* 27: 1–23.

——. (2005). *Addicted to Oil: America's Relentless Drive for Energy Security.* London, New York: I. B. Tauris.

Sagoff, M. (1997). "Do We Consume Too Much?" *Atlantic Monthly* 80–96.

Simmons, M. (2005). *Twilight in the Desert: The Coming Saudi Oil Shock and the World Economy.* New York: John Wiley.

Snil, V. (2003). *Energy at the Crossroads: Global Perspectives and Uncertainties.* Cambridge: MIT Press.

Solow, R. (1974). "The Economics of Resources or the Resources of Economics." *American Economic Review* 64(2): 1–14.

Stevens, P. (2005). "Oil Markets." *Oxford Review of Economic Policy* 21(1): 19–42.

Stiroh, K. (2002a). "Are ICT Spillovers Driving the New Economy?" *Review of Income and Wealth* 48(1): 33–57.

——. (2002b). "New and Old Economics in the New Economy." In *Economic Policy Issues of the New Economy,* ed. H. Siebert. Berlin: Springer.

Summers, R., and A. Heston. (1991). "The Penn World Table (Mark 5): An Expanded Set of International Comparisons, 1950–1988." *Quarterly Journal of Economics* 106(2): 327–368.

Temple, J. (2002). "The Assessment: The New Economy." *Oxford Review of Economic Policy* 18(3): 241–264.

Thompson, G. (2004). "The U.S. Economy in the 1990s." In *Where are National Capitalisms Now,* eds. J. Perraton and B. Clift. Basingstoke, UK: Palgrave.

Thompson, P., and T. Taylor. (1995). "The Capital-Energy Substitutability Debate." *Review of Economics and Statistics* 77(3): 565–569.

Tilton, J. (1989). "The New View of Minerals and Economic Growth." *Economic Record* 65: 265–278.

——. (ed.). (1990). *World Metal Demand: Trends and Prospects.* Washington, DC: Resources for the Future.

——. (2003). *On Borrowed Time? Assessing the Threat of Mineral Depletion.* Washington, DC: Resources for the Future.

Triplett, J., and B. Bosworth. (2003). "Productivity Measurement Issues in Services Industries: 'Baumol's Disease' Has Been Cured." *Federal Reserve Bank of New York Economic Policy Review* 9(3): 23–33.

Williams, E., R. Ayres, and M. Heller. (2002). "The 1.7 Kilogram Microchip: Energy and Material Use in the Production of Semiconductor Devices." *Environmental Science and Technology* 36(24): 5504–5510.

World Resources Institute. (2000). *Weight of Nations: Material Outflows from Industrial Economies.* Washington, DC: World Resources Institute.

Appendix: Data Sources

Real GDP (1995 USD, PPP basis): OECD, *Economic Outlook*, data CD-ROM.

Total Primary Energy Supplies: OECD/IEA, *Energy Balances of OECD Countries*, various issues.

World and U.S. metals consumption data from U.S. Geological Survey database, available at http://minerals.usgs.gov/minerals/.

Industrial production, producer prices deflators, and metals price indices: IMF, *International Financial Statistics*, various issues; except U.S. metals prices (and world steel prices): from USGS database. Industrial countries' industrial production and producer prices indices used as proxies for world values.

Resource depletion: World Bank, World Development Indicators online database.

Imports by origin: IMF, Direction of Trade Statistics online database.

Canadian sectoral energy consumption data: Natural Resources Canada, available at http://oee.nrcan.gc.ca/corporate/statistics/neud/dpa/home.cfm?text=N&printview=N.

U.K. sectoral fossil fuel energy consumption data: Office of National Statistics Environmental Accounts, available at http://www.statistics.gov.uk/CCI/nscl.asp?ID=6805.

U.S. sectoral fuel consumption data: Energy Information Administration, available at http://www.eia.doe.gov/emeu/mer/consump.html.

Canadian and U.K. sectoral output data: OECD STAN data CD-ROM; U.S. sectoral output data: Bureau of Economic Analysis Industry Economic Accounts, available at http://www.bea.gov/bea/dn2.htm.

Nonrenewable Exhaustible Resources and Property Taxation

Selected Observations

By C. LOWELL HARRISS*

ABSTRACT. In this paper, Professor Harriss connects the historic debate about using up nonrenewable resources and its impacts on future generations with the need to help finance government in less disruptive ways. He explains the Georgist program of taxing "economic rents" either directly by a rent tax or indirectly by a severance tax.

The use of natural resources—gifts of nature—to serve the general public must have persuasive appeal. Some things Nature created. No one can plausibly claim the benefits, the fruits, on the basis of having created the asset. The community, defined somehow, can take the fruits without endangering the source. Although much explanation and defining are called for, the principle seems clear.

This paper concerns nonrenewable, exhaustible resources. When used, they are used up—natural gas, for example. Problems of definition do arise, such as metal that becomes usable scrap. These have relevance to points made later. The reality before us at the moment is that people are consuming resources that are irreplaceable. Nothing users now pay goes to renew the source.

*C. Lowell Harriss is Professor Emeritus of Economics, Columbia University. He is the author of many books and articles. His teaching specialties include the economics of taxation. The views he expresses here are his own views and not necessarily those of any organization with which he is associated.

American Journal of Economics and Sociology, Vol. 65, No. 3 (July, 2006).

Should the interests of future generations get more recognition than they do today? This brief note links that question with some involving the financing of government and with challenges articulated persuasively in the late 19th century by Henry George. His followers, "Georgists," have continued the discussion. There are no clear lines defining "Georgism" and "Georgist." Yet whatever their origin, there are elements that not merely deserve respect but that, even more, can be a guide to improving public policy in the years ahead.

I

Rents

THE TERM *RENT* has a variety of accepted usages. The concern here is "economic rent." It differs from the kind of payment made when one rents a car. Think of economic rent as payment for the use of a resource whose existence does not depend upon the payment. Land. The original surface of the Earth, with a very few exceptions, is Nature's creation. One pays, perhaps heavily, to get a parcel in one location, one use, rather than another; but the payment does not create the area. Houses or trucks or autos exist because someone paid to get them produced. But not the products of Nature. Lines are not always clear. Some portion of other payments may be for a scarcity element and properly termed "rent-like"—outstanding athletes, for example. We do make payments, whatever the designation, that do not, over the long run, go to create, or renew, the source—some portion of human skill that is the creation of Nature.

Economic rents that are directed to paying for government services will not reduce the source as other taxes can do. This principle can be utilized where conditions permit. Property taxation in the United States represents use of the principle. Americans could be better off by financing more of (local) government by reducing the taxes on structures and other manmade capital and relying more fully on taxes on land, on the economic rent of land.

Part of what may be attributed to land represents underground nonreproducible resources. Some of what the owners of land get corresponds to economic rent, in the pure sense. And some of what the landowner obtains is the result of capital and effort required to extract the mineral. This return may be treated in business accounts as profit, interest, and labor expense, or something else—probably some combination. Accounting systems do not identify nor measure pure economic rent.

The amounts are certainly far different from what the available measurements suggest. Prof. Mason Gaffney in an unpublished manuscript identifies more than a dozen sources of economic rent related to land that are ignored in customary accounts, for instance, owner-occupied business property.

In economic rent there is a potential source of government revenue of unique nature—unique in that taking it in taxes would not induce the distorting and depressing effects of equivalent taxes on labor and capital, effects known as "excess burden" in that they reduce real income by amounts greater than the government receives in taxes. Substituting economic rents for some of the taxes on capital and labor would thus reduce the excess burden, the deadweight loss produced by revenue raising. Net benefit for the economy, a true economic surplus—the principle seems beyond question! Yet public discussion can focus on ability to pay with the intuitive conclusion that all are equal. But a dollar of land value and a dollar of structure value will differ significantly in their economic effects.

II

Justice

GEORGIST WRITINGS frequently place emphasis on "justice" without clear definition. The general sense is a more equal sharing of what Nature has provided (as distinct from what humans have created). Who among us, looking around a city, can deny that there have been some large unearned increments in land prices over the years? Who will not have felt that Nature put Persian Gulf oil in locations not well related to those who can use it? There are disparities that one can say, with confidence, do not relate rationally to contributions (production) of the recipient—"unjust." But which? How great? How have they come into existence?

One can, and one probably should, condemn many actions in the past that determined the ownership of land and other natural resources. Inheritance has continued this ownership. Some Georgists have advocated that an arrangement wrong from its origin will produce results to be condemned today and indefinitely. Sometimes, such reasoning is used to justify wide-scale changes now. But at its best, such argumentation must be used with utmost caution and reserve, and for the most part rejected. What exists today does result from the past—yesterday and probably centuries of choices and decisions. Who today can be qualified to evaluate the myriads of actions of the past? Today's practical opportunities are to decide as well as possible for the future. Nonrenewable resources present alternatives somewhat like those of land rent and some that are quite different.

Markets today cannot reflect the choices of the future. Twenty or 50 years in the future, what evaluations would consumers place on natural gas burned now or on the alternatives that are reflected in today's decision-making processes? Of course, some actions now do embody more than trivial concerns for the future—capital formation, for example. And use now of exhaustible resources reflects more or less deliberate rejection of interests and concerns about future users. Some things are renewable, such as forests, perhaps rather quickly while others take longer. Some, such as petroleum, are used up today. Our descendants will not have them.

III

Present Versus Future

Is THE PRESENT cheating the future? This question deserves explicit attention. The worth of the accumulation of knowledge must vastly exceed the worth of the nonrenewable resources being consumed. Present market prices must reflect the views of buyers and sellers today about the future. Yet market prices at present cannot incorporate the valuations of persons who will be affected. People adjust as life moves along. Conditions now reflect decisions in the past that will continue to exert influence, for good and for ill, indefinitely. As parents and as participants in businesses, as voters and as politicians, we human beings do take account somewhat of the using up of natural resources. If recognition now is judged, somehow, to be made quite short-sightedly, what investments might be available to improve results?

IV

Property Taxation

TAXATION IS AN AGENCY or instrument of collective action. Every community in the country uses property taxation. It is as old as American government. What exists differs widely from place to place. Relatively few areas have exhaustible resources, minerals presumably embodied in land prices and, to varying extent, in assessed valuations. Where such assets (values) exist, the community can get revenue from the resource. For petroleum products and other minerals whose prices are determined in large (probably world) markets, the land tax burden will fall on the owner of the land. The amounts that will have been capitalized in the price at which present owners bought land may be lost in the mists of time or decided in a purchase yesterday. For sand and gravel and any other locally-oriented resource, the quantity produced and thus the price can respond to a property tax on the natural resource element; but this would be a quite unusual case.

Property taxation has substantial merit as a revenue source for local government. And on a scale used typically to evaluate revenue sources, a tax on resource rents would rank high. Whether the resources are renewable or exhaustible would not be of much significance for such evaluations. But the merits of this source hardly pertain to local government—except perhaps to help finance some government spending directly related to the mineral, such as schools for the children of employees.

The property tax requires discretion and judgment in administration. Assessment of the worth of what is underground must usually reflect the reality that there are no fully reliable market transactions. Local assessors can often get good data on many sales of single-family houses. Mineral properties do not have comparable markets.

Owners, who have reason to wish for underassessment, may

control sources of evidence. And, of course, knowledge of what is underground can be incomplete and its worth in the future highly uncertain. A local assessor's skill may be no match for that of the owner's representative. Yet government can mobilize capacity to value using all the evidence ingenuity can acquire.

And property taxation is not the only possible instrument.

V

Severance Taxation

Several states impose severance taxes. They are more appropriate than property taxes for getting revenue from natural resources. They may be a percentage of price at some stage in the process. Or the tax may be a money amount: so much per ton of coal or thousand cubic feet of natural gas. Who ultimately pays such a tax? Conditions differ and relate to broad market processes and contractual relations. Burdens are widely diffused beyond state borders while also affecting in-state property values.

Severance taxes are imposed for good reasons by units of government that are larger than localities. As a practical matter, that means states. True, state boundaries were not set to help in achieving rational allocation of natural resource revenues. But there seems nothing better. (The federal government does not use the term *severance tax* for what it receives from leasing lands, including offshore.)

VI

Concluding Comments

NATURAL RESOURCES PRESENT greater potential for financing government than has yet been realized. And as other papers in this volume indicate, there are numerous problems and reasons for study. The reality of exhaustion cannot be denied; some when used will not be replaceable. Common sense says: "Take care for the future." But how?

The record gives little (no) reason for confidence in longer-run forecasts. True, any exhaustible resources that is being used can be used up. Yet other things will also be going on.

For exhaustible resources one might argue, with good reason, for using part of the revenues for a trust fund for future benefits. Alaska, in contrast, makes large distribution of the annual revenues from oil to current residents.

Greater use of land values to pay for government, the Georgist principle, will (albeit somewhat indirectly) enlarge the role of natural resources.

Lessons for Economic Reform Based on Pennsylvania's Experiences with the Two-Tiered Property Tax

By ROBERT ANDREW PETERS*

ABSTRACT. Although economic theory indicates that the imposition of a two-tiered property tax system facilitates urban revitalization, localities in most states have not been authorized to institute a two-tiered property tax. The authority to implement such a tax is partially determined by a state constitution's uniformity and equal protection clauses and tax rate ceilings. An analysis of these provisions reveals 23 states may establish a two-tiered tax, but implementation in 20 of the states must await the passage of state-enabling legislation. Because of the dearth of experience in enacting legislation and the absence of literature that provides guidance for securing its passage, the politics of enacting Pennsylvania's 1998 statute are assessed. The case study clearly indicates that enabling legislation enjoys bipartisan support as well as the backing of urban and rural representatives. However, the legislation's fate is primarily determined by the composition of local electorates and the political power of farm lobbies.

I

Introduction

URBAN REVITALIZATION EFFORTS HAVE TRADITIONALLY FOCUSED ON infrastructure improvements, industrial development bonds, tax abatements, and urban renewal grants. Over time, these forms of targeted aid have been supplemented by programs that benefit multiblock sections of

*The author is at the School of Public Affairs and Administration, Western Michigan University, 1903 West Michigan Avenue, Kalamazoo, MI 49008-5440; phone: (269) 387-8938; e-mail: robert.peters@wmich.edu. The author expresses his gratitude to the reviewers and journal editor. Their comments were invaluable in improving the article.

American Journal of Economics and Sociology, Vol. 65, No. 3 (July, 2006).

downtown business districts, such as enterprise zones, downtown improvement districts, and tax increment financing. Although these initiatives have been adopted throughout the country, there is an option that has not been widely replicated: the two-tiered property tax.

Its infrequent use belies the fact that a two-tiered property tax offers significant incentives for urban redevelopment. Instead of imposing a uniform millage[1] on all of a parcel's elements, the two-tiered tax levies a higher tax rate on land than on improvements to land. Whenever a two-tiered tax replaces a single-rate property tax or whenever the tax rate differential between land and improvements is increased, the revisions raise the relative price of holding land and lower the relative cost of improvements. It is these changes that reduce the financial benefits of holding land for speculative purposes and provide incentives for improving the land. The two-tiered tax, therefore, generates incentives for redevelopment and a more efficient use of land.

Even though local governments may decide to institute this type of levy, legal and political factors may discourage its use. The equal protection and uniformity clauses of state constitutions determine whether or not a two-tiered tax is permissible. In the event that the levy is constitutional, local adoption cannot proceed unless the state government first enacts statutes authorizing its use. One of the key factors affecting the probability of enacting a state statute is the composition of local electorates and the political influence of lobbyists who represent their interests.

I have organized this article as follows. In Section II, I describe in greater detail the two-tiered tax's impact on redevelopment. This discussion is followed by Sections III and IV, where I review state constitutional provisions and the impact of social interests and lobbying groups on the adoption of state enabling legislation, respectively. Section V provides the foundation for an analysis of the Pennsylvania General Assembly's decision to extend two-tiered taxing authority to Pennsylvania's boroughs.[2] A summary of the findings and implications for other jurisdictions are contained in Section VI.

Pennsylvania is the focus of this study because of the frequency with which the legislature has dealt with the issue of two-tiered taxes in the state. Two-tiered taxes were originally placed on the agenda by Pennsylvania-born economist Henry George, who proposed to eliminate property taxes on improvements to land. Although the state legislature did not adopt George's notion of a land-value tax, legislation was enacted in 1913 that authorized cities of the second class (Pittsburgh and Scranton) to establish a two-tiered property tax within their boundaries. Since authorization was limited to these two cities, other jurisdictions could not legally institute a two-tiered tax under the original enabling statute. However, when the issue resurfaced during the 1950s, statutory authority to adopt the two-tiered property tax was extended to third-class cities. Nine school districts were added to the state statute in 1995 and, three years later, the General Assembly approved legislation that permitted boroughs to also implement the tax. Given the frequency with which Pennsylvania's legislature has considered, and enacted, enabling legislation, there is ample opportunity to examine the historical record for evidence of variables that influenced the legislature's behavior.

II

Economic Development Incentives Provided by the Two-Tiered Property Tax

SINCE PROPERTY TAXES ARE BASED UPON the value of both the land and the improvements to land, construction or renovation projects on landed property generate higher property values and a concomitant increase in annual property tax burdens. The economist Henry George proposed to eliminate the negative implications of taxing improvements by shifting the entire levy so that it would burden only land. Throughout the past century, the Pennsylvania General Assembly has authorized second- and third-class cities, boroughs, and nine school districts to use a hybrid of the traditional property and land-tax systems. Instead of shifting the entire tax burden to land, state statutes permit these entities to impose a higher millage on land than on improvements.

Table 1 offers a simple numerical example to illustrate the manner in which a switch from the traditional, one-rate tax to a two-tiered property tax can affect outcomes. The table's numbers are based on

Table 1

Property Tax Options for a Hypothetical City, by Millage Rates, Millage Ratios, and Revenues

Property Tax Option	Land Millage	Improvement Millage	Millage Ratio	Revenue Generated by Land	Revenue Generated by Improvements	Total Revenue
Original	24	24	1:1	2.4 Million	21.6 Million	24.0 Million
First	50	21	2.38:1	5.0 Million	19.0 Million	24.0 Million
Second	62.5	25	2.38:1	6.25 Million	22.5 Million	28.75 Million

the assumption that the total assessed value of city property is $1,000,000,000. Of that amount, it is assumed that land accounts for 10 percent of the aggregate ($10,000,000) and the initial tax rate is 24 mills. Annual revenues generated by the tax on land and improvements are $2.4 million and $21.6 million, respectively. If the tax rate for land is raised to 50 mills and total tax revenues are held constant, then the millage rate for improvements can be reduced to 21 mills. The resulting ratio of tax rates for land and improvements rises from 1:1 to 2.38:1. Due to the increase in the relative price of land and decline in the relative price of improvements, property owners spend fewer resources on land itself and expend more funds on improvements to the land. The adoption of a revenue-neutral two-tiered property tax, therefore, expands the number of construction and renovation projects and stimulates economic development (Bourassa 1987; Brueckner 1986).[3] In urban areas, greater construction and renovation activity tends to yield an increase in housing density. It should also be noted that whenever alterations in tax policy raise the present value of immediate development above the present value of holding vacant land, Bourassa (1992) and Douglas (1980) contend that speculation declines and vacant land is developed. The resulting fill-in development and expansion of existing buildings enhance the city's tax base, eventually increase the density of city development, and reduce the demand for suburban growth (DiMasi 1987).[4]

The benefits of the two-tiered tax are compromised whenever adoption is accompanied by an attempt to increase total property tax revenues. Consider the following example in Table 1: if the hypothetical city replaced the original 24 mill tax with a two-tiered structure of 62.5 mills for land and 25.0 mills for improvements to land, then total revenue would rise from $24 million to $28.75 million and the ratio of millage rates would increase from 1:1 to 2.38:1. Since the tax rate ratios for the first and second two-tier tax systems are identical, both systems generate incentives to shift resources from land to improvements (e.g., construction and renovation projects). However, the larger total tax burden imposed by the second example raises the after-tax cost of owning city property. These higher costs reduce the rate of return for city properties in relation to surrounding jurisdictions. It is this change in the relative rate of return that leads investors to reduce their funding of city projects and expand monies flowing to the more profitable suburban areas. The siphoning of investment from the city to the suburbs continues until the increase in demand for suburban properties is reflected in higher prices and the rising prices reduce the suburban rates of return until the returns on suburban and city properties equalize. At this point, a larger share of the benefits from implementing a two-tiered tax accrues to the city and the difference in investment spending generated by each of the two-tiered tax systems declines.

The benefits of a two-tiered property tax are augmented by its capacity to limit the revenue hemorrhage of cities confronting population and economic decline. Under the traditional single-rate property tax, owners of abandoned factories and commercial buildings have an incentive to minimize their tax burdens by seeking lower assessed values for their structures. They usually do not seek reductions in the assessed value of land because lower assessments suggest the parcels are less desirable for development and, therefore, are less marketable (Sullivan 1997; Witte and Bachman 1978). Reductions in the assessed value of land, in other words, adversely affect the probability of selling an abandoned factory or commercial space, which is an undesirable outcome for the property's owners. Since the two-tiered property tax derives a larger portion of its revenues from land than the traditional one-tiered tax, and since land values are more stable than the value of structures, the adoption of a two-tiered tax minimizes the magnitude of revenue losses that accompany a declining population and economic base (Spossey 1997). A more stable revenue source enhances the probability of maintaining the city services and infrastructure that are necessary for stemming the loss of population and jobs.

III

Constitutional Provisions Affecting Implementation

THE UNIFORMITY AND EQUAL PROTECTION CLAUSES OF STATE CONSTITUTIONS determine the degree to which property tax burdens can be shifted from improvements to land. State courts have consistently interpreted these provisions to mean that assessments and tax rates must be uniform within each property classification. Given these legal precedents, a locality cannot impose one millage rate on land and a lower rate on improvements unless the state constitution authorizes the use of these two classifications or the legislature can exercise sufficient latitude in defining property classifications.

An examination of state constitutions reveals tremendous variations in the legislatures' authority to establish property classifications. Twenty-seven state constitutions do not permit the legislature to establish property classifications or partial exemptions beyond those delineated by the constitution. Given that Louisiana is the only state in which land is defined as a separate category and that courts overturn any attempt to legislatively augment the constitutionally defined categories, the two-tiered property tax is unconstitutional in all of the 27 states except Louisiana. Louisiana's capacity to shift property tax burdens from improvements to land is constrained by a provision that limits the assessment ratio to 10 percent for residential land and improvements and 15 percent for all other property except public service property (Louisiana State Constitution, Article VII, Section 18).[5]

Of the remaining 23 states listed in Table 2, 18 constitutions incorporate uniformity clauses that permit state legislatures to define property classifications. The constitutions of the final five states do not include a uniformity clause. As a result, property tax administration is governed by the equal protection clauses of the state and national constitutions. When this factor is combined with state court declarations that the constitutional requirements of the uniformity and equal protection clauses are identical, the five state legislatures are provided the same latitude to classify property as the 18 states. It is therefore evident that 23 state constitutions permit the legislature to enact statutes giving a specified set of local governments the authority to establish a two-tiered property tax.

There are, however, additional constitutional constraints on the degree to which local governments in these 23 states are permitted to shift property tax burdens from improvements to land. The tax revolt of the late 1970s and 1980s produced state constitutional amendments that impose ceilings on property tax rates. In the State of Oregon, for example, a 1987 amendment stipulates that the total

Table 2

State Constitutional Provisions Relating to the Classification of Land and Improvements, by Region, 1997

Northeast	South Atlantic	North Central	South Central	Mountain	Pacific
27 states in which land and improvements are not permissible classifications:					
Maine	Florida	Illinois	Alabama	Idaho	California
N. Hampshire	Georgia	Indiana	Arkansas	Nevada	Washington
New Jersey	S. Carolina	Kansas	Louisiana	Utah	
	W. Virginia	Michigan	Mississippi	Wyoming	
		Missouri	Tennessee		
		Nebraska	Texas		
		Ohio			
		Wisconsin			
23 states in which land and improvements are permissible classifications:					
Connecticut	Delaware	Iowa	Kentucky	Arizona	Alaska
Massachusetts	Maryland	Minnesota	Oklahoma	Colorado	Hawaii
New York	N. Carolina	N. Dakota		Montana	Oregon
Pennsylvania	Virginia	S. Dakota		New Mexico	
Rhode Island					
Vermont					

Source: Constitutions of the 50 states.

property tax rate for local government and public schools shall not exceed 25 mills in 1991–1992 and gradually declines to a maximum of 15 mills in 1995–1996 and thereafter. The provision was repealed in 1997 and was replaced by an amendment that dictates further reductions in property tax burdens (Oregon State Constitution, Article XI, Sections 11 and 11a). In the absence of a constitutional amendment that raises the ceilings on property tax rates, the capacity to shift property tax burdens from improvements to land is severely constrained.

If the preceding elements do not produce barriers to implementing two-tiered taxes, then the final, essential component is the enactment of state-enabling legislation. The passage of legislation is an inherently political activity in which lawmakers represent their districts' interests. Since all interests do not exercise similar levels of influence, the following two sections assess the degree to which the composition of the local electorates and the political power of their lobbyists affected the extension of two-tiered property taxing authority to Pennsylvania's boroughs.

IV

Interest Group Characteristics and Political Power

AN INDIVIDUAL'S BELIEFS AND PREFERENCES are not developed in a vacuum. They are a function of society's norms and beliefs that are the foundation for defining what is valued and preferred (Berger and Luckmann 1966; Cox 2001). These special meanings are what influence the strategies of policy entrepreneurs. For example, the stereotypes that dominated congressional deliberations on AIDS prevented policy entrepreneurs from altering the conceptual discourse or setting new parameters for evaluating spending policy. Due to the characteristics ascribed to homosexual males and those who inject illegal drugs, funding for prevention, research, and treatment was significantly smaller than would have been the case if the disease had initially infected groups that society views in positive terms (Jennings 1999). The influence of interest group characteristics was reinforced when public conceptions about the infected populations expanded to include those who were defined in positive terms (heterosexuals, children, and those who do not abuse drugs). At this point, the nature of the debate shifted and additional monies were provided (Pollock 1994). It is therefore apparent that societal views of interest groups provide insights into the agenda-setting process, policy formulation and design, and legislative behavior (Czech and Krausman 2001).

The impact of society's perceptions is magnified when these perceptions are commingled with an interest group's political clout. Even though these variables can be measured on an interval scale, the presentation is simplified when they are defined on nominal scales and placed on a two-by-two table. Table 3 indicates the upper-left quadrant is occupied by advantaged groups that are politically powerful and defined in positive terms. Legislators are predisposed to advocate policies that benefit these groups because they are able to generate the votes, contributions, or political participation that improve a lawmaker's reelection prospects. The politician's electoral fate is also enhanced by public reaction to policies that benefit advantaged groups. Since the groups are defined in positive terms, the public approves of government bestowing additional benefits on them

Table 3

The Political Power of Target Populations, by Their Social Construction

	Social Construction	
Political Power	Positive	Negative
	Advantaged	**Contenders**
Strong	Elderly	Large Unions
	Veterans	Big Business
	Dependents	**Deviants**
Weak	Children	Criminals
	Mothers	Drug Addicts

Source: Schneider, and Ingram (1993).

(Schreiber and Ingram 1993). Advantaged groups, therefore, exercise sufficient power to wrest more than their fair share of benefits from government, and the public views the outcome as just.

A prime example is Social Security and Medicare. The political power of elderly citizens enables them to claim a larger share of public resources than is the case for children. The record also reflects the public's propensity to view the elderly as deserving. Congressional debates consistently define an expansion of benefits as just compensation for the elderlys' lifetime of hard work and sacrifice that built the country and made it great. Due to these elements, as well as the number of citizens who have elderly parents or grandparents and expect to grow old themselves, a Harris poll (1998) revealed that 67 percent of adults felt the balance of federal spending for the programs benefiting the elderly and children was about right. Even in the mundane world of allocating state funds for medical education, Knott and Weissert (1995) conclude policy inducements are skewed in the direction of advantaged groups.

The oversubscribing of benefits, however, creates a legitimation crisis for government. It must explain why "democracies concentrate wealth and power in the hands of a few rather than the many" (Schreiber and Ingram 1993: 339). Policymakers attempt to address this issue by linking policies that benefit advantaged groups to universal, instead of special, interests. More specifically, the rationale for these initiatives typically focuses on the group's link to achieving important public purposes, such as national defense and economic competitiveness (Schreiber and Ingram 1993).

The political power and group characteristics that generate an oversubscription of benefits also contribute to a reticence to impose burdens. When confronted with the possibility of increased costs, advantaged groups marshal their forces in opposition to the measures, and the general public is sympathetic to their cause. The dynamics that generate an oversubscribing of benefits, therefore, produce an undersubscribing of costs (Schreiber and Ingram 1993).

Deviants occupy the opposite corner of Table 2. Since this category encompasses groups that wield minimal political power, they have little control over the formal agenda or the design of policies and programs that affect them. When this element is combined with society's attribution of negative characteristics, policymakers are prone to impose burdens or punishments even when these approaches are less effective than bestowing benefits (Schneider and Ingram 1993).

The third aggregate is composed of dependents, such as children and the handicapped, who are portrayed in positive terms but are not politically powerful. Although their virtuous characteristics make it politically expedient for policymakers to appear sympathetic to the groups' interests, the low rates of political participation and clout produce insufficient incentives for lawmakers to adequately address their needs. The resulting gap between the groups' needs and publicly provided resources is minimized through mechanisms such as stigmatizing the receipt of benefits, providing insufficient outreach programs, and raising the costs of participation. The political environment, therefore, yields benefits that are undersubscribed and burdens that are oversubscribed (Schneider and Ingram 1993).

The final group is composed of contenders who are politically powerful but are defined in negative terms. Since interests such as wealthy individuals, labor unions, and large corporations are actively engaged in the political process, they possess sufficient political clout to secure resources. However, the negative characteristics attributed to these interest groups produce negative reactions to the provision of clearly defined, visible benefits. Lawmakers, therefore, balance the groups' political clout and the public's negative perceptions by granting benefits that are evident to the contenders, but not to the general public. On the burden side of the ledger, regulations give rise to the impression that burdens are being inflicted when, in reality, the initiatives impose minimal, negative effects. Both sets of strategies enable legislators to minimize the political ramifications of granting benefits to, and restricting the regulatory burdens borne by, politically powerful groups that are defined in negative terms (Schneider and Ingram 1993).

Given the impact on the policy-making process of characteristics attributed to interest groups and their political clout, the following section utilizes these variables, along with constitutional and statutory constraints, to assess Pennsylvania's efforts to expand the range of governments authorized to establish two-tiered property taxes.

V

The Politics of Enacting Enabling Legislation

THE LEGISLATIVE RECORD INDICATES THAT LOCAL GOVERNMENTS AND THE FARM LOBBIES were the only interest groups that were attempting to influence the enabling legislation's development and composition. Local governments were actively pursuing a liberalization of state law that would permit a wider array of jurisdictions to adopt the two-tiered property tax. The farm lobbies opposed passage because the owners of developed parcels could use the legislation to shift part of their property tax burden to farmers. Since the outcome was clearly affected by each group's political power and characteristics, the following subsections examine these variables and the manner in which they were manifested in the legislative process.

A. Characteristics and Political Clout of Local Governments

Since local governments are in closer geographic proximity to the people than state or national governments, they are generally perceived as the most responsive to citizen needs and the most efficient providers of goods and services. A Council for Excellence in Government poll (1999) found that 51 percent of the respondents felt connected to local government. This figure compares to 41 percent for state government and one-third for the federal government.

The preference for government that is physically closest to the people is also demonstrated by the question of whether people trust government to do what is right just about always or most of the time. Thirty-five percent of the respondents trusted local government a great deal or quite a lot. The figures for the state and national governments were 33 percent and 21 percent, respectively (Center for Excellence in Government 1999). Similar results were generated by a 1997 Gallup poll, but a Pew Research Center for People and the Press survey (1998) found a difference of 7 percentage points between those who had a great deal of confidence in their local and state governments. When the preceding outcomes are combined with the percentage of respondents who trust small business (56 percent) and large corporations (22 percent), the Council for Excellence in

Government (1999) concluded that the "preference for small, local institutions over large, national ones is not unique to government."

The positive image and political clout generated by these attributes are augmented by local governments' role as partners in the delivery of services; localities administer a variety of programs that are partially funded by the state and national governments. State legislators, therefore, consult with their local counterparts in order to minimize the possibility of instituting policies that may harm local government or generate a backlash by local constituents. Due to the political power emanating from local government's position in the federal system and the public's positive views of local government, these entities are defined as an advantaged group.

There is, however, a caveat to including local governments on the list of advantaged groups. Variations in the political, economic, historical, cultural, socioeconomic, and financial composition of localities translate into differences in local interests and positions on impending legislation. A locality confronting a declining population and stagnating economy will support a different set of state economic development initiatives than a high-growth area that is experiencing labor shortages. Whenever these differences are reflected in divergent positions on legislation, the influence of local governments is reduced. It is, therefore, evident that local governments' impact on the legislative process is affected by the extent to which local interests are dispersed or concentrated (Peters 1994).

B. Local Government's Impact on Pennsylvania's Legislative Process

One hundred years of Pennsylvania legislative history indicate that support for two-tiered property tax enabling legislation has been bipartisan and geographically dispersed. In 1913, the Pennsylvania General Assembly used its constitutionally provided latitude (Constitution of the Commonwealth of Pennsylvania, Article VIII, Section 1) to authorize second-class cities (Pittsburgh and Scranton) to enact two-tiered property taxes (*Senate Journal*, 1959 session: 3533). A half-century passed before similar authority was extended to third-class cities. Since the latter statute was ratified without a floor debate in either chamber and there was only one legislator who voted against the bill, the record does not divulge the rationale for restricting eligibility to third-class cities or the roles of interest group characteristics and political clout in determining the bill's fate and composition.

Although no legislation was introduced between the mid-1950s and mid-1970s, each of the 34 subsequent bills replicated the language of the third-class city statute. More specifically, the bills authorized jurisdictions to "levy separate and different rates of taxation for (borough, city, or township) purposes of all real estate classified as land, exclusive of the buildings thereon, and on all real estate classified as buildings on land. When real estate taxes are so levied, the rates shall be determined by the requirements of the (jurisdiction's) budget as approved by (the governing body)." During the 1990s, this language was supplemented by (1) a prohibition against implementing two-tiered property tax rates that would generate more revenues than the maximum, permissible millage for land and buildings, and (2) a stipulation that tax rates for land and structures must be uniform within each classification. The former acknowledges the public's desire to limit tax burdens, while the latter recognizes constraints imposed by the constitution's uniformity clause.

As was the case with the enactment of legislation for third-class cities, the legislation was supported by members of both parties. Table 4 indicates that 63.8 percent of the 34 bills were cosponsored by Republicans and the total number of co-sponsors who represented urban or rural districts was virtually identical. Within each party, the data indicate that a majority of the Republican co-sponsors represented rural constituencies, while a majority of the Democratic co-sponsors were from urban districts.

An analysis of the geographic distribution of House seats indicates that the preceding pattern is a function of each party's power base rather than philosophical differences among factions of the two parties. Even though the urban districts within the City of Philadelphia and its surrounding suburbs were an important source of Republican House seats, statistics for the 1993–1994 session indicate the party derived 63.3 percent of its House seats from rural districts. This figure is slightly higher than the percentage of Republican co-sponsors who represented these areas.

Table 4

Party Affiliation of House Districts Sponsoring Township Legislation, by Population Density

	Population Density		
Party Affiliation	Urban	Rural	Total
Republican	81 42.2%	111 57.8%	192 100%
Democrat	71 65.1%	38 34.9%	109 100%
Total	152 50.5%	149 49.5%	301 100%

Chi-square = 14.65.
Relationship is not statistically significant at the 0.01 level of significance.

The Democratic Party, on the other hand, derived a minority of its legislative seats from rural districts in the state's traditional coal mining and steel manufacturing regions, while 63.8 percent of its 1993–1994 House members represented the cities and suburbs of Philadelphia and Pittsburgh. When this percentage is compared to the share of Democratic co-sponsors who represented urban districts, it is apparent that the urban Democrats' share of House seats approximates the share of co-sponsors.

The rationale for the legislation's popularity is clearly evinced in the legislative record. According to information gleaned from the House's 1992 school finance debate, the legislation's appeal to urban lawmakers was partly a function of the tax's capacity to stabilize local tax bases. The school finance bill, which included a provision authorizing school districts to implement a two-tiered tax, was sponsored by a representative whose district was located in the heart of Pennsylvania's steel manufacturing region. When foreign competition brought about the closure and razing of steel mills in her district, the tax base and revenues of the city and school district were depressed. Since the implementation of a two-tiered property tax supplied the city with the means to counter its revenue hemorrhage and eliminate its annual budget deficit, local officials sought to secure similar benefits for the local school district. The urban districts, therefore, defined the passage of school district enabling legislation as an essential ingredient in their efforts to restore the school district's fiscal health and generate sufficient funds to defray the cost of the new teachers' contract (*House Journal*, 1992 session: 882).

The base of support was augmented by the belief that property taxes should not penalize people who maintain and/or enhance their property. For example, State Senator Wozniak (1998) stated: "[I]t's absurd that you have to pay taxes when you improve your home. You own the home, and if you want to improve it, you should not be penalized. Usually, the land stays the same, and if you tax the land, most everyone is treated in a fair manner. The tax rate should not have anything to do with bettering the place that you call home." This statement is similar to Senator McGinnis's 1959 pronouncement that "if you rob a chicken coop, they will fine you once, but if you build one, they will fine you every year" (*Senate Journal*, 1959 session: 3533–3534).

Support among rural and suburban legislators was also based upon the belief that the two-tiered tax would encourage a more efficient use of urban land. It was believed that the use of a two-tiered tax would generate fill-in development that would reduce the demand for farmland and minimize urban sprawl.

The record, therefore, indicates that legislators were sympathetic to local governments' desire to stabilize their tax bases and to ameliorate the financial burdens and adverse environmental impacts associated with urban sprawl. Given the variety of legislators who sponsored legislation and the rationales that supported their positions, the probability of local governments speaking with one unified voice was maximized. Under these circumstances, localities possessed the attributes of an advantaged group—they were defined in a positive manner and their political power was not diluted by disagreements concerning the legislation's preferred fate.

C. Characteristics and Political Clout of Farmers

However, local governments' capacity to secure the passage of two-tiered tax enabling legislation was inhibited by the second advantaged group composed of farmers. Logic suggests larger groups are able to exert more political pressure than smaller groups. However, the influence of the latter group is magnified by the minimization of the free-rider problem. More specifically, the members of smaller interest groups, such as farmers, are more likely to "know each other, monitor each other, and encourage each other to pull his [sic] weight" (Matsusaka 1995: 149). Each group member, in other words, is aware that the group's success may depend on his or her actions, whereas, in a larger group, an individual's actions are unlikely to be a decisive factor in the outcome (Matsusaka 1995). The political influence of the farm lobby is manifested in the heavy subsidization of agriculture in industrial countries where the sector is relatively small, and the imposition of heavy taxes in many developing countries where the sector is large (Zablotsky 1995).

Agriculture's political clout is buttressed by the belief that family farmers are stewards of the land and, therefore, advance the goals of environmental protection and the preservation of open space (Browne et al. 1992). It is this belief that entices environmental groups and open space advocates to join the battle on some agricultural issues. Since Czech and Krausman (2001) define political power as the number of interest groups supporting a particular cause, the addition of these nonfarm groups enhances the political clout of the agricultural lobby. The political power generated by these issues is also impacted by the popular appeal of preserving farmland and open space. Although surveys from the late 1990s did not focus on the preferences of Pennsylvania residents, data from the adjoining state of New Jersey reveal that 72 percent of the respondents felt farmland and open space preservation was very important (Star Ledger/Eagleton-Rutgers Poll 1998).

When compared to local governments, it is also apparent that agriculture's capacity to influence state policy is not diluted by a profusion of issues affecting its interests. The smaller and more focused agenda enables the agriculture lobby to concentrate its resources on a limited range of issues and thereby maximize its influence. Due to the preceding elements, farmers are defined as politically powerful.

Agriculture's capability to secure benefits is buttressed by the positive characteristics society associates with family farms. Even though the number of family farms has been declining and agribusiness accounts for a growing share of agricultural output, Americans have retained a romanticized, positive view of farmers. This view emanates, in part, from a long-established belief that farmers are the most vigorous, independent, and virtuous citizens. It is also claimed that, due to the ownership of land, the interests of farmers, their communities, and freedom are intertwined. The farmers' connection to the land and means of production yields additional stereotypes that include an interest in preserving social stability, developing a sense of community, and serving as a repository for values such as honesty, integrity, hard work, charity, stewardship, self-reliance, and independence. Farming also provides an opportunity to be one's own boss and is an alternative to working for wages. Consequently, it provides a safety valve for the American economy and a means of preserving individual liberties. Lobbyists for American agriculture, as well as lawmakers who advocate for bills that are beneficial to farmers, have relied on these positive views of family farms to preserve old programs and advance agricultural interests (Browne et al. 1992). Due to agriculture's political clout and society's positive views, farmers clearly are an advantaged group for which benefits are oversubscribed and costs are undersubscribed.

D. The Farm Lobbies' Impact on Pennsylvania's Legislative Process

It was readily apparent during the 1992 House school finance debate that the Pennsylvania General Assembly was reluctant to undermine efforts to preserve open space or impose costs on the advantaged group of farmers. Representatives Freeman (D-Northampton), Godshall (R-Montgomery), and Heckler (R-Bucks) contended that granting two-tiered taxing authority to school districts would undermine the state's efforts to maintain farmland and open spaces. Whenever, in more specific terms, the homeowners in a taxing jurisdiction outnumber farmers, the former group could use their numerical advantage to establish a two-tiered property tax that would dramatically raise the millage rate for land. The heavier levy on land would reduce the farmers' profit margin and thereby enhance their incentive to sell farmland for residential development (*House Journal*, 1992 session: 876–877).

In order to eliminate the potential negative impact on farmers and open space preservation efforts and to enhance the probability of final passage, the bill's sponsor introduced an amendment that authorized only nine urban school districts to implement a two-tiered tax. By adopting this amendment, the legislature limited the use of the two-tiered tax to older urban cores, "where it is most needed, without harming open space and prime farmland" (*House Journal*, 1992 session: 882). The amendment also provided an opportunity for the legislature to "take up the rest of it when we have more time to reflect and understand its potential impact" (*House Journal*, 1992 session: 882).

The legislature's reluctance to harm agricultural interests resurfaced during the 1993 legislative session. A bill authorizing boroughs to impose a two-tiered tax was reported by the House Local Government Committee. Following the bill's rereferral to the House Appropriations Committee, it was reported without amendment, and the House voted 133–166 for final passage. Table 5 indicates that the vote was partisan: slightly less than 9 percent of the House Democrats opposed final passage, while a majority of the House Republicans voted against the bill. When population density is added to the equation, it is apparent that 61.1 percent of the urban Republicans supported passage, while 70.5 percent of the rural Republicans opposed the bill.

Seventeen of the Republicans who opposed final passage also cosponsored identical legislation during the next legislative session. Of the seven Republicans who responded to written inquiries concerning the apparent discrepancy, there was unanimous support for the concept of two-tiered property taxes. However, all of the respondents stated that the 1993 floor debate raised concerns about the legislation's potential negative impact on farmers. Instead of supporting a bill that could be injurious to agricultural interests, the legislators

Table 5

1993 House Vote on HB 2532, by Party Affiliation and Region

	Republican		
House Vote	Urban	Rural	Total
Yea	22	18	40
	61.1%	29.5%	41.2%
Nay	14	43	57
	38.9%	70.5%	58.8%
Subtotal	36	61	97
	Democrat		
House Vote	Urban	Rural	Total
Yea	62	31	93
	96.9%	81.6%	91.2%
Nay	2	7	9
	3.1%	18.4%	8.8%
Subtotal	64	38	102
Total	100	99	199

elected to err on the side of caution and oppose final passage. When subsequent inquiries revealed that boroughs located within their legislative districts did not encompass farmland, the possibility of shifting property tax burdens from homeowners to farmers was eliminated (Brown 1997; Gerlach 1997; Godshall 1997; Hershey 1997; Lynch 1997; Phillips 1997; Stern 1997). This information eliminated concerns about the adverse effects on their districts' farmers as well as the barrier to subsequently co-sponsoring and supporting enabling legislation for borough governments.

The enabling legislation was referred to the Senate Local Government Committee on October 4, 1994 and reported without amendment on November 14. As was the case in the House, the Senate rereferred the legislation to the Appropriations Committee on

November 21, nine days before the end of the 1994 session. Due to time limitations and the volume of end-of-session business, the Senate leadership did not bring the bill to the floor for consideration or a vote (Gardner 1997). Evidence concerning the potential impact of agricultural interests on Senate deliberations, therefore, was not provided by the 1994 session.

During the ensuing legislative session, 28 of the 39 Republican co-sponsors and 2 of the 4 Democratic co-sponsors represented rural areas. The decision to co-sponsor the borough bill and the first-class city bill did not represent a betrayal of their districts' agricultural interests because the co-sponsors did not represent any of the boroughs that encompassed farmland. Those who owned homes in boroughs represented by the bill's co-sponsors or in the City of Philadelphia, therefore, could not adversely affect their districts' agricultural interests by shifting part of the property tax burden to land.

The introduction of legislation in both legislative chambers and the progression of enabling legislation during the previous session suggested that a borough bill should have been approved within a relatively short period of time. This assumption was not supported by events. Due to concerns about the potential adverse tax implications for farmers who owned land within the boundaries of several boroughs, all of the two-tiered property tax bills languished in committee and died at the end of the session.

Concerns about farmland preservation and the legislation's negative impact on farmers resurfaced during the following session's Senate debate. Since the state's boroughs could not be separated into classifications, it was not feasible to replicate the school finance approach of excluding classifications of governmental entities that included farmland within their borders. The Senate, therefore, opted for an amendment that established the property tax categories of farmland and nonfarmland, and stipulated that a higher tax rate could only be imposed on the latter category. The provision also defined farmland as any tract that is actively devoted to agricultural use as defined in Section 3 of the Act of June 30, 1981 (P.L. 128, no. 43), the agricultural area security law. Additional protection for agricultural interests was provided by a nonseverable clause. Due to this clause, the entire statute would be voided by a judicial finding that the subclassification of land violated the constitution's equal protection or uniformity clause.

The House Local Government Subcommittee on Boroughs struggled with the constitutionality of the subclassification of land. Prior state court rulings indicated that property classifications are constitutional as long as (1) they are not capricious, arbitrary, or unreasonable, (2) they are based on the nature or use of the property, (3) they exhibit a fair and substantial relation to the objective(s) of the authorizing legislation, and (4) the ratios of assessed value to market value and tax rates are uniform within each classification. Since the subclassification of land is intended to preserve agriculture and this goal is embodied in several state statutes as well as the state's constitution, the provision addressed the first three criteria (House Local Government Subcommittee on Boroughs 1997). Although the fourth requirement is primarily a function of implementation, the legislation supports the imposition of uniform assessment ratios, assessment methodologies, and tax rates for each classification. Legal precedent, therefore, suggested that the legislation would pass constitutional muster.

The subcommittee was also concerned about the financial implications for farmers. As was the case in the Senate, subcommittee members were troubled that residential development could use the two-tiered tax as a vehicle for shifting part of its property tax burden to the farmers. Although the establishment of farmland and nonfarmland subclassifications was intended to eliminate opportunities for shifting property tax burdens to agriculture, there was offsetting testimony that farmers in other countries benefited from a two-tiered tax that did not subclassify land. If the positive effects could be replicated in Pennsylvania, then Pennsylvania's farm preservation efforts could be buttressed without the subclassification of land. Due to a lack of clear guidance concerning the merits of subclassification, the subcommittee chair requested the Pennsylvania State Farm Bureau and the Pennsylvania Grange to examine the issues and decide which option they would support. When both organizations deemed that it was in the farmers' best interests to retain the land subclassification amendment (Armstrong 1998), the subcommittee elected to go along with the farm organizations' decision and retain the amendment. This decision, in conjunction with the amount of time the bill languished in subcommittee, indicates the committee members were reluctant to impose burdens on the advantaged group of farmers.

Following resolution of the constitutional and agricultural issues, the subcommittee sent the bill to the full committee, which reported the bill on May 6, 1998. Because of calendar constraints imposed by House Chamber renovations, the House did not approve the bill until November 16. Eight days later the bill was signed by the governor.

VI

Conclusion

The two-tiered property tax's contribution to urban revitalization emanates from the millage rate differential between land and improvements on the land. Whenever the implementation of the tax lowers the millage for improvements and raises the millage for land, there is a commensurate reduction in the tax disincentive for constructing, renovating, and expanding structures. At the same time, there is an increase in the relative cost of holding land. These changes spur urban redevelopment by encouraging investment in structures and the development of vacant parcels. The benefits associated with the adoption of a two-tiered property tax are maximized when the change from a single-rate tax to a two-tiered tax is revenue neutral. If this constraint is violated, then investment will be impeded until the after-tax costs and rates of return on investments are equalized between the jurisdiction imposing the tax reform and the surrounding region. It should also be noted that the two-tiered property tax enables jurisdictions confronting population and/or economic decline to stabilize their tax base and provide the services that retard or reverse the downward spiral.

The legality of implementing a two-tiered property tax is determined by a state's constitution and statutes. At the present time, only 23 constitutions permit the legislature to establish those property classifications that are essential to the implementation of the two-tiered property tax. Dillon's Rule[6] stipulates that the localities in these 23 states cannot levy the two-tiered tax unless authorization is granted by state statute. If the state statutes are mute on the subject or prohibit its use, then local officials cannot adopt the tax until the state enacts the requisite enabling legislation. In the event that enabling legislation is passed, then the extent to which localities can shift tax burdens from improvements to land is affected by the constitution's limits on maximum property tax rates.

An analysis of Pennsylvania's constitution and statutes does not reveal any barriers to implementation. Article VIII, Section 1 permits the state legislature to define land and improvements as distinctive property tax categories. Since Article IX does not set a ceiling on property tax rates, there are no constitutional limits on the extent to which tax burdens can be shifted from land to improvements.

Local adoption of the two-tiered property tax therefore is dependent on the enactment of state-enabling legislation that is influenced by the political power and characteristics of participating interest groups. One of the interest groups that participated in the recent liberalization of Pennsylvania's two-tiered property tax statute was local government. This entity is defined as an advantaged group because of its members' geographical proximity to its constituents and its role as partner in the delivery of services. The polls consistently find that citizens trust and feel a closer connection to entities such as local government that are smaller and physically closer to them than the state or national counterparts. Since localities partner with the state and national governments in the delivery of goods and services, state legislators are reticent to approve bills that adversely affect localities or that may generate opposition among local interests.

The political clout of local governments was also augmented by the legislation's universal appeal, which minimized the fissures in local government ranks. Urban localities favored the incentives for economic development, while suburban and rural officials were attracted by a potential reduction in urban sprawl and the associated decline in financial and environmental costs. The tax was especially popular in areas that were confronting the exit of major employers. In these instances, the tax provided a means to stabilize the local property tax base and thereby preserve the resources that are essential to maintaining a beleaguered jurisdiction's infrastructure and competitiveness.

However, the positive characteristics and political clout of local governments were not sufficient to overcome concerns about the legislation's potential negative effects on farmers and the preservation of open space. The Pennsylvania General Assembly continually struggled with the possibility that inhabitants of residential developments could use their numerical advantage to shift part of their property tax burden to farmers. Legislators were also concerned that higher property tax burdens would increase the number of farmers who choose to sell their land and undermine the state's efforts to preserve open space. The gravity of these concerns clearly affected the disposition of legislation authorizing school districts and boroughs to levy a two-tiered property tax. In the former instance, only nine urban school districts were permitted to utilize a two-tiered tax. Enabling legislation for the Commonwealth's boroughs was stymied in subsequent legislative sessions until the House Local Government Subcommittee on Boroughs (1) developed a method for protecting farms within nine of the Commonwealth's 962 boroughs and (2) gained assurances from the farm lobbies that they did not object to the amendments. It is therefore apparent that the fate of two-tiered tax enabling legislation was dependent on eliminating from the bills jurisdictions containing farmland or gaining the acquiescence of agricultural organizations such as the Farm Bureau and the Grange.

Since farmers in other countries are also politically powerful, the necessity of placating farmers and supporting open space preservation efforts could be replicated in other parts of the world. However, the experiences of Australia and New Zealand highlight the need to add one caveat to the Pennsylvania experience. In each of these cases, the entire property tax burden was shifted to land and the change was accompanied by the introduction of steeply progressive tax rates. There was, however, one important distinction—the policies of these two countries were intended to "break up large landholdings and redistribute wealth, rather than . . . controlling urban sprawl, stimulating construction, or controlling land speculation" (Wyatt 1994: 2). Thus, it is apparent that the interests of farmers and open space advocates can dash the hopes of local governments unless there is an overriding national interest in curtailing the wealth and power of agricultural interests.

In an era where globalization and cost cutting have contributed to the closure of major employers in many localities, the two-tiered tax has the unique capacity to stabilize the property tax base and furnish the resources that are essential to maintaining a jurisdiction's competitiveness. It also produces incentives for the redevelopment of downtown business districts and the enhancement of housing stocks. Before the benefits of the tax can be realized, significant legal hurdles must be overcome. The constitutions in 27 states must be amended so that land and improvements to land can be defined as permissible property tax categories. In each of these states and in most of the remaining 23 states, the legislature must also enact enabling legislation that authorizes localities to adopt the tax. Throughout the legislative process, the characteristics and political clout of the participating interest groups will impact the legislation's composition and fate. The Pennsylvania case clearly indicates that agricultural interests and the desire to preserve open space will trump the desires of local governments. Consequently, the probability of enacting enabling legislation will be maximized by limiting the authorization to jurisdictions that do not include farms within their boundaries.

Notes

1. A millage is the term that defines the property tax rate. Since one mill generates one dollar of property tax revenue for each $1,000 of assessed value, real estate with an assessed value of $50,000 will generate $50 of revenue for each mill of property taxes.

2. Boroughs occupy a middle ground between city and township governments with per capita revenues, expenditures, and net municipal debt that are greater than townships but less than cities. Boroughs are typically governed by a seven-member council and a weak mayor. One-third of Pennsylvania's boroughs are located in urbanized areas (64 in the Philadelphia Urbanized Area, 124 in the Pittsburgh Urbanized Area, and 143 in smaller urbanized areas) and an additional 126 boroughs are classified as urban but located outside of these urbanized areas. The remaining 508 boroughs (52.5 percent of the Commonwealth's boroughs) are classified as rural (Governor's Center for Local Government Services 2000).

3. The numerical examples focus only on the effects of property tax policy. Even though they do not assess the ramifications of expending the revenues, it must be noted that changes in a jurisdiction's spending priorities magnify or diminish the incentives to improve land and thereby influence the long-term density of development.

4. The literature has examined the two-tiered property tax's impact on the distribution of tax burdens, property improvements, and urban redevelopment, but has not assessed the degree to which fill-in development has affected urban sprawl. Consequently, constraints on urban sprawl arising from the implementation of a two-tiered property tax have been based on conjecture and mathematical modeling rather than an analysis of Pennsylvania cities.

5. The judicial branch consistently differentiates between the classification of property for purposes of imposing differential tax burdens and classification for purposes of calculating cash or market value. In the 27 states that are constitutionally prohibited from imposing different tax burdens on various classifications of property, the courts nevertheless uphold a legislature's authority to establish property classifications for the purpose of estimating the properties' cash or market value. The rationale for the apparent discrepancy is the mechanisms markets utilize to establish a property's market value. The criteria for determining the market value of residential property is not the same as the criteria for establishing the value of industrial property. As a result, it is not possible to establish the fair or equalized market value unless the state government defines multiple classifications and assessment formulas that are appropriate for each classification.

6. Dillon's Rule is a legal doctrine that is named for Justice John Dillon. According to Dillon's Rule, local governments can exercise only those powers that are explicitly granted to them (e.g., the power to levy specific taxes), clearly implied by the explicitly granted powers, or essential to "meeting the declared objectives and responsibilities of local government" (Nice 1987: 138).

References

Articles, Books, and Surveys

Berger, Peter L., and Thomas Luckmann. (1966). *The Social Construction of Reality: A Treatise in the Sociology of Knowledge.* New York: Doubleday.

Bourassa, Steven C. (1987). "Land Value Taxation and New Housing Development in Pittsburgh." *Growth and Change* 18(4): 44–56.

——. (1992). "Economic Effects of Taxes on Land." *American Journal of Economics and Sociology* 51(1): 109–114.

Browne, William P., Jerry R. Skees, Louis E. Swanson, Paul B. Thompson, and Laurian J. Unnevehr. (1992). *Sacred Cows and Hot Potatoes: Agrarian Myths in Agricultural Policy.* Boulder, CO: Westview Press.

Brueckner, Jan K. (1986). "A Modern Analysis of the Effects of Site Value Taxation." *National Tax Journal* 39(1): 49–58.

Council for Excellence in Government. (1999). *America Unplugged: Citizens and Their Government.* Available at http://excelgov.org/usermedia/images/uploads/PDFs/America_Unplugged_full_report.pdf.

Cox, Robert Henry. (2001). "The Social Construction of an Imperative: Why Welfare Reform Happened in Denmark and the Netherlands but Not in Germany." *World Politics* 53(3): 463–498.

Czech, Brian, and Paul R. Krausman. (2001). *The Endangered Species Act: History, Conservation Biology, and Public Policy.* Baltimore, MD: Johns Hopkins University Press.

DiMasi, Joseph A. (1987). "The Effects of Site Value Taxation in an Urban Area: A General Equilibrium Computational Approach." *National Tax Journal* 40(4): 577–590.

Douglas, Richard. (1980). "Site Value Taxation and the Timing of Land Development." *American Journal of Economics and Sociology* 39(3): 289–294.

Governor's Center for Local Government Services, Department of Community and Economic Development. (2000). *Borough Council Handbook,* 10th ed. Harrisburg, PA: Department of Community and Economic Development.

The Harris Poll. (1998). *The Remarkable Lack of Intergenerational Conflict as to How Government Spending Should be Divided Between Services for Young and Old.* Available at http://www.louisharris.com/harris_poll/index.asp?PID=183.

Jennings, M. Kent. (1999). "Political Responses to Pain and Loss: Presidential Address, American Political Science Association, 1998." *American Political Science Review* 93(1): 1–13.

Knott, Jack H., and Carol S. Weissert. (1995). "Predicting Policy Targets Through Institutional Structure: Examining State Choices in Increasing the Production of Primary Care Physicians." *Administration and Society* 27(1): 3–24.

Matsusaka, John G. (1995). "The Economic Approach to Democracy." In *The New Economics of Human Behavior.* Mariano Tommasi and Kathryn Ierulli, eds. New York: Cambridge University Press.

Nice, David C. (1987). *Federalism: The Politics of Intergovernmental Relations.* New York: St. Martin's Press.

Peters, Robert A. (1994). "The Politics of Enacting State Legislation to Enable Local Impact Fees: The Pennsylvania Story." *Journal of the American Planning Association* 60(Winter): 61–69.

Pew Research Center for the People & the Press. (1998). *How Americans View Government: Deconstructing Distrust.* Available at http://people-press.org/reports/print.php3?Page ID=593.

Pollock, Philip H., III. (1994). "Issues, Values, and Critical Moments: Did 'Magic' Johnson Transform Public Opinion on AIDS?" *American Journal of Political Science* 38(2): 426–446.

Schneider, Anne, and Helen Ingram. (1993). "Social Construction of Target

Populations: Implications for Politics and Policy." *American Political Science Review* 87(2): 334–347.

The Star-Ledger/Eagleton Poll. (1998). *Public Coming and Goings: Want Open Space Preservation/Transportation Improvement? HELLO—Willing to Pay for Them? GOOD BYE.* Available at http://slerp.ruters.edu/releases/119–1.pdf.

Witte, Ann D., and James E. Bachman. (1978). "Vacant Urban Land Holdings: Portfolio Considerations and Owner Characteristics." *Southern Economic Journal* 45(October): 543–558.

Wyatt, Michael D. (1994). "A Critical View of Land Value Taxation as a Progressive Strategy for Urban Revitalization, Rational Land Use, and Tax Relief." *Review of Radical Political Economics* 26(1): 1–25.

Zablotsky, Edgardo Enrique. (1995). "The Process of Government." In *The New Economics of Human Behavior.* Mariano Tommasi and Kathryn Ierulli, eds. New York: Cambridge University Press.

Correspondence and Interview

Armstrong, State Representative Thomas E. Pennsylvania General Assembly. Phone Interview, June 1, 1998.

Brown, State Representative Teresa E. Pennsylvania General Assembly. Correspondence, September 15, 1997.

Gerlach, State Senator James W. Pennsylvania General Assembly. Correspondence, October 20, 1997.

Godshall, State Representative Robert W. Pennsylvania General Assembly. Correspondence, September 17, 1997.

Hershey, State Representative Arthur D. Pennsylvania General Assembly. Correspondence, September 22, 1997.

Lynch, State Representative Jim. Pennsylvania General Assembly. Correspondence, September 19, 1997.

Phillips, Merle H. Pennsylvania General Assembly. Correspondence, September 19, 1997.

Stern, State Representative Jerry A. Pennsylvania General Assembly. Correspondence, September 17, 1997.

Wozniak, State Senator John N. Pennsylvania General Assembly. E-Mail Correspondence, February 4, 1998.

Legal and Legislative Documents

Constitution of the State of Louisiana.

Constitution of the State of Oregon.

Legal and Legislative Documents: Commonwealth of Pennsylvania

Constitution of the Commonwealth of Pennsylvania, 1873.

Constitution of the Commonwealth of Pennsylvania, 1968.

Pennsylvania General Assembly, House of Representatives. *Legislative Journal.* 1992 session, 876–882.

Pennsylvania General Assembly, House. *Hearings of the House Local Government Subcommittee on Boroughs.* October 2, 1997.

Pennsylvania General Assembly, Senate. *Legislative Journal*. 1959 session: 3533–3534.
Pennsylvania House Bill 2517. 1983 session. Printer's Numbers 3641.
Pennsylvania House Bill 438. 1993 session. Printer's Numbers 489, 1614, 1828, and 1982.
Pennsylvania House Bill 2846. 1993 session. Printer's Number 3776.
Pennsylvania House Bills 3089–3091. 1993 session. Printer's Numbers 4260–4262.
Pennsylvania House Bill 2532. 1993 session. Printer's Number 3238.
Pennsylvania House Bill 1258–1261. 1995 session. Printer's Numbers 1418–1421.
Pennsylvania House Bill 554–557. 1997 session. Printer's Numbers 617–620.
The Pennsylvania Manual. 1975 session. Harrisburg, PA: Department of General Services.
The Pennsylvania Manual. 1983 session. Harrisburg, PA: Department of General Services.
The Pennsylvania Manual. 1993 session. Harrisburg, PA: Department of General Services.
The Pennsylvania Manual. 1995 session. Harrisburg, PA: Department of General Services.
The Pennsylvania Manual. 1997 session. Harrisburg, PA: Department of General Services.
Pennsylvania Senate Bills 1491–1497. 1975 session. Printer's Numbers 1863–1869.
Pennsylvania Senate Bills 2014–2020. 1977 session. Printer's Numbers 1174–1180.
Pennsylvania Senate Bills 214–219 and 222. 1979 session. Printer's Numbers 215–220 and 223.
Pennsylvania Senate Bills 362–366. 1981 session. Printer's Numbers 365–369.
Pennsylvania Senate Bill 580. 1995 session. Printer's Number 604.
Pennsylvania Senate Bill 211. 1997 session. Printer's Number 940.

Panel Discussion

Gardner, Jack. (Speaker). (1997). *Tax Reform Forum*. (Videotape Recording). Chambersburg, PA: 10,000 Friends of Pennsylvania.
Spossey, City of Washington (PA) Mayor Anthony. (Speaker). (1997). *Tax Reform Forum*. (Videotape Recording). Chambersburg, PA: 10,000 Friends of Pennsylvania.
Sullivan, Dan. (Speaker). (1997). *Tax Reform Forum*. (Videotape Recording). Chambersburg, PA: 10,000 Friends of Pennsylvania.

A Simple General Test for Tax Bias

By Mason Gaffney*

Abstract. The paper infers the biasing effects of taxes from their differential effects on the present values of rival uses for given tracts of land. After-tax wage rates, interest rates, and commodity prices are exogenous, hence not affected by taxes, which are therefore all shifted to land rents and values. The effects are differential among rival uses, hence change their ranking in the eyes of the landowner-manager. Most taxes downgrade the highest use into a lower use, inducing quantum leaps away from higher and better uses into lower and worse uses. The paper uses forestry as an allegory for all land uses. It compares yield taxes, property taxes, income taxes, and site value taxes. It finds that a change from the first three to the site value tax would induce quantum leaps from lower to higher uses of land.

The method here is to infer the biasing effects of taxes from their differential effects on present values of rival uses for land. A local tax jurisdiction is an open economy. Our simplifying premise is that arbitrage equalizes all *after*-tax rates of return on new investing, at levels determined in world capital markets. Labor is free to come and go, and product prices are set in world markets. Given those premises, all taxes are shifted to land, the only factor fixed in an otherwise open economy; tax jurisdictions are defined as fixed areas of land.

Using these premises lets us devise a simple test for tax neutrality.[1] Treat net present value derived from a land improvement as a residual, and impute this residual value to land. Find algebraically the ratio of after-tax land value to before-tax land value. If the ratio is simply (1 – t) (where *t* is a tax rate), the tax is neutral—the highest and best use of land after tax is the same as that before tax.[2] The ratio (1 – t)

*The author is a Professor of Economics at the University of California, Riverside, CA 92521; email: migaffney@surfcity.net. Many of his writings are available at http://www.masongaffney.org.

American Journal of Economics and Sociology, Vol. 65, No. 3 (July, 2006).

is independent of any parameter the landowner controls. The tax base on marginal land must be zero, lest the land be sterilized.[3]

The simplicity of this technique allows for complexity in the applications, without losing any threads in tangles of detail. We can analyze or just inspect many parameters in the ratio to find what specific avoidance maneuvers a tax will induce and to estimate what excess burdens will result. In this paper, we analyze effects on substitution of capital for labor and for land, including effects on capital turnover and frequency of site renewal. We analyze differential effects on different grades or qualities of land. We can also show how to find revenue-neutral tax rates, when tax A is substituted for tax B. We can point toward dangerously snowballing "Laffer-curve effects" and how to minimize them by selecting more neutral kinds of taxes.

The present study uses timber culture as an example because this enables a simple analysis, along with continuous grounding in reality. Timber is a good allegory for all other forms of investment. It occupies 32 percent of the private land area of the nation and is weighty in its own right (Daugherty 1995). This short paper does not treat other kinds of capital explicitly, but does explain a simple means of modifying the analysis to do so. The writer has published the relevant mathematics elsewhere (Gaffney 1976a, esp. Appendix I).

One distinguished commentator, Gordon Tullock, has suggested orally that this is Georgist tax theory restated. He is partly right, partly wrong. The findings are consistent with Henry George's ideas about the neutrality of taxing land values. However, George had no capital theory except an error-ridden one that no one cares to remember, while the present paper deals mostly with durable capital.

I

Harvest or Yield Tax

"YIELD" TAXES ARE IMPOSED on the harvest value of timber ("stumpage"), net of harvest costs, but gross of up-front capital costs. The tax rate is flat, at rate t. The taxable event is timber harvest. Yield taxes are widely believed to be neutral because the growth rate of stumpage after tax is the same as it is before tax. Our analysis is more comprehensive, however, considering the whole investment cycle, and finds a heavy bias. First, we set up the model:

S = site value from discounted cash flow (DCF) absent taxes
R = revenue from "stumpage" (sale value net of harvest costs) at maturity (year "*m*")
m = maturity (years from planting to harvest)
i = relevant interest and discount rate
t = tax rate applied to the base "R" after *m* years
P = planting cost, year zero

One may incorporate intermediate costs and revenues in the model without disturbing it, either by compounding them forward to year *m* (where they are commensurable with "R"), or by discounting them to year zero (where they are commensurable with "P"). This lets us analyze cycles of timber culture in their totality, unbound by the simple case where all costs are incurred at time zero, and all revenues come at one other point in time. Better yet, this is the "simple means" wherewith one can generalize the model from timber to any other kind of capital improvement, whatever its time pattern of costs and revenues.

Site value (S) is the present value of timber less its planting cost (P). That residual value is imputed to the site. To make it hugely more general and useful, and only marginally more intricate, we assume the investment cycle to be repeated every *m* years, in perpetuity. That accounts for the "–1" in the denominator of Equation (1).

$$S = \frac{R - Pe^{mi}}{e^{mi} - 1} \tag{1}$$

Equation (1) is Faustmann's formula for "Soil Expectation Value," widely discussed in the literature. It is derived by discounting the numerator not just once, but as an infinite chain repeatable in perpetuity (Gaffney 1957; Scott 1987, and works there cited).[4] An advantage of this model is to dispense with any arbitrary limit on the time horizon; it lets us treat capital turnover and replacement.

To show the effect of a yield tax, let:

$$\sigma = \text{Site value after Yield Tax}$$

$$\sigma = \frac{R(1-t) - Pe^{mi}}{e^{mi} - 1}. \tag{2}$$

By inspection, since P is not deductible, there is a leverage effect in the tax: it falls harder on marginal investments. It induces entrepreneurs to abort marginal investments on all land, and all investments on marginal land, causing an "excess burden." By assumption, this excess burden does not result from forward shifting to consumers with elastic demand, nor from backward shifting to suppliers of capital or labor. It results from "downward" shifting to land. It changes what now appears to the owner as the highest and best use of land, after tax.

To measure the bias, we find σ as a fraction of S:

$$\sigma/S = 1 - t\left[1 + \frac{P/S}{1 - e^{-im}}\right]. \tag{3}$$

Equation (3) is smaller than $(1 - t)$, except when $P = 0$. Simple inspection of the algebra now lets us identify several kinds of bias. Equation (3) is highly sensitive to the parameters P, S, i, and m. Equation (3) is a decreasing function of P, and an increasing function of S, m, and i. Thus the yield tax biases landowners against intensive planting (high P) and against shorter cycles (low m), and against marginal land (low S). It also magnifies the advantage of those with strong financing (low i) over those with weak credit. The last force will act toward fostering higher concentration of ownership.

Taxes on marginal land are greater than zero. Equation (3) may easily become zero or negative, meaning land will have no use at all (without adapting the parameters to avoid taxes). If after-tax land value is zero or less, land-time has no value to the owner, and there is no economic reason to restock land. The combination would lead to a bias in favor either of nonuse or of "volunteer" regeneration, where P is held at zero. This comes at the cost of deferring m and lowering R, possibly to zero. Bias is a maximum against marginal land (low S) and, by clear inference, against marginal increments of P and R to all land.

Table 1

Values of σ/S from Equation (3), where $i = 0.07$; $t = 0.40$

	P/S = >	0.2	0.5	0.75	1	1.5
m	$1 - e^{-im}$	—	—	—	—	—
5	0.30	0.33	−0.07	−0.40	−0.73	−1.4
10	0.50	0.44	0.20	0.00	−0.20	−0.60
15	0.65	0.48	0.29	0.14	−0.02	−0.32
20	0.75	0.49	0.33	0.20	0.07	−0.20
50	0.97	0.52	0.39	0.29	0.19	−0.02
60	0.99	0.52	0.40	0.30	0.20	−0.01

Table 1 displays the bias by numerical example, using i = 0.07 and t = 0.40. The 40 percent yield tax rate is chosen because it is the revenue-neutral rate corresponding to a 1 percent property tax rate, as explained and calculated later.

To avoid taxes, landowners are induced to move from the upper right toward the lower left in Table 1; in other words, from high P/S and low m to low P/S and high m. The resulting loss of net present value before tax is a measure of excess burden. Just how far each landowner will move depends on a host of particulars far too numerous to treat in the small compass here. The point is that the tax introduces a powerful arbitrary bias, acting in predictable directions.

The landowner subject to yield taxes is in the same position as a share tenant. Modern work on share tenancy, following Gale Johnson and Stephen Cheung (1969), also stresses the logical counterpart: crop-sharing motivates tenants to take up land without limit. Private landlords big enough to dominate their markets use their bargaining power to prevent that by limiting the land they mete out to each tenant; but the fisc has no such power over private landowners. A byproduct of yield taxation is, therefore, a tendency toward reenforcing concentration of ownership of forest land.

Many forest outlays come well after time-zero: examples are thinning, pruning, fire and pest control, and timber stand improvement (TSI). Each such outlay is a separate investment cycle of shorter life than *m*. Its resulting revenue is the increment it generates in total R. Each such cycle would be punished by a yield tax in terms of its own short life, not the entire tree lifecycle of *m* years. The bias against such outlays is, from Table 1, obviously extra heavy.

II

Property Tax on Standing Timber

PROPERTY TAXES ARE IMPOSED ANNUALLY on a base equal to the assessed market value of standing timber, starting when timber is planted. The base is not the value of timber for immediate harvest, which may be nil for some years. It is its investment value to a buyer who will hold it until maturity. The tax rate is flat, at rate *p*. The taxable event is owning timber on the annual tax date.

S, R, m, i, and P are as before.

p = property tax rate
θ = site value after property tax on timber (not land)

θ is the value that satisfies Equation (1) when we add *p* to *i*:

$$\theta = \frac{R - Pe^{(i+p)m}}{e^{(i+p)m} - 1}. \qquad (4)$$

By inspection of Equation (4), the effect of the property tax on timber is the same as the effect of raising the discount rate by the amount of the tax.

Finding θ as a fraction of S, we get what appears to be a complex expression; but we will simplify it greatly:

$$\theta/S = \frac{e^{im} - 1}{e^{(i+p)m} - 1} - (P/S)\frac{e^{pm} - 1}{e^{pm} - e^{-im}}. \qquad (5)$$

Equation (5) looks fierce, but may be tamed by tabulating its two coefficients. Better yet, they are complements, reducing them to one. We name the first coefficient, Ω. Thus, Equation (5) may be rewritten:

$$\theta/S = \Omega - (1 - \Omega)P/S \qquad (5A)$$

Table 2

Values of θ/S from Equation (5A), where $i = 0.07$; $p = 0.01$

	P/S = >	0.2	0.5	0.75	1	1.5
m	Ω	—	—	—	—	—
5	0.85	0.82	0.78	0.74	0.70	0.63
10	0.83	0.80	0.75	0.70	0.66	0.58
15	0.80	0.76	0.70	0.65	0.60	0.50
20	0.77	0.72	0.66	0.60	0.54	0.43
50	0.60	0.52	0.40	0.30	0.20	0.00
60	0.55	0.46	0.34	0.21	0.10	−0.13

Ω, in turn, may be tabulated, and varies within narrow limits. Needed values of Ω are given in Table 2, Column 2. Thus Equation (5), despite its complex first impression, becomes docile and tractable.

Table 2 displays values of θ/S when i = 0.07, and p = 0.01. It is comparable with Table 1 to expedite comparing the effects of property taxes and yield taxes. The 40 percent yield tax rate (t) is revenue-neutral with a 1 percent property tax rate (p), as we calculate and explain below.

Tax avoidance induces landowners to move from the lower right toward the upper left in Table 2; in other words, from high P/S and high m to low P/S and low m. Like the yield tax, the property tax induces less application of capital. However, the taxes differ in their effects on long versus short cycles. Yield taxes favor longer cycles; property taxes favor shorter ones.

Note, though, that the property tax bias is weaker than the yield tax bias, and much weaker when m is low. This finding refutes "conventional wisdom" about the catastrophic effects of the property tax on standing timber. The property tax is not without sin, but neither is it the most counterproductive tax. Its biases are considerably abated by a double capitalization effect. Like all taxes, it lowers site value, but it also lowers the value of standing timber itself, thus tempering the tax burden. The literature sometimes recognizes the first effect, but hardly ever the second.

A useful byproduct of that model is to determine what yield tax rate is revenue neutral when yield taxes are substituted for property taxes on timber (as they now are in most states). No one rate can be revenue neutral everywhere because this will depend on the value of m. For an example, let $m = 50$. This would apply on the West Coast, where rotations are much longer than in the southeast. Now we ask: "What value of t makes $\sigma = \theta$ when $p = 0.01$?" This is revenue neutral in the sense that the present value of taxes is the same in either case.

It does not adjust for taxpayer avoidance reactions, which limits its generality, and overstates revenue from both taxes, but especially from yield taxes. Tables 1 and 2 indicate that bias is stronger under a yield tax than under a property tax, so avoidance behavior will be correspondingly more extreme. Thus, the comparison made here is probably too favorable to a yield tax.

The calculation is greatly simplified, without significant loss of accuracy, by setting $P = 0$.[5] Now we have $\sigma = \theta$ when:

$$S(1 - t) = S(\Omega)$$
$$t = 1 - \Omega = 1 - 0.60 = 0.40. \tag{6}$$

Thus, for revenue neutrality, a yield tax rate of 40 percent is needed for each 1 percent cut in the property tax. At such a high rate, there would be a severe Laffer-curve effect: a higher rate bringing in lower revenues. This effect might be so strong that no yield tax rate, however high, could replace property tax revenues.

In most states, however, the yield tax rate is much lower. In California, the rate is capped at 2.9 percent, levied in lieu of a 1 percent property tax rate. This entails not just a change in the tax base, but a near approach to tax exemption. The low tax levy makes yield taxation popular with forest interests and accounts for the support they give it. It conceals the severe excess burden that yield taxes would impose at revenue-neutral rates.[6]

The revenue-neutral value of t is the simple complement of Ω. Table 2 shows that Ω is sensitive to m, but only moderately so. This means that even where rotations are 15 years instead of the 50 years used in Equation (6), it still takes a yield tax rate of 20 percent to be revenue neutral. That is lower than 40 percent, but still consistent with our basic finding that very high yield tax rates are required for revenue neutrality.

In addition, there are other taxes to consider. The yield tax levied in lieu of a property tax induces foresters to lower both the amount of P and its *frequency* as well. Lower and less frequent P also means lower and less frequent harvests, where most payrolls are generated—and taxed. Payroll taxes, income taxes, sales taxes, gasoline taxes, and all other activity-based taxes come along less often and in lesser amounts. If we summed all taxes generated in forests, and in ancillary activities, yield taxes higher than the 40 percent shown would be needed for revenue neutrality. Again, rates this high would cause a heavy Laffer-curve effect such that revenue neutrality might never be attained.

Thus, the tax bias shown is more than just "allocational," or microeconomic. It is also a bias against aggregate employment on the nation's fixed stock of land. Yield taxes make timber culture less labor using, more land using, and more capital using (in the Austrian sense of longer investment cycles). Timber culture, in the present analysis, is an allegory that applies to all investments of whatever kind, so its implications are general and macroeconomic. Tax biases like those analyzed here affect every parcel of land subject to taxation. The writer has developed this theme elsewhere (Gaffney 1976a, 1976b).

III

Tax on Income from Property

THE EXTREME INTERTEMPORAL BIAS of the yield tax, and some of its bias against P, are abated by letting the grower deduct P from the tax base. It makes a great difference *when* the grower may deduct it. For a tax on all property income (from both timber and site), let him or her deduct it at maturity, *m*. Let:

π = Site value after tax on property income
r = Corresponding tax rate on property income

$$\pi = \frac{R(1-r) - Pe^{im} + rP}{e^{im} - 1} \tag{7}$$

$$\pi/S = 1 - r(1 + P/S). \tag{8}$$

Equation (8) does away with explicit intertemporal bias: *m* does not appear in it. However, Equation (8) retains a bias against high values of P and low values of S.[7] For $m > 30$ years or so, the gradient of bias is only negligibly less than that shown in Table 1 for the yield tax. Comparing Equations (8) and (3) makes clear the reason why. Equation (3) approaches Equation (8) as *m* approaches infinity, and, in practice, as *m* exceeds 30 or so.

There is implicit intertemporal bias, too, because the bias against P tends to lower and/or defer the application of P. The cost of holding land is lowered during the downtime of land, the time between harvesting one rotation and planting the next. Less investment in P, when it does occur, also gets the next crop off to a slower start (e.g., if the owner seeds instead of planting nursery stock, and even more so if he or she waits for volunteer regeneration).

The absence of explicit intertemporal bias gives added force to the bias against P. The landowner has no avoidance route to soften the tax impact except to lower P. Further, the deduction of P means that the revenue-neutral tax rate is higher than for the yield tax. The higher rate, of course, leads into Laffer-curve effects, somewhat offsetting the efficiency gain that comes from letting owners deduct start-up costs.

IV

Tax on Net Site Rent or Net Site Value

THE TAX BASE MAY BE NARROWED to land, in two different ways:

a. With an income tax, by deducting P at the front end (expense it), instead of capitalizing it for later deduction;
b. Assess land directly, and levy a property tax based on site value.

A. *Expensing P*

Let Γ = Site value after tax on yield, when P is expensed;
u = corresponding tax rate

$$\Gamma = \frac{R(1-u) - Pe^{im} + uPe^{im}}{e^{im} - 1} = S(1-u)$$

$$\Gamma/S = 1 - u. \tag{9}$$

Here, at last, we have a kind of tax neutrality. Equation (9) is independent of P, S, *i*, or *m*. Note, also, that we have not destroyed the tax base. The usual objection to expensing is that it is equivalent to tax exemption. So it is, for those items that are expensed. However, land purchase is not deducted nor expensed (except extra-legally, which is another story).[8] Only P is expensed, while income imputable to the site remains fully taxable. Furthermore, taxable site income is enhanced by the benefit that inures from letting foresters expense P. Under our premises, this benefit lodges in higher site rents.

The rate must be raised a great deal to maintain revenues. There is no explicit Laffer-curve effect, but a problem with this tax is moral hazard. The grower, when he or she expenses P, essentially thereby becomes the manager of capital supplied by the Treasury. The grower then owes the Treasury a high fraction of gross sales (R). The temptation to "fudge" might be too high for practical administrative control. There also is the problem of what to do when the investing firm has no outside income against which to deduct the expense of P. Marketing of excess expenses is conceivable, but the unpopularity of "safe harbor" provisions helped kill the investment tax credit (ITC) in the early 1980s.

Further, during the downtime of land after harvest, this method offers a carrot but wields no stick (has no income effect or cash-flow effect), compared with the next method.

B. A Property Tax Limited to Site Value

Let:

$$\varphi = \text{Site value after tax on site value}$$
$$w = \text{Corresponding tax rate on } \varphi$$
$$T = w\varphi$$
$$T/i = \text{Present value of future taxes in perpetuity}$$
$$\varphi = S - T/i = S - w\varphi/i. \qquad (10)$$

Since φ is both the tax base, and the after-tax value, it appears on both sides of Equation (10). This is the classic phenomenon of land tax "capitalization."[9] One resolves the apparent "dilemma" albegraically by collecting terms. Doing so, the property tax rate is simply added to the capitalization rate, a routine procedure among professional appraisers and assessors.

$$\varphi = \text{Si}/(\text{i} + \text{w}) \qquad (11)$$
$$\varphi/\text{S} = \text{i}/(\text{i} + \text{w})$$

Equation (11) is independent of P or *m* or S. It is allocationally neutral. That means that the rate may be raised to any high level without imposing an excess burden, and without any self-defeating Laffer-curve effect. Equation (11) is also an increasing function of *i*. This means the effect of the tax is to weaken the advantage of buyers with strong financing (low *i*) over those with poorer access to credit.

Note a counterintuitive feature of land tax capitalization in Equation (11). The tax rate, *w*, may rise above 100 percent without destroying the tax base, φ. A higher tax rate lowers the base, but this is not a Laffer-curve effect because the base falls in lesser proportion than the rate rises, resulting always in higher collections. You have to apply a higher rate, but you keep raising more revenue, and you do not abort any investing or producing, even on marginal land.

Popular political rhetoric about Laffer-curve effects (now also called "dynamic revenue forecasting") lumps all taxes as though they were homogeneous. Some apparently cool-headed economic analysis, regrettably, does the same. This is careless and misleading. It keeps us from analyzing the structure of different taxes. Some taxes have powerful Laffer-curve effects; some have weak effects; some have none at all.[10] To understand and remedy our revenue predicament, we need to take account of these basic differences.

There is no moral hazard problem with the site-value tax ($w\varphi$), as there is with the tax on land income ($u\Gamma$). The landowner is not managing any of the Treasury's capital; only his or her own. Accurate assessment of site value now becomes more critical in one sense, but in another sense not at all. Accurate assessment is highly desirable, for obvious reasons of distribution, social morale, and revenue. However, William Vickrey often pointed out that inaccurate land assessments will not bias land use, our present subject, as long as they are not functions of the use or ownership of land.

From Equation (11), site rent (Si) is now divided between the landowner and the fisc in the proportion that i bears to w. To maintain constant revenues, w must vary in proportion to the general level of market i. This is a feature of all property taxes, owing to the capitalization effect. It may appear tricky on paper, but tax collectors have coped with it over several thousand years of property tax history.

How about revenue neutrality? Suppose a state exempts standing timber from the property tax, and compensates by raising site taxes. What value of w is required to make $\varphi = \theta$ when $m = 50$ and $p = 0.01$? Simple mathematics gives us a start on answering this question, but will have to be interpreted.

$$\text{For } \varphi = \theta: \quad i/(i + w) = \Omega - (P/S)(1 - \Omega). \tag{12}$$

Solving Equation (12) for w:

$$w = i\left[\frac{1}{\Omega - (P/S)(1 - \Omega)} - 1\right]. \tag{12A}$$

Let $P/S = 0.2$, $i = 0.07$, $m = 50$, $p = 0.01$. Then $w = 0.06$. This seems like a high jump, from 0.01 to 0.06, but it overstates what is normally needed. Bear in mind that in timber culture, the ratio of site value to the value of the "improvement" (i.e., timber) is very low in the last years before harvest. The value of capital in this model begins from the value P, at time zero, but then grows exponentially for m years. It rises to the value R at the end of the growth cycle. Solving Equation (1) for R, and given the parameter values just posited, $R = 194P$ (or 39S). Thus, in the mature years of timber, its value dwarfs the site value. In applying this model to other kinds of land improvements, the revenue-neutral value of w would be much lower than 0.06.

Likewise, where growth cycles are shorter, as with Southern Yellow Pine, values of m are much lower, resulting in higher values of Ω. These in turn give much lower values for w, because w is supersensitive to Ω. At $m = 15$, for example, $\Omega = 0.80$, and $w = 0.02$. The Appendix gives additional reasons why, in practice, the revenue-neutral rate of w is generally much lower than 0.06. The most important of these is that the site tax has no excess burden, while the tax on timber does.

Another feature of the site tax is that it picks up speculative elements of land value derived from uses other than timber culture. In many regions, these values are much higher than values derived by discounting future timber harvests alone. Taking this into account, a lower rate is revenue neutral.

The premise of Equation (11) is that assessments be based on market value. In many states, in fact, timberland owners enjoy preferential assessment of land. Under California law, for example, a state agency controls assessments, applying a legislated formula that is structured so as to ensure valuations below market (*California Revenue and Tax Code*, Section 434.5, analyzed in Gaffney 1995). The California Code also prescribes that valuations be derived from timber culture alone. Under these constraints, the revenue-neutral tax rate would have to be higher, but the "high" rate would be only nominal since it is applied to assessed values that are well below market values. The de facto rate would still be what Equation (12A) shows. An open and above-board system would, of course, use true assessments that follow the market.

V

Summary

YIELD TAXES, PROPERTY TAXES on capital, and income taxes all impose substantial excess burdens on timber culture and, by extension, on all land uses. They sterilize marginal land completely, and abort marginal increments of capital and work on all land. To abate problems of the income tax, we may allow expensing capital outlays (other than land purchase). To abate problems of the property tax, we may exempt timber and raise the rate on site value.

Notes

1. Some analysts prefer to treat rates of return after tax (RORAT) as the residual, and the criterion of neutrality. We do not enter that thicket here. For those preferring the RORAT approach, the writer has run such a test elsewhere (Gaffney 1967). The results were broadly consistent with those presented here.

2. The ratio might also be $1/(1 + t)$, or some equivalent, as in Equation (11), and be neutral.

3. A zero tax on marginal land implies a zero tax for the marginal investment on *all* land, a requisite for neutrality.

4. A simple derivation is to begin with $(P + S)e^{mi} = R + S$, and solve for S.

5. Readers may confirm this by setting the rate of 0.40, arrived at in Equation (6), in the full equation in its complex form.

6. Another factor, in *Realpolitik*, is the insurance against double taxation such as might occur if an owner were to pay property taxes for many years and then be faced with a newly enacted yield tax.

7. This is the factor omitted by Thomson and Goldstein (1971) in their defense of the neutrality of income taxation.

8. Many tax proposals now bruited, like the "flat tax" of Hall and Rabushka, would allow expensing of land purchase. Thus far, however, nothing like this has been enacted.

9. This is a puzzle or paradox for neophytes, but mathematically trivial, and routinely used in the trade by appraisers and assessors.

10. Considering income effects, wealth effects, and liquidity effects, the tax may actually raise production: a reverse Laffer-curve effect. These points are important, but beyond the scope here.

References

Cheung, S. N. S. (1969). *The Theory of Share Tenancy.* Chicago: University of Chicago Press.

Daugherty, Arthur. (1995). "Major Land Uses in the U.S." Agricultural Economics Report #723, Economic Research Service, USDA, September.

Gaffney, Mason. (1957). "Concepts of Financial Maturity of Timber and Other Assets." Agricultural Information Series #62, North Carolina State College, Raleigh.

——. (1967). "Tax-Induced Slow Turnover of Capital." *Western Economic Journal* Sept.: 308–323.

——. (1976a). "Capital Requirements for Economic Growth." In *U.S. Economic Growth from 1976 to 1986: Prospects, Problems and Patterns* (vol. 8). Joint Economic Committee, U.S. Congress. Washington, DC: GPO.

——. (1976b). "Toward Full Employment with Limited Land and Capital." In *Property Taxation, Land Use and Public Policy.* Ed. Arthur Lynn, Jr. Madison: University of Wisconsin Press.

——. (1970–1971). "Tax-Induced Slow Turnover of Capital." Unabridged in five ports. *American Journal of Economics and Sociology* 29(1): 25–32; 29(2): 179–197; 29(3): 277–287; 29(4): 409–424; 30(1): 105–111.

——. (1995). "Property Tax Reform in the Big Picture." Paper delivered at the Conference on Land, Wealth, and Poverty, The Jerome Levy Institute, Annandate-on-Hudson, NY, November 3.

Johnson, Gale. (1950). "Resource Allocation Under Share Contract." *Journal of Political Economy* 58: 111–123.

Scott, Anthony. (1987). "Faustmann, Martin." In *The New Palgrave* (vol. 2). Ed. John Eatwell et al. London: Macmillan Co.

Thomson, Procter, and Henry Goldstein (1971). "Time and Taxes." *Western Economic Journal* 9: 27–45.

Appendix: Taxable Capacity of the Site Value Base

FOR P/S > 1, Equation (12A) seriously overstates the required value of *w* because the excess burden of any tax other than the site tax is very great on marginal land (low S, high P/S). As a purely mathematical exercise, Equation (12A) indicates that with some parameters (low values of S and low values of Ω), no value of *w*, however high, would be high enough to be revenue neutral. This, however, is economically impossible within our "physiocratic" premises, where taxable surplus and land rent are synonymous. A very high rate of *w* will tax away all the land rent there is, which is the entire taxable surplus that could be collected by any tax.

Five reasons explain the high values of *w* yielded by Equation (12A). First is tax capitalization, as explained above. Second is the narrow base. Site value (S) may exceed planting costs (P) at time zero, and normally does, but by the time of harvest (*m*), the value of timber will have grown to many times S.

Third is that our premise in finding θ originally is to ignore taxpayer avoidance maneuvers. This results in overstating θ relative to φ by assuming that landowners continue to grow timber just as though their trees were not being taxed—even though the tax makes some land values negative (Table 2). In fact, this deadweight loss or excess burden drives some land out of use altogether, and lowers P invested on all lands. It is a maximum where S is low and Ω is low (which is when *m* is high). These are the very conditions that are required to make the denominator of Equation (12A) approach zero. Thus, Equation (12A) ceases to give accurate values for the required *w* when it appears, in terms of simple mathematics, to give sky-high values of *w*.

Fourth is that applying the tax p to standing timber induces owners to shorten rotations, cutting timber earlier, and thus eroding the tax base. This is another dimension of the deadweight loss or excess burden.

Fifth, a purely practical matter, is that hardly any state has ever assessed saplings at full market value in practice, as premised in the mathematics. The practice has been to overlook green timber until a few years before maturity, so the property tax on standing timber has yielded less revenue than it would if practice followed theory and law.

Financing Transit Systems Through Value Capture

An Annotated Bibliography

By JEFFERY J. SMITH and THOMAS A. GIHRING*

ABSTRACT. Much of the literature on value capture reports empirical findings on the incidence of rising land values related to distance from a transit stop following the installation of rail transit improvements. This annotated bibliography shows that the elevated value effects of transit access are well documented. The authors maintain that it is now time for transit/land-use research to move from hypothesis *testing to practical applications of value capture.* Longitudinal models can help predict land-value increments over a period of time, yielding estimates of the total capturable revenues that would support the debt financing of transit improvement projects.

Most mass transit systems are financed not by riders but largely by subsidies from local public general funds. Those subsidies necessitate either higher taxes, reduced spending on other public services, or both. Does mass transit need to be so highly subsidized? An underutilized source of funding involves the recovery of site value gains associated with properties near to transit stations. Were transit systems

*Jeffery J. Smith is the President of the Geonomy Society, a group of academics and activists that provides information about the impacts of the flow of natural rents on economies, societies, and the environment. This IRS 501(c)3 organization publishes *The Geonomist*, 3508 SE Madison St., Portland, OR 97214, http://www.geonomics.org; e-mail: jjs@geonomics.org. Thomas A. Gihring is an international planning consultant based in Seattle. He previously taught graduate urban planning, and has undertaken several studies in land-value taxation and value capture. He is currently working on the reform of planning laws and practice in Bosnia-Herzegovina and property tax reform in Oregon. E-mail: tagplan@comcast.net. An earlier version of this appeared on the website of the Victoria Transportation Policy Institute, www.vtpi.org, and was reprinted by the Urban Land Institute.

American Journal of Economics and Sociology, Vol. 65, No. 3 (July, 2006).

to use the economic value that they themselves generate, transit agencies would not have to rely so heavily on public subsidies. Is it possible to finance a mass transit system through a combination of fare box revenues and the capture of land-value increments within transit corridors? Nobel laureate economist William Vickrey claims that mass transit operators could keep fares at a minimum, even at zero during off-peak hours, by funding the bulk of operating costs as well as development costs from site rents (see Reference 13).

Transit riders, by not paying the full operating cost, are not the only free riders. Another group consists of the landowners who are in a position to capture windfall profits by virtue of owning land near newly installed rail transit stations. When a transit station opens, the owners' land values rise, so presently they are in a position to sell their advantageous locations at a higher price. Were owners to relinquish these unearned value increments, they would incur no loss on their original capital investment. But retaining this speculative gain is actually receiving a publicly created benefit. Unearned value increments are a public windfall that neither they nor any individual alone produced. Hence, the public sector is justified in recapturing at least some of this "betterment"—the value of the services rendered by installing the transit improvements. Earlier in our Western culture, this form of economic justice was more clearly understood; indeed, the words *own* and *owe* were once one word at a time when owners owed rent to their lord or king. Today, we frequently see the justice of government's compensation of owners for a property "taking"; we are now beginning to relearn the justice of compensating society for a "giving."

Value capture is the appropriation of land-value gains resulting from the installation of special public improvements in a limited benefit area. It is a betterment levy, based on ad valorem assessments of ordinary property taxes, and is similar in conception to development exactions and impact fees. The aim is to finance all or part of the costs of local transportation projects. Based on the "benefits received" rationale for public taxation, it proposes to recapture what is essentially publicly created value. Unlike building value, which derives from private capital investments, land value represents the speculative dimension of real estate. Thus, value capture is a variation of an unearned increment tax, and is based on the premise that property owners benefiting from a government-conferred locational advantage should pay some portion of the cost of public improvements from which the added value is originally derived.

Having stated the claim of fiscal surety and the principle of recovering land-value increments, is value capture practicable in real-world settings? R. T. Meakin notes that Hong Kong's rail transit system receives no subsidy (see Reference 72). All costs, including interest on bond indebtedness, are paid from land rents derived from development in station areas. W. Rybeck estimated the added land values sequential to the development of Washington, D.C.'s Metro, and found a surplus of incremental land value (see Reference 23). D. Riley, who studied the London tube extension, also found that surplus land values were generated (see Reference 22).

To date, the bulk of the published studies related to the subject of value capture focus on the impacts of transit facilities on property values, with data obtained from properties within transit corridors. N. Tideman and R. J. Borhart both maintain that this limited perspective underestimates the full impact (see References 24 and 33). Potentially, a greater revenue base would be available to transit agencies if rising land values within an entire metropolitan region were appropriated through a general land-based property tax. The land tax also engenders cost-reducing incentives. Land acquisition comprises a substantial portion of the capital costs associated with constructing public transit facilities. This cost could be effectively reduced if ground rents were collected. That is, when the public sector captures incremental land values through the general property tax or through special levies on land within transit corridors, less value remains for private owners to capitalize into purchase price paid by successive buyers.

Localities could adopt additional forms of "green" taxation such as congestion pricing and vehicle emission permit fees to help fund transit systems. Other possible revenue sources include joint development agreements and the leasing of publicly owned sites near stations.

In the past, real estate developers built transit systems as an associated amenity, and recouped the capital costs from the sales of developed sites (still a common practice in Japan). Such profits from land residuals are commonplace in the private sector, but could reasonably be extended to the public domain—where local government covers the financial risks and the costs of building transit systems. Robert Cervero holds that a central element of joint development amounts to a quid pro quo, whereby private developers' benefits from transit accessibility are capitalized into higher rents and occupancy rates, and transit agencies' capital funding is enhanced through cost-sharing mechanisms (see Reference 35).

Most studies of value-capture financing for transit focus on U.S. cities, where low-density development and auto dependency prevail. Studies have begun to emerge from developing countries where denser cities and a more even modal split can be found. Mobility is vital to a nation's prospects for development. As Lewis and Williams point out, affordable transit availability offers cash savings to householders (see Reference 71). One of the best examples of mass transit is found in Curitiba, Brazil. Colombia and Uruguay have built roads using value capture; Mexicali has replaced its conventional property tax with a land tax. Some of the authors cited in the fourth section of this bibliography ("Lessons from Developing Countries") note that while progressive legislation may be on the books, the practical means of capturing site values for transit projects is hampered by inadequate land registration records and lagging assessments.

This bibliography does not contain most of these older studies of transit-induced site values. In 1994, Huang compiled a bibliography on value-capture financing. The RICS Policy Unit recently compiled a review of literature on public transport and land values. Diaz, Lewis and Williams, Pickett and Perrett, the Transportation Research Board, and the U.S. Subcommittee on the City prepared earlier compilations. Some of the studies in this our bibliography do cite updates of older studies. If any of these or the more recent studies that we site here have influenced policymakers to employ value-capture financing methods, we are not aware of them.

Included in this bibliography are 76 annotated titles under the following topical headings:

- Financing public transportation
- Prospects for cost recovery

- Effects of transit facilities on property values
- Lessons from developing countries

I

Financing Public Transportation

1) Buchanan, M. *Urban Transport and Market Forces in Britain*, Anglo-German Foundation for Study Industrial Society, London, 1988, pp. 211–219. Available from: AGFSIS, 17 Bloomsbury Square, London, England.

The report features sections on buses, trains, and roads. The application of market forces and competition may decrease the public cost of transport and decrease traffic congestion in the United Kingdom. Thus far, policies addressing market forces have been confined largely to bus service, where deregulation has produced little change in service levels. Although public savings have been realized in large urban areas, the tendering process has led to major increases in county council costs and public transport staff. Market forces have not been effectuated in the railway system in the same way; large subsidies are still required. Tighter financial targets, the disposal of surplus land, and the subcontracting of work have all been undertaken, as have improvements in administration. Construction of new railways is being funded in part by the consequent increase in land values, an example being the London Docklands Railway. Four methods are discussed: allocating the subsidy to specific purposes; paying the subsidy via a third party; separating the operation of railways from the provision and maintenance of infrastructure; and privatization. Methods to commercialize road infrastructure include urban parking management, the financing of new highway construction from tolls, and road pricing. (From Transport and Road Research Laboratory in TRIS Database under "Taxing Property Values for Transit.")

2) California Tax Data. *California Codes, Public Utilities Code, Sections 33000–33020*, 100 Pacifica #470, Irvine, CA 92618. With reference to the *Draft 2004 Countywide Transportation Plan*, Los Angeles County.

The California Public Utilities Code authorizes the creation of

Special Benefit Districts in the vicinity of rail rapid transit stations. In 1985, the Southern California Rapid Transit District formed two benefit districts to help fund construction costs of the Segment I of the Metro Red Line. The funding plan included $130 million in revenues from the two districts to cover 9 percent of the capital costs of transit infrastructure and neighborhood amenities. The law requires nexus and proportionality between the assessments and locational benefits received. Districts contain zones based on distance from the station up to a half-mile radius in areas outside of the Los Angeles central business district. Assessment fees are graded according to the proximity of each zone to the station.

3) Gihring, Thomas A. "Applying Value Capture in the Seattle Region." ***Journal of Planning Practice & Research*** **16(Winter 2001): 307–320.**

The "geo-bond" financing mechanism features the capture of land rent as distinct from other assessment devices that include the value of building improvements. Using the Broadway station area of Sound Transit's proposed LINK light rail line, the author employs a model simulating the tax effects of (1) a land-value tax as an alternative to the conventional property tax, and (2) a land-value gains tax, within the half-mile radius transit benefit district. The land-value-taxation (LVT) produces the desired development incentive effects, as it shifts the burden off buildings in this "main street" retail setting. The gains tax targets the surplus land value: the difference between the annual assessed land-value increase and the amount of general property tax paid. Over a 12-year period, the gains tax would support a bond principal amount of about $24 million. Sound Transit estimates the construction of the station and street improvements (excluding right-of-way acquisition) at $80 million.

4) Hagman, Donald G., and Dean J. Misczynski, eds. ***Windfalls for Wipeouts: Land Value Capture and Compensation.*** **Planners Press, American Planning Association, Chicago, IL, 1978 (funded by the U.S. Dept. of Housing and Urban Development).**

Special Assessment Districts (SAD) by local governments, once used extensively, fell out of favor during the Great Depression. Yet by the 1970s, the tool was making a comeback. In 1913, Los Angeles,

Oakland, Portland, and Kansas City raised 20 percent of their budgets from SADs. When the Depression wiped out land value, civic bonds became difficult to pay off and lost their ratings. Then, by 1972, cities with populations over 100,000 that had SADs in effect (about 5 percent of all local jurisdictions) funded an aggregate 12 percent of their budgets through this method. With regard to the use of the land-value tax (LVT), the editors questioned the effectiveness of Pittsburgh's experience in shifting the property tax rate from buildings to locations, citing a 1973 Price Waterhouse study (written before the rate differential was increase to 6:1, land to improvements). Nevertheless, the solid results from using the LVT for developing Waikiki Beach, Hawaii were also noted.

5) Hayashi, Yoshitsugu. "Issues in Financing Urban Rail Transit Projects and Value Captures." ***Transportation Research. Part A: General*** **23A (January 1989): 35–44.**

In Japan, urban rail transit projects are suffering from cost burden due to the current financing system's dependence on borrowed money from loans and bonds that are repaid mainly by fares. The transit fund cannot bear increased expenditures from accelerated construction demand and the rising cost of land acquisition. This paper reexamines the financing system and analyzes the possible means of raising revenues. From the viewpoint of the benefits principle, the author examines the imbalance between those who bear the costs and those who receive the benefits, using Japanese examples.

6) Higginson, Martin. "Alternative Sources of Funding." ***Public Transport International*** **48 (September 1999).**

The author cites several transit systems. Copenhagen, Denmark, is funding a line to a new suburb by selling off public land in the Orestad business district, privatizing development, and collecting more property tax revenue from the rising land values. The rail project was scheduled to open in 2002.

7) Howard, Jane A. ***Strategies to Implement Benefit-Sharing for Fixed-Transit Facilities.*** **Series Report from Transportation Research Board, National Research Council, National Cooperative Transit Research & Development Program No. 12, 1984–1985.**

A Local Improvement District is a special property assessment to pay for capital improvements benefiting a defined area. In Portland, Oregon, it is designed to collect some site rent (attributed increases in land values) for the purpose of funding transit-related improvements such as street paving, streetscape amenities, and trolley stanchions. As the result of a required vote, downtown owners unanimously approved the LID and are assessed by square footage of land (excluding buildings), with greater weight given to frontage within 100 feet of transitways. The LID is paying off $1.5 million in bonds over 20 years. That amounts to more than a quarter of the $5.5 million total project cost.

8) Ito, M. *Establishing New Measures to Construct New Railroad Lines.* JTERC Reports 11, Japan Transport Economics Research Centre, Tokyo, Japan, **March 1989.**

This study examines the New Joban Railway Line in the northeastern area of Tokyo. It estimates land values of properties along the corridor, with and without the rail line, and calculates the resulting increment. Methods of ensuring that a region receives an adequate return on its investment are discussed. Included are (1) local taxes for a Railroad Construction Fund; and (2) reduction of station construction costs, either by setting up a trust company to construct a combination of station retail outlets, or by making the developer or local companies responsible for some of the costs. For rural areas, the author recommends a system of integrated development, ensuring that development of residential, educational, and cultural facilities along the line keep pace with rail construction. Also included are suggested methods by which problems of acquiring railway land can be overcome. (See IRRD 857359 in Transport Research Laboratory on TRIS Database, "Taxing Property Values for Transit.")

9) Ridley, T. M., and J. Fawkner. *Benefit Sharing: The Funding of Urban Transport Through Contributions from External Beneficiaries*. Report from the 47th Congress, International Union of Public Transport, 19 avenue de lîUruguay, B-1000 Brussels, Belgium, 1987.

"Specific improvement assessments" funded the first 35 kilometers of Milan, Italy's Metro. The special levy is assessed on properties within 500 meters of stations. This form of LID had raised 36 billion lire, but following its initial success the levy was replaced by a real estate transfer tax that feeds into the local general fund.

10) Shinbein, Philip J., and Jeffrey L. Adler. "Land Use and Rail Transit." ***Transportation Quarterly*** **49(3) (1995): 83–92.**

In a case study of Orange County, New York, a case is made for shifting transit subsidies from the present system of general taxes to land-value taxes. Shinbein and Adler argue that it is realistic to think of self-financing transit improvements from LVT. Joint development programs coupled with permissive zoning to encourage high-density "pocket communities" near transit stations increase land values that can be recaptured to pay for capital costs of rail infrastructure.

11) Strathman, James G., and Kenneth J. Duecker. "Regional Economic Impacts of Local Transit Financing Alternatives: Input-Output Results for Portland." ***Transportation Research Record*** **1116 (1987).**

This study ranks several taxing methods for funding transit. The one found least likely to distort economic activity is the gasoline tax, followed closely by the property tax. The least desirable method of raising revenues is a higher onboard fare, followed by a payroll tax. Taxes on income, parking, and sales produce moderate distortion effects. An input-output model is used to estimate the change in sectoral output that would result from a transfer of resources to transit operations, using all seven financing alternatives. The range in total sectoral output change is substantial.

12) Transportation Research Board, Transportation-Technology Management, Inc. ***Financing for Capital Investment: A Primer for the Transit Practitioner.*** **TCRP Report 89, Transit Cooperative Research Program, 2003.**

This federally funded report includes a comprehensive explanation of capital sources and financing mechanisms for local transit projects. It makes no mention of value capture or the use of land-value increments as a revenue source.

13) Vickrey, William S. "The City as a Firm." In ***The Economics of Public Services*****. Eds. M. S. Feldstein and R. F. Inman (London: MacMillan, 1977), pp. 334–345.**

The author offers a modern extension of the land rent theories of

Henry George, setting the problem of land-value taxation and financing of public services in a general equilibrium framework. "The aggregate of the land rents generated by the urban agglomeration produced by the existence of activities with economies of scale within the city will equal the subsidies required to enable these activities to sell their output at prices equal to their respective marginal costs."

14) Walther, E., L. A. Hoel, L. J. Pignataro, and A. K. Bladikas. *Value Capture Techniques in Transportation: Final Report, Phase One.* Report No. DOT-T-90-11, Office of the Secretary of Transportation, of the United States May 1990.

The authors provide an overview of the potential use of value-capture techniques. Included is a general set of criteria for state and local officials to evaluate the applicability of value capture to specific funding situations. A series of techniques in communities of various sizes is provided, along with a decision support methodology based on a set of 63 indicators to evaluate specific value-capture proposals. Techniques include: special assessment districts, donations, negotiated investments, public/private partnerships.

II

Prospects for Cost Recovery

15) Allen, W. Bruce. "Value Capture in Transit." *Journal of the Transportation Research Forum* 28(1) (1987): 50–57.

This case study in south metropolitan Philadelphia offers an interdependent set of models of modal choice, station choice, and travel savings using the economic law of market areas. These models (1) spatially separate auto users from transit users, (2) spatially separate the users of station A from the users of station B, and (3) spatially connect the locii of all points where the user saves an equal amount of money by using transit over auto. All of these models yield hyperbolas that bend around the stations on the line. The station choice model is tested using auto access data for all suburban stations of the Philadelphia-Lindenwold high-speed line for a morning rush hour (recording 13,000 observations), and assumes the station chosen most often from any given location is the preferred station. The savings model is tested by postulating that residential sales price is a function of the characteristics of the property, the neighborhood, distance from the CBD, and savings (using over 1,300 real estate transactions from 1980). Each dollar's worth of daily savings is found to add $443 to the value of the property. The benefit to nontransit census tracts (less congestion and shorter travel times) was not added in; if it were added, savings would be 30 percent higher. Without it, $4,581 could be captured per single-family home. Within the transit census tracts, this adds up to $279.5 million, or 117.9 percent of the construction cost of the Lindenwold Line, the right-of-way of which did not need to be purchased. Buying the land and constructing bridges would have raised the cost to $820 million, of which captured land rent could have paid one-third (unless all rent were captured, which would drop land's price to zero). In order for the costs to be borne by the beneficiaries, land value should be captured at the time it is created, that is, between the announcement of a new improvement and its actual opening.

16) Alterkawi, Mezyad. "Land Economic Impact of Fixed Guideway Rapid Transit Systems on Urban Development in Selected Metropolitan Areas: The Issue of the Price-Distance Gradients." Ph. D. dissertations, Texas A&M University, 1991. Stock No: 91-33904 University Microfilms International.

This study noted Toronto's experience with the Yonge Street Subway. Property tax revenues in the vicinity of the transit line increased an average of $5 million annually. This compares favorably with the $4 million annual cost of servicing the subway's capital financing bonds.

17) Anas, Alex. *The Effects of Transportation on the Tax Base and Development of Cities*. Report for the U.S. Dept. of Transportation, April 1983.

Transportation improvements and investments change zone-to-zone travel times and costs. This researcher's model forecasts changes in land values. The forecasts are determined annually and by small geographic zones in a metropolitan area. The Chicago application shows that under 1970 conditions, capitalized land-value changes are nearly 36–40 percent of the capital cost of rail rapid transit proposals then proposed for Chicago's southwest side. Similar calculations for bus systems appear more promising. Anas suggests a one-time lump-sum property assessment rather than an increase in the land tax rate, since that latter would lower "site values." This would lower selling price, while the value remains the same (what buyers are willing to pay: price plus tax).

18) Batt, H. William. "Value Capture as a Policy Tool in Transportation Economics: An Exploration in Public Finance in the Tradition of Henry George." ***American Journal of Economics and Sociology*** **60(1) (2001): 195–228.**

This study shows how value capture could have been used to finance a nine-mile portion of the New York State interstate highway system. The added increment of land value attributed to the Northway sector amounted to 11 times that of the cost of right-of-way acquisition and road and bridge construction. Batt concludes that the windfall gains in land value that fell to private landowners could easily have paid off the bonds issued to build the project. Furthermore, the added taxes from value-capture assessments in the highway corridor removes the invitation to landholders to speculate on their sites. Batt argues that pricing approaches to the recovery of costs of government services is more attractive than conventional "command and control" police power approaches.

19) Hack, Jonathan. ***Regeneration and Spatial Development: A Review of Research and Current Practice*****. IBI Group, Toronto, 2002.**

This paper provides specific examples of how, and to what degree, urban transit investment (principally light rail) has stimulated urban regeneration and created private opportunities for private sector investment in transit corridors, notably around transit stations. The case studies provided are derived from a review of research to date that showcases recent examples of LRT investment in Europe and North America. See Appendix for a summary of observations.

20) Nakagawa, D., and R. Matsunaka. ***Funding Transport Systems: A Comparison Among Developed Countries.*** **Pergamon, 1997.**

The authors repeat the findings of Tsukada and Kuranami (see Reference 75) that in Japan private railroads manage real estate within rail corridors, and thereby enhance profits.

21) Nathanson, Phyllis J., and Gary Booher. *Survey of Joint Development and Value Capture Activity in Selected Metropolitan Areas.* City of Los Angeles Planning Department, 1983.

Among several systems noted in this survey, Miami's Metrorail raised sufficient site rent to cover 25 percent of its total capital cost ($116 million).

22) Riley, Don. *Taken for a Ride: Trains, Taxpayers, and the Treasury.* Centre for Land Policy Studies, 4 O'Meara Street, London SE1 1TE, 2001.

London's Jubilee Underground extension cost £3.5 billion, raising the nearby land's rental value by £1.3 billion. Public collection of 25 percent of that increase would pay off the Jubilee in 20 years. In the vicinity of Edinburgh, Scotland, developers are co-funding a new line on old right-of-way.

23) Rybeck, Walter. *Transit-Induced Land Values: Development and Revenue Implications*. Report published in *Commentary*, Council on Urban Economic Development, October 24, 1981, pp. 23–27.

In his report to Congress, this former staff member to Sen. Paul Douglas noted that Washington, D.C.'s Metro, after some $3 billion in expenditures by 1981, was 40 percent complete and had generated over $2 billion in additional land value. By January 2001, after $9.5 billion in capital expenditures, the completed system had generated between $10 and $15 billion in new land value.

24) Tideman, Nicolaus. "Integrating Rent and Demand Revelation in the Evaluation and Financing of Services." In *Does Economic Space Matter?* Eds. Hiroshi Ohta and Thisse Jacques-Francois (London: MacMillan, 1993), pp. 133–150.

Taking into account more than just the property selling price, this researcher considers how a transportation project changes the returns to land, labor, and capital, compared to the project's costs: (1) the increase in privately collected rent—in other words, the increase in the selling price (and lease value) of land; (2) the increase in taxes on land; (3) the decrease (more usual than an increase) in its value because capital can't be moved (as the land rose in value but buildings fell in value); (4) the change in taxes on existing buildings; (5) the taxes on new buildings erected in response to the transportation improvements; (6) the cost of extra public services for the added buildings (unless there are user fees); (7) the extra tax revenue if there is a sales tax or a wage tax that reduces land values; (8) the savings in travel time if low fares reduce congestion; (9) reduced smog; and (10) the loss of human happiness from uncompensated personal adjustment to the change in the built environment. The sum of these 10 items is compared to the transportation system costs.

25) U.S. Congress, House Committee on Banking, Finance, and Urban Affairs, Subcommittee on the City. *New Urban Rail Transit: How Can Its Development and Growth-Shaping Potential Be Realized?* U.S. Government Printing Office, 1980.

From page 81: Burkhardt and Howard summarize historical evidence: "Major land value increases occurred in many station areas of New York City's expanding transit system in the early 1900s." From page 124: Donald Richmond states, "[t]he (Toronto Transit) Commission . . . experience . . . suggests that the long-term land-leasing program can completely recover land acquisition costs over a reasonable time period."

III

Effects of Transit Facilities on Property Values

26) Al-Mosaind, Musaad A., Kenneth J. Duecker, and James G. Strathman. "Light Rail Transit Stations and Property Values." *Transportation Research Record* 1400 (1993): 90–94.

Two forces are said to be at work in the relationship between proximity to light-rail transit stations and single-family home values. Proximity to LRT stations may improve the accessibility of residents to an urban area's central business district, and may also result in transportation cost savings. These effects are evidenced in higher property values. In metropolitan Portland, Oregon, two distance models to East Side MAX stations were compared. The first showed a positive capitalization in sale prices for homes within a quarter-mile walking distance. This effect was equally felt for all homes within that distance zone. The second model found that property values were estimated to decline with increasing distance to a station. However, a clear price gradient was not evidenced. The authors conclude that nuisance effects produced by the station may play a role in reducing the potential benefits of close proximity to nearby homes. On average, the total contribution of proximity to stations adds nearly 10.6 percent to home prices.

27) Anas, A., and Regina Armstrong. *Land Values and Transit Access: Modeling the Relationship in the New York Metropolitan Area: An Implementation Handbook*. Report No. FTA-NY-06-0152-93, U.S. Federal Transit Administration, Office of Technical Assistance and Safety, Springfield, VA (National Technical Information Service), September 1993.

This article presents findings of a multiyear study of the relationship between land values and transit access in the New York area, as precursor to capturing this value for public transit. Initiated as an element of the Third Regional Plan for the New York/New Jersey/Connecticut region, the results serve as a research prototype for transit systems throughout the United States. Two economic models are presented that predict shifts in land values within the region and at a parcel scale in relation to transit stations. "The total benefits of reducing wait times on transit equal \$3.7 billion (\$1.57/trip). Taxing the producer surplus increases would raise \$100 million/yr, enough to finance a doubling of the number of trains (an unknown cost)."

28) Armstrong Jr., Robert J. "Impacts of Commuter Rail Service as Reflected in Single-Family Residential Property Values." *Transportation Research Record* 1466 (1994): 88–97.

The effects of regional commuter rail service on single-family residential properties in metropolitan Boston, Massachusetts, are examined. The study area encompasses 1,920-square kilometers, with a total 1990 population of over 630,000, and uses records of 451 property sales. Controlling for structure and site attributes, as well as location and amenity variables, the regression model reveals a clear association between regional access to the Boston central business district provided by the rail service and appreciated property values. Residences located in communities having a rail station have a market value that is approximately 6.7 percent greater than residences in unserved communities.

29) Barker, William G. "Bus Service and Real Estate Values." *68th Annual Meeting of the Institute of Transportation Engineers, Toronto, Ontario, 1998* (available from ITE, 1099 14th Street, NW, Washington, DC 20005-3438).

Real estate developers and lending institutions are not willing to base investments on the location of easily changed bus routes. However, the availability of local bus service does increase the value of at least some urban real estate.

30) Benjamin, John D., and G. Stacy Sirmans. "Mass Transportation, Apartment Rent and Property Values." *Journal of Real Estate Research* 12(1) (1996): 1–8.

This study examines the effects of transit access, measured in ground distance to the nearest Washington, D.C. Metrorail station, on residential rent levels. From over 250 observations of 81 apartment complexes, the authors find that rents decrease by 2.4 percent to 2.6 percent for each one-tenth mile in distance from a Metrorail station. Distance does not have a significant effect on occupancy rates. Rent levels play an important role in property values because any positive or negative influence on rent will in turn affect a property's market value.

31) Bernick, M., R. Cervero, and V. Menotti. *Comparison of Rents at Transit-Based Housing Projects in Northern California*. Working Paper 624, University of California at Berkeley, Institute of Urban and Regional Development, 1994.

"Rents at the BART housing projects are higher than those of nearby projects."

32) Bollinger, C., K. Ihlanfeldt, and D. Bowes. "Spatial Variation in Office Rents Within the Atlanta Region." *1996 TRED Conference, Lincoln Land Institute, Cambridge, Mass*., Georgia State University, Policy Research Center, July 1998.

This is a hedonic rent study of office buildings in the Atlanta area from 1990 to 1996. Part of the rent differences among office buildings is due to differences in wage rates, transportation rates, and proximity to concentrations of office workers. The convenience of face-to-face meetings facilitated by office agglomerations is also reflected in office rents, providing evidence that agglomeration tendencies continue to be important in explaining office concentrations, despite the ability of information technology designed to reduce the need for some such contacts.

33) Borhart, Robert J. *Corridor Reservation: Implications for Recouping a Portion of the "Unearned Increment" Arising from Construction of Transportation Facilities.* Final Report, Virginia Transportation Research Council, Charlottesville, VA. Series title: VTRC; 94-R15, 1994.

Increases in land rents show up in higher property taxes, not only in property selling prices. The author quotes President Franklin D. Roosevelt supporting value capture.

34) Cambridge Systematics. *Economic Impact Analysis of Transit Investments: Guidebook for Practitioners*. TRB Report 35, Transit Cooperative Research Program, Transportation Research Board (http://www.trb.org), 1998.

This comprehensive guidebook describes various technical methods for measuring the economic impacts of transit investments, including changes in adjacent property values. It also includes a summary of research findings on the increases in property values found around BART stations in the San Francisco Bay area. See Appendix for a summary of property value impacts. Tables 9.6 to 9.10 list 15 studies dating from 1970 to 1996 that calculate the premium effect of transit investments, measured in unit area of property.

35) Cervero, Robert. "Rail Transit and Joint Development: Land Market Impacts in Washington, D.C. and Atlanta." *Journal of the American Planning Association* 60(1) (1994): 83–94.

In addition to public-private cost sharing and the lease revenues derived from commercial space in rail stations, joint development projects generate more fare revenues as they stimulate more transit trips. This study examines how transit investments affect office market indicators. Evidence shows that joint development projects create measurable land-value increases and other associated benefits. Among five dependent variables studied, office rent levels are most closely correlated with transit factors, especially ridership. Other benefits associated with transit centers are low vacancy rates, higher absorption rates, and larger office building size. In conclusion, urban rail transit will significantly benefit land use and site rents only if a region's economy is growing and supportive programs such as permissive zoning are in place.

36) Cervero, Robert. "Transit-Based Housing in the San Francisco Bay Area: Market Profiles and Rent Premiums." *Transportation Quarterly* 50(3) (1996): 33–49.

This study focuses on apartment rents rather than housing prices. In the proximity of three BART stations selected, most residents occupied multi-unit complexes of 20–60 units and formed single-car households comprised of young adult singles or couples without children, with above-average incomes. Within two of the station areas, housing rents were found to be 10–15 percent higher than the average rents in nonstation areas. In the third (Richmond) station area, no rent premium was found. In his guarded conclusion, Cervero states that "in theory, the existence of a rent premium for multi-unit projects suggests value capture mechanisms (e.g., forming benefit assessment districts) could be used to help finance rail systems," adding without explanation: "although this is very difficult to implement in practice." It is not clear what prompted this caveat—the implied difficulty of political acceptance, or perhaps legal constraints to the formation of taxing districts or bond-financing mechanisms.

37) Cervero, Robert, and Michael Duncan. "Benefits of Proximity to Rail on Housing Markets: Experiences in Santa Clara County." *Journal of Public Transportation* 5(1) (2002): 1–18.

Hedonic price models show that nearness to light rail and commuter rail stops substantially add value to residential parcels. Large apartments within a quarter-mile of LRT stations command land-value premiums as high as 45 percent. Such market profits provide a potential source of local revenue from value-capture programs.

38) Cervero, Robert, and Michael Duncan. "Transit's Value Added: Effects of Light Commercial Rail Services on Commercial Land Values." Presented at TRB Annual Meeting, 2002 (available at http://www.apta.com/info/briefings/cervero_duncan.pdf).

This study models the value effects of proximity to light rail and commuter rail stations, as well as freeway intersections, in Santa Clara County, California. Substantial capitalization benefits to commercial retail and office properties were found, on the order of 23 percent for a typical commercial parcel near an LRT stop, and more than 120 percent for commercial land in a business district within a quarter-mile of a commuter rail station.

39) Cervero, Robert, Christopher Ferrell, and Steven Murphy. "Transit-Oriented Development and Joint Development in the United States: A Literature Review." ***Research Results Digest*** **52, Transit Cooperative Research Program (October 2002).**

This is a comprehensive review of literature on transit-oriented development (TOD). Topics include: definition of TOD, agency roles, impacts and benefits on land markets, supportive policies and regulations, the use of value-capture financing, and station area design supportive of TOD. The authors suggest that transit boards might share in the land-value benefits derived from proximity to transit by participating in joint development as well as value capture.

40) Chen, Hong, Anthony Rufolo, and Kenneth Dueker. "Measuring the Impact of Light Rail Systems on Single Family Home Values: An Hedonic Approach with GIS Application." ***Transportation Research Record 1617*****, TRB, National Research Council, Washington, DC (1998).**

Proximity to transit stations accounts for a 10.5 percent home price differential. This confirms the findings of Al-Mosaind et al. (see Reference 26). They conclude that the positive effects outweigh the negatives.

41) Damm, David, Steven Lerman, Eva Lerner-Lam, and Jeffrey Young. "Response of Urban Real Estate Values in Anticipation of the Washington Metro." ***Journal of Transport Economics and Policy*** **(September 1980): 315–335.**

The authors draw conclusions from reviews of earlier studies of value-capture financing, showing that in response to new transit lines, land values are enhanced in centers of concentrated activity and in predominantly undeveloped areas. Their Metro case study demonstrates that the values of retail properties are highly sensitive to proximity to transit stations. This suggests that retail areas are better suited for value-capture policies.

42) Diaz, Roderick B. "Impacts of Rail Transit on Property Values." Commuter Rail/Rapid Transit Conference, Toronto, Ontario, American Public Transit Association, 1999.

The author summarizes recent North American studies examining the impact of 12 rail projects, including both heavy rail and light rail. Several variables contributing to positive and negative changes in property values are identified. In Miami, home values near stations increased by up to 5 percent (Gatzlaff 1993). In Toronto, nearby home value increases averaged $2,237 (Bajic 1983). In general, proximity to rail increases accessibility, which is the primary factor in rising property values. http://www.apta.com/info/online/diaz.pdf (from "Rail Transit and Property Values" in *Information Center Briefing* 1(March 2001), at http://www.apta.com/info/briefings/briefings_index.htm).

43) Dunphy, Robert T. *The Cost of Being Close*. ULI Working Paper 660, Urban Land Institute, October 1998.

In southern California, real estate consultant Larry Netherton compared examples of comparable housing for sale at different distances from a central business area. Buyers would have to travel another 15 to 30 minutes to trim $10 to $15 per square foot off the price of a house. In Orange County, two similar upper-end housing projects were compared, one near major employment, retail, and cultural centers, and the other 20 miles away from employment centers. The closer units sold for an average of $599,400, and the distant units sold for $320,000—a difference of about $280,000, or $14,000 per mile, or $11,200 per minute of extra commute time. In more distant Riverside County, the closer project was priced at $214,900, while a same-sized, similar house 20 miles farther out sold for $141,900. The differential here was $73,000 total, or $3,600 per mile, or $2,400 per minute of extra commute time.

44) Fejarang, R. A. "Impact on Property Values: A Study of the Los Angeles Metro Rail." Transportation Research Board 73rd Annual Meeting, January 1994.

In a city such as Los Angeles, value impacts can be caused by regional as well as local behavior. Did the announcement of Metro Rail impact property values? The announcement involved a consortium of federal, state, and local funding propositions that began in 1983 and were legislated in 1988. The period studied was from 1980 to 1990, during which plans became actualized. That is, investments were secured and rail transit was under design and construction, but not yet available for riders or for rider-dependent shopping. Isolating exogenous variables was accomplished at both macro and micro levels. Using a pretest-posttest control group, property values following the period of actualization were found to be significantly different from prior values. Property values near rail lines were found to be significantly different from property values located a distance (from Transport Research Laboratory).

45) Garrett, Thomas. *Light Rail Transit in America: Policy Issues and Prospects for Economic Development.* Federal Reserve Bank of St. Louis (http://www.stlouisfed.org), 2004.

A hedonic pricing model applied to residential property values in St. Louis found that average home values increased $140 for every 10 feet closer proximity to a MetroLink rail transit station, within a radius of 1,460 feet. A home situated 100 feet from a station had a price premium of $19,029, representing a 32 percent increase in average property value. Garrett's study also noted an increase in values beyond the maximum radius, but noted that other factors not related to station proximity and not included in the model, such as traffic volumes on nearby streets, are probably responsible for this effect. His analysis did not investigate transit impacts on commercial property values, which he surmises would also experience positive effects.

46) Gatzlaff, Dean H., and Mark Smith. "The Impact of the Miami Metrorail on the Value of Residences Near Station Locations." *Land Economics* 69(1) (1993).

Miami Metrorail began operations in the mid-1980s in a recently burgeoning city with a sprawling land-use pattern. The 20 miles of rail line traverse the city center, extending to the more affluent north and the poorer south. Both of the extensions were planned in the path of existing built-up neighborhoods where it was hoped that redevelopment would occur. For a study methodology, the researchers chose to examine home resales, those having sold more than once over the past 18 years. They looked for price changes after the announcement of the new line. The anticipated presence of transit service did perceptibly lift nearby site values, but only in the south-side neighborhoods, where recent capital investments had been activating the redevelopment market. Compare this study with the downtown Miami study of retail sales increases (Reference 59). The larger study of Miami's rail system is also cited in References 21 and 42.

47) Goodwin, Ronald E., and Carol A. Lewis. *Land Value Assessment Near Bus Transit Facilities: A Case Study of Selected Transit Centers in Houston, Texas.* Southwest Region University Transportation Center, Houston, TX, 1997.

Site values in the Houston region were falling due to shrinking incomes and diminished incomes. However, values fell less near bus stops than they did in more distant locations.

48) Gruen, Aaron. *The Effect of CTA and Metra Stations on Residential Property Values: Transit Stations Influence Residential Property Values*. Report to the Regional Transportation Authority, June 1997.

By improving accessibility, lessening congestion, and reducing household transportation costs, transit service adds value to residential locations. Observing 96 Chicago-area Chicago Transit Authority (CTA) and Metra stations, Gruen used hedonic modeling supplemented by a literature review and interviews with realtors and other experts on local market conditions. More important than the presence of a transit station is the perception of neighborhood desirability. Still, the proximity of transit does positively affect property values. The price of a single-family house located 1,000 feet from a station is 20 percent higher than a comparable house located a mile away. Realtors in both the affluent suburban West Hinsdale station area and the gentrifying Logan Square area on Chicago's northwest side point out that prices have been increasing and that these locations increasingly appeal to younger, higher-income professionals, many of whom commute via CTA or Metra to downtown Chicago. Apartment properties located closer to train stations tend to realize higher rents and occupancy levels than comparable apartments less conveniently located (http://www.ggassoc.com/ from "Rail Transit And Property Values." *Information Center Briefing* 1 (March 2001), at http://www.apta.com/info/briefings/briefingsindex.htm).

49) Huang, W. *The Effects of Transportation Infrastructure on Nearby Property Values: A Review of the Literature*.

Working Paper 620, Institute of Urban and Regional Development, Berkeley, CA, 1994.

The effect of the presence of transportation infrastructure on distant lot values is small, but there are many distant lots; therefore, the hedonic method may underestimate incremental site rents. Furthermore, it may be a mistake to regard as exogenous the values attributed to other amenities that developers add in response to accessibility-induced value.

50) Kay, J. H., and G. Haikalis. "All Aboard." ***Planning*** **66(October 2000): 14–19.**

In Dallas, DART has shown what a modern city driven by the private sector can accomplish with rail transit. Property values around transit stations have jumped by approximately 25 percent since DART began operation in 1996. However, Dallas's extensive land area complicates transit's contribution to the regional transportation system. In a sidebar, Haikalis describes New Jersey's new Hudson-Bergen Line. (Available from: APA, 122 South Michigan Avenue, Suite 1600, Chicago, IL 60603-6107, TRIS Database: "Taxing Property Values for Transit.")

51) Knaap, Gerrit, Lewis Hopkins, and Arun Pant. ***Does Transportation Planning Matter? Explorations into the Effects of Planned Transportation Infrastructure on Real Estate Sales, Land Values, Building Permits, and Development Sequence*****. Research Paper, Lincoln Institute of Land Policy, 1996.**

This study observed property values in the Westside LRT corridor in Washington County, suburban Portland, Oregon. The study compared values prior to construction with values at the beginning of LRT operations. Values of parcels located within one-half mile of the line were found to decrease with distance from the stations but rise with distance from the rail line between stations. Thus, the opposite affects of accessibility and nuisance were deduced.

52) Landis, John, Robert Cervero, Subhrajit Guhathukurta, David Loutzenheiser, and Ming Zhang, Rail Transit Investments. ***Real Estate Values, and Land Use Change: A Comparative Analysis of Five California Rail Transit Systems*****. Monograph 48, Institute of Urban and Regional Studies, University of California at Berkeley, July 1995.**

This study measured ground distance to BART stations in Alameda and Contra Costa Counties, California. The authors found that 1990 single-family home prices declined by $1 to $2 per meter distance from a BART station. They did not find a significant impact on home values based on proximity to CalTrain commuter rail stations, although houses within 300 meters of the CalTrain right-of-way sold at a $51,000 discount. No increase in value around commercial/industrial stops was found, but the authors note that commercial property observations encounter significant data measurement problems.

53) Lewis-Workman, Steven, and Daniel Brod. "Measuring the Neighborhood Benefits of Rail Transit Accessibility." ***Transportation Research Record*** **1576 (1997): 147–153 (Transportation Research Board, http://www.trb.org).**

The authors found that within a one-mile radius from the Pleasant Hill Rail Station in the Bay Area, average home prices decline by about $1,578 for every 100 feet distance from the station. In the area within a one-mile radius from the Forest Hills, 67th Avenue, and Rego Park rail stations, average home prices decline by about $2,300 for every 100 feet distance from the station.

54) Nelson, Arthur C. "Effects of Elevated Heavy-Rail Transit Stations on House Prices with Respect to Neighborhood Income." ***Transportation Research Record*** **1359 (1992): 127–132.**

Nelson studied the impact of elevated rail stations on single-family homes in Atlanta. Within its low-value neighborhoods, Nelson found that a transit stop raises residential values. Curiously, high-value communities installing a transit stop appears to lower site values—by nearly the same amount. Both the affluent northern and the poorer southern sectors contribute equally to ridership, although the south's contribution includes a higher percentage of the workforce. Owner occupancy rates were almost equal in both north and south. The observed impact on property prices: in the south sector, the farther from a station, the lower the home's value; in the north, the reverse was found. Total negative effects in the north were $9 million; total positive effects in the south were $10 million.

55) Nelson, Arthur C. "Transit Stations and Commercial Property Values: A Case Study with Policy and Land-Use

Implications." ***Journal of Public Transportation*** **2(3) (1999): 77–95.**

Nelson develops a theory of commercial property value with respect to both transit station proximity and the role of policies that encourage commercial development around transit stations without discouraging such development elsewhere. He applies this theory to sale of commercial property in Atlanta's "Midtown," located 1 kilometer (.6 mile) north of the downtown edge. Midtown is served by three heavy rail transit stations operated by the Metropolitan Atlanta Transit Authority (MARTA). To encourage transit-oriented development near MARTA stations, the city waives parking requirements and floor area ratio restrictions. Commercial property values are affected positively by both access to rail stations and by policies that encourage more intensive development around those stations. Using ground distance to a MARTA station as a measure of access, a citywide analysis finds that price per square meter falls by $75 for each meter of linear separation. Average prices rise by $443 for a location within special public interest districts.

56) Parsons Brinkerhoff. "The Effects of Rail Transit on Property Values: A Summary of Studies." Research carried out for Project 21439S, Task 7. NEORail II, Cleveland, OH, February 27, 2001.

This paper summarizes the results of several previous studies in tabular form. The authors note that varying methodologies make it difficult to compare results. Nevertheless, it is clear that in most cases access to transit systems is valued by property owners. Rail's influence on residential values is demonstrated more clearly than on commercial uses; however, influence on commercial values appears to vary by: (1) how much accessibility is improved, (2) the relative attractiveness of locations near stations, and (3) the strength of the regional real estate market.

57) Pickett, M. W., and K. E. Perrett. ***The Effect of the Tyne and Wear Metro on Residential Property Values.*** **Supplementary Report 825, Transport and Road Research Laboratory, Crowthorne, Berkshire, UK, 1984.**

Three different methods of analysis are performed on the data collected. Results show an average increase of £360 (1.7 percent) in the value of properties near Metro stations during the four-month period surrounding the date on which each section of line opened. In reference to related studies, Dvett et al. found a small but significant positive effect on the value of single-family dwellings at three of the six BART station areas studied. Lerman et al. found that distance from Washington Metro stations influences property values, the value rising as the opening date nears and falling if the opening is delayed. The Regional Commission in Atlanta found an associated increase in industrial property values.

58) Price Waterhouse Coopers. "Review of Property Value Impacts at Rapid Transit Stations and Lines." Technical Memorandum 6, Richmond/Airport—Vancouver Rapid Transit Project, April 3, 2001.

The authors review transit impact studies from selected cities across North America. The reviewers find a positive relationship between property values and station location, but also a possible negative impact on single-family homes along the line due to nuisance impacts. Four research reports are summarized: (1) "Transit Case Studies for the City of Hillsboro, Oregon," (2) "Transit Benefits 2000 Working Papers," (3) "Light Rail Transit Impacts in Portland, Oregon," and (4) "Impact of the Vancouver, BC Skytrain on Surrounding Real Estate Value."

59) Richert, Thomas M. *Economic Impacts of Automated People Mover Development in Commercial Centers*. Advanced Transit Association, 1999.

After one year of operation of the APM, retail sales in downtown versus the greater metro region grew in Denver by 8 percent, in St. Louis by 4 percent, and in Miami by 1 percent (where patronage of downtown commercial space had been lagging historically). Higher retail sales translate into higher site values.

60) Rice Center for Urban Mobility Research. *Assessment of Changes in Property Values in Transit Areas.* Urban Mass Transit Administration, Houston, TX, 1987.

This is a summary of earlier findings from Toronto, Baltimore, Denver, San Diego, and San Francisco. Some transit centers showed a 100 percent to 300 percent increase in commercial site values. In Atlanta, 61 percent of the businesses within 500 feet of a transit stop reported increased sales.

61) Ryan, S. "Property Values and Transportation Facilities: Finding the Transportation-Land Use Connection." ***Journal of Planning Literature*** **13(4) (1999): 412–427.**

Ryan reviews empirical studies of the relationship between the presence of transportation facilities—highways, heavy rail, and light rail transit systems—and property values. Inconsistencies in findings from this literature over the past several decades are explained. For example, results vary based on whether researchers measure accessibility in terms of travel time or travel distance. Measuring distance yields mixed results in property value effects. Measuring time yields the expected inverse relationship between access to transportation facilities and property values. The delineation of study areas also influences the direction of effects. This study offers a new interpretation of the transportation facility-property value literature, improving the ability to measure relationships and to anticipate land-market responses to transportation facilities.

62) Sedway Group. "Regional Impact Study." Commissioned by Bay Area Rapid Transit District (BART), July 1999. In Transit Resource Guide, ***Rail Transit and Property Values,*** **American Public Transportation Association, 1666 K Street NW, Washington, DC.**

This is a review of studies of the benefits associated with BART service, measured in residential and office property impacts. Reported single-family home values fell by $3,200 to $3,700 for each mile distance from a BART station in Alameda and Contra Costa Counties. Apartments near BART stations were found to rent for 15 percent to 26 percent more than apartments more distant from BART stations. The average unit land price for office properties also decreased as distance from a BART station increased, from $74 per square foot within a quarter-mile of a station to $30 per square foot at locations exceeding half a mile.

63) Voith, Richard. ***Transportation, Sorting, and House Values in the Philadelphia Metropolitan Area.*** **Working Paper No. 90-22, Federal Reserve Bank of Philadelphia, Economic Research Department, 1990.**

Households prefer residential locations close to their jobs. Employees who work in central business districts choose census tracts served by commuter rail and tend to own fewer cars. Holding all other factors constant, the value of commuter rail service is capitalized into house value, adding an extra $5,716 or 6.4 percent. In the aggregate, the accessibility premium in real estate values for the Philadelphia suburban area is in the order of $2.43 billion. At a discount rate of 10 percent, residents with rail service should be willing to pay $243 million a year to keep the system, whether or not they use it. Despite the increasing decentralization of the region, over half of the residents of the metropolitan area have a direct interest in the quality of public transportation and economic health of the CBD, regardless of workplace.

64) Weinberger, Rachel R. *Commercial Rents and Transportation Improvements: Case of Santa Clara County's Light Rail.* WP00RW2, Lincoln Institute of Land Policy, 2001.

In Santa Clara County, California, property owners sued the county, claiming losses in value from the nearby light rail. To determine the actual effect of the light rail facility on property values, Weinberger examined commercial property rents, comparing accessibility to transit and to highway as determinants of rent, and analyzed the effects over time. Controlling for other factors, properties within a half-mile of light rail stations were found to command almost 15 percent more rent. Highway access, being ubiquitous, offers no particular locational advantage. As the transit system matured, nearby properties accrued greater benefits. But, in times of high demand, so did all other locations command higher rents.

65) Weinstein, Bernard L., and Terry L. Clower. *The Initial Economic Impacts of the DART LRT System.* Center for Economic Development and Research, University of North Texas, July 1999.

Values of properties adjoining Dallas's DART light rail stations grew 25 percent more than similar properties not served by the rail system. Proximity to stations appears to be an economic advantage for most classes of real estate, especially Class A and C office buildings, and commercial strip retail outlets. Average occupancy rates for Class A buildings near rail stations increased from 80 percent in 1994 to 88.5 percent in 1998, while rents increased from an average $15.60/sf to $23/sf. Commercial strip retailers near the stations experienced a 49.5 percent gain in occupancy and a 64.8 percent improvement in rental rates (http://www.dart.org/economic.htm; from "Rail Transit and Property Values" in *Information Center Briefing* 1 (March 2001), at http://www.apta.com/info/briefings/briefings_index.htm).

IV

Lessons from Developing Countries

66) Calvo, Christina M. *Options for Managing and Financing Rural Transport Infrastructure*. Technical Paper No. 411, World Bank, Washington, DC, 1998.

Berkshire, England, successfully privatized the maintenance of roads. Calvo suggests applying this model to developing countries, where central governments are often hierarchical and indifferent to rural areas. "If increases in land value are captured mainly by the local elite or by outsiders, however, there will be little motivation for mass participation in the project."

67) Cervero, Robert, and Bambrang Susantono. "Rent Capitalization and Transportation Infrastructure Development in Jakarta." *Review of Urban and Regional Development Studies* 11(1), Department of City and Regional Planning, University of California, Berkeley, CA, 1999.

Freeway off-ramps raised the rents of nearby offices in Jakarta, Indonesia. Thus, value capture would be fair, but the method is not feasible because land ownership and values are not registered; furthermore, owners can buy off tax collectors.

68) Dalvi, M. Q. "Value Capturing as a Method of Financing Rail Projects: Theory and Practice." From the 7th CADATU Conference: Urban Transport in Developing Countries, New Delhi, India, 1996.

Hong Kong's Mass Transit Railway Corp. chooses to not sell land but co-develop it. Property rental income financed about 22 percent of MTRC's operating cost in 1993.

69) Heggie, Ian G. "Financing Public Transport Infrastructure: An Agenda for Reform." Proceedings of Seminar M, PTRC Summer Annual Meeting, P327, Planning and Transport Research and Computation (International) Co., 1989.

In the developing world, value generated by a transport system can be significant. However, often cadastre records are missing or lack information on registered owners and the value of parcels. Furthermore, better-off owners "are often influential local politicians."

70) Leinbach, L. R. "Transport and Third World Development: Review, Issues, and Prescription." ***Transportation Research. Part A, Policy and Practice*** **29(5) (1995): 337–344.**

This study notes several studies that provide solid empirical evidence of the positive benefits of rural roads. Nevertheless, land tenancy is often a major determinant of who benefits; if land is unevenly distributed, the land-poor will receive little benefit. In urban areas, the amount and location of transportation infrastructure affects labor accessibility, which in turn affects the manufacturing sector and economic productivity. The author emphasizes the importance of fairly allocating capital investments to provide maximum benefit to the greatest number of people. The problem is that typical market demand models lead to an undifferentiated view of benefits and costs to households. Author's suggestion: A general citizens' dividend derived from the excess revenues of a high land-value tax on property owners might provide a solution to the maldistribution of wealth, or effective demand.

71) Lewis, David, and Fred L. Williams. ***Policy and Planning as Public Choice: Mass Transit in the United State*** **(Hampshire, U.K.: Ashgate, 1999).**

In this book's preface, the authors state: "The public realizes $5 in cash savings for each tax dollar invested in transit services." This general benefit is achieved by the external cost-cutting effects of reduced private car usage, reduced traffic congestion, and the relatively low cost of transit riding by consumers. On page 141 is a chart clearly showing the correlation between transportation mobility and national wealth. The implication is that policymakers having the aim of lifting people out of poverty must take seriously the goal of affordable mobility, which transit offers.

72) Meakin, R. T. "Hong Kong's Mass Transit Railway: Vital and Viable." In ***Rail Mass Transit for Developing Countries. Proceedings of the Conference Held in London on 9–10 October, 1989*** **(London: Telford, 1990). pp. 125–143.**

Meakin notes that in Hong Kong the transit system receives no subsidy. All of its costs, including interest, are derived from rents from land development. "Discussion" by J. Faukner notes that the World Bank requires mass transit (but not roads) to be self-financing, and that lenders should minimize environmental impacts.

73) Ortiz, Alexandra. "Economic Analysis of a Land Value Capture System Used to Finance Road Infrastructure: The Case of Bogota, Colombia" Ph.D. diss., University of Illinois at Champaign-Urbana, 1996.

Starting in 1926, the city of Bogota has charged property owners the anticipated rise in site value ("valorization") before road construction has begun. Revenue from these charges declined in the late 1980s as assessments fell behind and as poor landholders could not afford even the lagging assessments. A 1992 valorization had collected 80 percent of its target by mid-1995. Presumably, the city made up the difference with other taxes, since new roads did get built. What worked for roads could work for transit as well. Ortiz concludes that pre-emptive betterment charges for infrastructure would not be needed if a general land tax were working well. Colombia has a municipal land tax rate of 1 percent and a national rate of 2 percent, plus a land gains tax of up to 50 percent, yet land is registered at only 20 percent of its value.

74) Prest, A. P. *Transport Economics in Developing Countries* (Praeger, 1969).

Prest relates how Uruguay has historically funded roads from land rent. In 1928, the country set up its Permanent Fund for Development and Farm-to-Market Roads, financed by taxes on gasoline, tires, and land value—prorated by distance from the road. Even at a very low rate, 0.125 percent to 0.65 percent, the land-value tax funded one-third of the road construction budget. However, assessments did not keep pace with rising land values, and confusion arose when proximity to more than one road entered into the reckoning. Hence, the LVT fell into disuse.

75) Tsukada, Shunso, and Chiaki Kuranami. "Value Capture with Integrated Urban Rail and Land Development: The Japanese Experience and Its Applicability to Developing Countries." Proceedings of Seminar M, PTRC Transport and Planning

Annual Meeting, University of Sussex, England, PTRC Education and Research Services, September 1990.

To win matching funds from the Japanese central government for planned urban rail systems, local governments must raise 35 percent of the construction costs. Some jurisdictions increase the property tax rate to raise the required revenue. Taxable value is determined by distance from the rail station and the city center. Another strategy is to develop fallow land along the rail corridor. One private rail line earns 18 percent of its total revenue from real estate (plus 54 percent from the railway and 28 percent from other businesses). The authors recommend that public transit agencies serve extant demand from riders, coincide construction with an economic upswing, cooperate with the private sector, commit themselves then deliver on their promises, and become competitive with other transport modes.

76) Walmsley, D., and G. Gardner. *The Economic Effects of Public Transport*. Transport Research Laboratory in TRIS Database: "Taxing Property Values for Transit," 1993.

Studies from Western Europe, North America, and various developing countries show how changes in the organization and financing of public transport affect patronage and urban development. Its general findings could apply, perhaps on a smaller scale, to other improvements in public transport such as busways. It considers funding from: (1) revenues, (2) taxation, (3) land-value capture, (4) advantages and disadvantages of assured funding, and (5) the involvement of private capital. Besides improving public conveyance, rapid transit systems can also improve the environment and the "image" of a city, as well as encourage new urban development and enhance safety. Bus transit deregulation in the United Kingdom illustrates how market disciplines can be applied to bus operation, and how privatization might affect public transport. The report offers recommendations for transport planners in Eastern European countries.

V

Additional Titles, Not Annotated

Day, Philip. *Land Value Capture.* Report to the Local Government Association of Queensland, February 1992.

Freeman, Mark J., and F. G. Price. "Value Capture: A Neglected Factor in the Funding of Transport Facilities." Reprint of paper prepared for the Annual Transportation Convention, C. 106, Paper 5D-10, Roads and Transport Technology, No. 687. Pretoria, South African Council for Scientific and Industrial Research, Division for Roads and Transport Technology, August 1989.

Hilling, David. *Transport and Developing Countries* (London: Routledge, 1996).

Hsu, Kuo-Wei. *The Impact of Mass Rapid Transit Systems on Land Values: Case Study, Taipei,* Chaoyang University, Taiwan, 1996 (available from: secret@mail.cyut.edu.tw).

Lawrence, Wai-chung Lai. "The Effect of MRT [Mass Rapid Transit] on Land Values Rekindled." *Journal of Property Valuation & Investment* 9(2) (1991). MCB University Press Limited.

Marwick, Peat. *Fiscal Impact of Metrorail on the Commonwealth of Virginia*. KMPG, Inc., November 1994.

Scheurer, Jan, Peter Newman, Jeff Kenworthy, and Thomas Gallagher. *Can Rail Pay? Light Rail Transit and Urban Redevelopment with Value Capture Funding and Joint Development Mechanisms.* Institute for Science and Technology Policy, Australia, 2000.

Tisato, P. "A Comparison of Optimization Formulations in Public Transport Subsidy." *Rivista Internazionale di Economia dei Trasporti* 27 (June 2000).

United Nations Centre for Human Settlements. *Comparative Modal Efficiencies in Urban Transport, With Reference to Developing Countries, Volume I.* Report No. HS/236/91E, Mass Public-Transport Modes and Sustainable Development, 1991.

VNI Rainbow Appraisal Service. *Analysis of the Impact of Light Rail Transit on Real Estate Values.* San Diego: Metropolitan Transit Development Board, 1992.

Voith, Richard. "Changing Capitalization of CBD-Oriented Transportation Systems: Evidence from Philadelphia, 1970–1988." *Journal of Urban Economics* 33 (1993): 361–376.

ZHA, Inc. *Amherst Corridor Alternative Analysis: Economic Development/Value Capture Study: Task IV, Analysis of Land Use and Development Activity.* For Niagara Frontier Transportation Authority, December 1988.

VI

New Directions in Transit Research

As this bibliography shows, a large part of the literature on the subject of transit financing is aimed at empirical evidence of the incidence of rising land values associated with distance from transit stops following the installation of rail transit improvements. Most of the research related to value capture conducted since 1980 alludes to but does not answer the question of whether special land assessments within transit station benefit areas might be a feasible method of financing transit systems. The importance of value capture is mounting due to urban planners' current interest in transportation-oriented development (TOD) in station areas. Here, the symbiotic relationship between improved transit ridership and the land-use efficiencies that derive from concentrated, mixed-use development offers an even better prospect for enhancing land-value increments to be captured. A land-based property tax is also thought to be an effective way to encourage private development around a public transit project.

The jury is in. After 25 years of empirical research, it appears the positive effects of transit access are confirmed. It is now time for transit/land-use research to move from hypothesis testing and toward the practical applications of value capture. We propose two means to accomplish this. The first is a shift in the research paradigm; the second is the dissemination of information to local public agencies.

1. It would be helpful to shift research methods from the use of static models to dynamic models, showing the effects of transit station access on property values over a period of time during which land-value increments accumulate. A perusal of the citations in this bibliography will show that the typical measure of effects is the percentage difference between the property price in nearby versus distant locations with respect to rail stations. But in actual fact, would not the differential vary over time? The research question could be reframed: How does the *growth* of property values in the vicinity of station areas compare with *growth* in areas within the same jurisdiction not served by rail

stations? The longitudinal dimension is useful because the research findings have direct relevance to the application of value-capture mechanisms that rely on a cost-recovery period.

Dynamic models can help predict land-value increments over a period of years, yielding estimates of the total capturable revenues that would support the debt financing of capital improvement projects. Rather than the expression of percentage differential in values, an appropriate measure might be comparative average annual growth rates. For example, the observed differential in property values following the construction of transit improvements may actually reveal a comparative growth rate of 8 percent in the wider service area versus a 12 percent growth rate within station areas (which could be circumscribed as potential transit benefit districts, or special assessment districts).

When projecting rates forward from the point of announcing a new transit line, the percentage differential is expected to increase, due to superior locational advantage, upzoning (allowable density increases), and mounting development activity. However, over a long time period, the annual growth rate differential may decrease and finally level off, causing land values to increase at a decreasing rate. Model refinements could include expected variations in growth rates over time, accounting for stages in the land-appreciation cycle: predevelopment, construction/full operation, to the mature market phase (RICS 2002). Basing their projections on such research findings in various regional transportation markets, transit planners would have the necessary data to put together capital financing with such mechanisms as increment bond financing for special assessment districts. This leads to the second means of moving research in a new direction.

2. Even after years of research and legislative initiatives, the manuals and primers produced by transit organizations rarely contain specific reference to value-capture financing. For example, *Financing for Capital Investment: A Primer for the Transit Practitioner* (Transportation Research Board 2003) outlines several conventional financing mechanisms and capital

> sources. These consist of various types of bonds, partnerships, and taxes (including tax increment financing and special assessment districts and lease transactions), but none of these mechanisms is linked to land-value increments as a revenue source. If the main body of research were to shift its emphasis to real-world applications of value capture, government-sponsored reports, manuals, and policy guides could be expected to devote greater attention to value-capture mechanisms.

The authors acknowledge the commendable work conducted by the RICS Policy Unit. The Royal Institution of Chartered Surveyors and the Office of the Deputy Prime Minister commissioned a study identifying the relationship between land use, land value, and public transport. Stage 1 of the study (review of literature) was undertaken by Jeremy Edge and Jamie Fox at ATIS REAL Weatheralls, by David Banister, Claudio de Magalhaes, and Stephen Marshall at University College London, and by Andrew Marsay at Symonds Group. Listing about 150 references, this review finds common elements and summarizes the evidence in tables. It is interesting that the authors note the scarcity of time-series analysis, and state that ideally data should be available from before the decision to build transit improvements is taken, and immediately after opening, as well as downstream. This is our principal finding as well.

References

RICS Policy Unit. (2002). *Land Value and Public Transport, Stage 1: Summary of Findings*. RICS Contact Centre, Coventry, UK.

Transportation Research Board, Transportation-Technology Management. (2003). *TCRP Report 89*. Transit Cooperative Research Program.

The Complex Taxonomy of the Factors

Natural Resources, Human Action, and Capital Goods

By FRED E. FOLDVARY*

ABSTRACT. Contemporary neoclassical economics has reduced factor analysis to two homogenous inputs, K and L. This excessive simplification has led to a deficient understanding of economic reality and a misunderstanding of concepts such as the producer surplus. This paper presents a taxonomy of the factors, including the complexity of natural resources. A better understanding of the role that factors play will enhance an understanding of economic reality and policy.

The classical categories of factors—land, labor, and capital goods—were recognized by Adam Smith and Jean-Baptiste Say and were a central element of classical economic thought. The neoclassical turn of the late 1800s melded land and capital goods into a homogenous variable, K. Textbooks still typically give *pro forma* recognition to the three factors but then proceed to ignore land in their applied topics, such as economic development, taxation, and macroeconomic policy. The near absence of land in mainstream economic analysis has been amply described (Foldvary 2005), but what has apparently also been lacking is an analysis of the complex taxonomy of the factors of production and its implication for policy.

*The author is a lecturer in economics at Santa Clara University. His book include *The Soul of Liberty, Public Goods and Private Communities*, and *Beyond, Neoclassical Economics*. Foldvary's research areas include real estate economics, social ethics and public finance.

American Journal of Economics and Sociology, Vol. 65, No. 3 (July, 2006).

The foundation for the taxonomy of the factors is the distinction between "nature" and nonnature. The relevant contrast to nature is what human beings produce, and therefore also the human action that produced the goods. The apt economic meaning of *nature* is therefore "everything that is prior to and apart from human action."

Human action was analyzed by Ludwig von Mises (1949) as purposeful behavior, persons acting to achieve ends. The definition above excludes as "nonnatural" any act that is consciously, deliberately, purposefully committed by a person. Nothing that human action does is "natural." The concept of an act as "unnatural" is thus meaningless for economics; phenomena are either natural or nonnatural.

Land in economics synonymously means "natural resources." Land can be divided into three categories: (1) space, (2) nonliving matter, and (3) biological natural resources.

Spatial land (land #1) in turn consists of (a) territorial space: the surface spatial soft-shell envelope at the earth's surface in which life is located, including the space holding the waters; (b) spectral space, in other words, frequencies of the electromagnetic spectrum; and (c) routes for satellites and other spacecraft.

Material natural resources (land #2) can be categorized according to the states of matter: (a) solid substances such as minerals and coal, oil in solid substances such as shale and tar sands, and ice; (b) liquid substances such as water and oil; (c) gaseous substances such as air and natural gas, as well as properties of gas such as the capacity to carry soundwaves; and (d) other states of matter such as plasma. The last category (d) exists, but has little economic significance.

Biological natural resources (land #3) include (a) living beings; (b) the genetic base of life; and (c) the ecological relationships among living beings, including the habitat.

I

Territorial Space

EACH OF THE THREE DIVISIONS of natural resources interacts differently with the other factors. Territorial space (1a) is, for practical purposes, a fixed resource. The earth does gain in volume and mass as meteors strike it, but the annual expansion is so tiny that it is irrelevant for economic analysis. As long as the earth exists, territorial space is, for economics, absolutely fixed in supply, as it can neither be created, destroyed, nor altered. Territorial space always remains land regardless of the matter or activity within some boundary.

Territorial space is the most important natural resource for human activity, as all activity must have a location, and productive locations are scarce. Although the fixed supply of this natural resource is often recognized in the economic literature, sometimes it becomes confused with capital goods, and land is claimed to be not really completely inelastic in supply. To understand the fixity of land, we must first clarify the other factors.

Human capital consists of the talents and capacities that human beings possess genetically and of an increase in a worker's productivity due to education, discovery, and innovation. When conceived, a human being is natural, but as the being develops during gestation and after birth, the outcome is also a product of human action, the ingestion of nutrients by the mother, the upbringing by the parents and guardians, and influences from peers and society, so that after conception, a human being is no longer a natural being and thus labor is not a natural resource.

Economic goods, synonymous with economic wealth, are resources, products, and services, excluding human capital, with a market value. A good has a market value if at least one arm's-length buyer will typically voluntarily purchase or trade the item.

Capital goods are goods that have been produced but not yet consumed. Following William Hutt (1974), economic consumption is the using up of economic value, i.e., reducing its market value. Consumer goods, loosely, are goods that typically become consumed within a short time after having been produced. Again following Hutt, production is the creation of economic value. Capital goods can be intangible, such as knowledge or reputation capital.

Capital goods are often defined as "the produced means of production" or "wealth devoted to procuring more wealth" (George [1879] 1975). Inventories are the "tools" and "means" used by retailers to provide services to customers, but the concept becomes murkier when households own the goods. An owner-occupied house is commonly recognized as a capital good providing housing services, just as it does to a tenant who rents from a landlord. But, likewise, an owner-occupant's car is a capital good, as is a can of corn on a shelf in a pantry. Household inventory is just as much a capital good as is inventory in a store that sells to customers.

Ancient capital goods could be considered as land for purposes of taxation, as the producers are long gone and the taxation of the long-ago clearing, leveling, and draining of the natural materials need not apply to recently produced goods. Land could also be treated as a residual for anything that is not capital goods and not labor; for example, garbage, having no positive market value, becomes land.

But the definition of capital goods as produced but not yet consumed does not imply that the goods are economic goods or wealth. The goods can have a zero market value, or a negative value, and may be items that people would pay to be rid of. Trash is therefore a capital good with negative value. Capital goods never revert to land. Fertilized soil is not a natural resource, not land. The capital goods that fertilize land, rather than becoming land, on the contrary, convert the land into a capital good.

With these meanings in mind, we can see that claims that territorial space is not fixed are false. Those who assert that land can be increased by clearing, draining, filling, and leveling have in mind a definition of "land" as the usable solid surface of the earth, rather than a natural resource. "Landfill" is a capital good, as is any improvement to a site. Indeed, as a natural resource, land cannot be improved. Thus, the supply of territorial space is not affected by any improvement that makes the space more useful for economic activity.

Another fallacy is the belief that territorial space is not purely natural because its price depends on the demand due to population, commerce, and civic works. But what is natural is the supply, the physicality of the three-dimensional space. Demand is necessarily that of human beings, and cannot be natural.

The characteristics of spectral space (1b above) are similar to those of territorial space. The frequencies of the electromagnetic spectrum and their properties, such as light, thermal radiation, and capacity to carry transmissions, are fixed in nature and cannot be altered by human action. Labor can provide only the texts that are carried by the spectrum. It has been claimed that the scarcity and thus the supply of the spectrum depends on technology, but this applies to territorial space as well. The concept of "supply" thus requires clarification.

The term *supply* has two meanings, which are usually not distinguished in economic analysis. Consider the shares of stock for a corporation. If no new shares are being issued, the supply-1 of shares means the total number of shares held. Supply-1 is fixed. Supply-2 consists of the number of shares offered for sale in the financial market. That supply behaves like the supply of produced goods, as, holding expectations and all else constant, more shares will be offered for sale at a higher price.

Likewise, the supply-1 of land means the quantity (measured, for example, as cubic meters or square meters) within some boundary. That supply is fixed. Supply-2 consists of plots of land offered for sale in the real estate market. Supply-2 of land is of course not fixed, and this supply curve can slope up, as with the supply-2 of shares of stock. Also, the supply of land for a particular use is supply-2, where offers to sell land increase with price.

Technology thus affects supply-2, not supply-1. Land-saving technology reduces the land needed for a particular use, such as farming, but does not affect supply-1. Likewise, technology that reduces the amount of frequency bandwidth needed for some use affects supply-2 only.

For goods currently being produced, the total supply-1 of inventory is typically offered for sale in the market, so in that case, supply-1 equals supply-2. This is the supply curve that is typically presented in textbooks and in economic analysis, where the distinction does not matter.

Spatial routes are not all fixed in supply-1. Airline routes can depend on the locations and size of airports, as well as the air-travel technology. These routes are thus produced by the technology of the aircraft and the production of landing places. Airline routes are capital goods. However, satellite orbits (land 1c) are natural, as they must orbit the earth in the limited space surrounding the earth. Some routes, such as geosynchronous orbits, stationary above a location on the surface, are even more limited. These routes are not a product of the locations of satellite signal transmission and reception.

All three types of spatial land are fixed in supply-1, and therefore the tapping of their economic rent for community or public revenue has no deadweight loss (i.e., excess burden). The second consequence of a fixed supply is capitalization. An increase in the demand for spatial land can increase only the value of that land, as it cannot affect the quantity.

But the works that increase value are capital goods, so the increase in the yield of the space is not natural. The return on land is termed "rent" in economics, and "rent" will refer here to only the return on the land factor. The rental of a capital good can be referred to as the "yield" or "rental" of the capital good. When a house is rented to a tenant, only the portion of the rental that pays for the space is rent, the rest being a yield of the capital goods and wages for the labor services of providing the real estate.

Thus, the rental of a site is actually a payment for three factors: rent, the yield of capital goods, and wages. Likewise, the increase in real estate rentals due to public goods such as streets, security, and public transit are yields of those capital goods, not of the natural land. Land rent excludes the rental added by nonnatural factors such as civic goods and security services.

The rent of a television frequency is that which one would pay for a new or abandoned portion of the spectrum. Once the spectrum is in use, its value may rise due to the activity of the enterprise using the spectrum, just as a location in a city might increase in value due to the fame of the firm occupying a site. These increases in value are a return to capital goods.

Given the natural features of a site and the capital goods attached to a site, a greater population can increase the demand to be located there, and thus raise the site values. As noted above, demand does not affect the characteristics of the supply of land, so the greater demand simply due to population is land rent and does not affect the natural characteristic of spatial land.

II

Speculation for Space

THE CHARACTERISTICS OF TERRITORIAL SPACE make speculation outcome different from that of commodities. Speculation is the purchase or short-sale of an item with the expectation of a favorable shift in supply or demand. Speculators buy futures contracts for wheat, for example, expecting the demand to rise or the supply to shrink. Land speculators expect the demand for land to rise due to new infrastructure or development or growing population or increasing commerce.

Commodity speculation makes the price of commodities in the present rise, which reduces current use, shifting more use to the future and reducing future shortages. But the quantities of land cannot be so shifted. Land rent can rise due to population movements to fringe areas with less productivity, increasing the rent of the more productive areas. Speculation extends and quickens this movement. Speculators tend to be those most optimistic about future prices, so they can push land prices higher than current use warrants. Higher costs for real estate reduce profits, so investment can fall, leading to lower demands for other goods and even to a recession.

If site owners have to finance civic goods and services, this reduces the gains from land speculation and reduces, if not eliminates, the destabilizing effects of spatial speculation.

III

The Nonproducer Surplus

SINCE NEOCLASSICAL ECONOMICS melds land into capital goods, it masks the income from nonlabor resources as returns to homogenous "capital" rather than explicitly separating land's share. Land rent cannot be so easily hidden in microeconomic theory, so it is assigned another name. This creates a contradiction that is seldom confronted in the economics literature.

In the models of perfect competition and monopolistic competition, the long-run equilibrium is zero economic profit. If marginal costs rise with increasing production, supply curves slope up, creating a producer surplus of price minus marginal costs. But the producer surplus is an economic profit. So how can there be no economic profit? The owners of firms do not receive it; therefore, the surplus must flow to the owners of factors. But if the labor market is atomistic, highly competitive with replaceable workers, then wages obtain only normal returns with no surplus. Likewise, competitive producers of capital goods also have zero profits.

The only other factor left is land, and since territorial space cannot expand, land rent can persist as economic profit. The economic cost of land is zero, and thus all land rent is a surplus. The producer surplus is thus site rentals, aside from the minor surplus of monopoly labor and monopoly capital goods. If labor has a common wage and all capital goods of a type are equal in price, then the differing costs of production are generated by differences in the productivity of locations. Site rentals thus absorb these greater productivities, a combination of land rent and the rentals of the capital-goods infrastructure tied to land.

If the civic goods that get capitalized into site values are not paid for by the site owners, the site owners receive a surplus for goods they did not pay for. The title holders of land are in that case nonproducers, and the surplus they receive as rentals should be called the "nonproducer surplus." Instead, neoclassical theory masks the rent of natural resources by calling it a "producer surplus," as though it was going to the producers, the workers, and owners of capital goods.

IV

Material Land

UNLIKE SPATIAL LAND, material natural resources can be altered by human action, and then they are no longer land. Solid natural substances such as minerals and coal (land 2a) are subject to depletion, since these were created in the past and, as far as is known, there is no new creation. Natural ice, however, can expand, as water turns to ice during the winter. These natural materials become capital goods when they are altered by human action.

Extraction alters minerals and coal by transporting them and by isolating them. Human action also is involved in exploration for these materials. But such action does not alter the physical capital good, which remains natural until physically changed in form or location, just as spotting an eagle with a telescope does not transform it into a domestic animal. What is affected by exploration is the yield from the extraction. Part of the yield is rent for the scarcity of the resource; part is a return on labor, including exploration; and part is a return to the capital goods used in the extraction and exploration.

The use of minerals and coal necessarily involves depletion, as they are nonrenewable resources. If there is no government intervention, a free market efficiently allocates the amount of extraction. Speculation in material land is similar to that in produced commodities, and while it can carry prices to a level not warranted by actual rents, it does not have the greater impact of speculation in territorial space, since the particular commodities have a narrower use.

The price of minerals such as copper is global, so the local infrastructure will not affect the local price of copper. Thus, while the public collection of spatial rent is market enhancing in avoiding capitalization subsidies to landowners, it does not hold this economic effect for material land. The collection of the land rent from material land can serve a moral purpose, but not an allocation purpose. Indeed, the taxation of extractions of coal and minerals needs to be carefully crafted to avoid inducing too fast or too slow extractions relative to the nontaxed market rate. There needs to be a combination of ex ante charges for leaseholds, current charges for extraction, and ex post charges based on the estimated economic profit. The firm also should be required to pay the social cost of environmental damage, which would induce it to prevent such damage if the cost of prevention were less than that of pollution.

For liquid material land (2b), namely, water and oil, there is a potential problem of dividing the extraction among separate owners of the pool. There needs to be an organization or agreement among the owners of the pool to avoid turning the resource into a common good that becomes too quickly exhausted.

Material land may be fixed in supply-1 like oil or renewable like water. The current literature on "peak oil" suggests that the supply-1 of oil will in the future decrease as ever more oil is extracted and few new sources are found. But there are also large quantities of oil that can be extracted from shale and tar sands, land 2a, which will substitute for liquid natural oil at higher prices. Uranium, as land 2a, will increasingly substitute for oil as a source for the generation of electricity, including the energy used in producing hydrogen for fuel cells.

The supply of water is to a great extent controlled by governments, which subsidize its provision to agriculture and allow large areas to become contaminated. The efficient provision of water prices is at the social cost, which includes the provision of sufficient amounts for fish and other wildlife preservation and extraction at sustainable levels, with pollution at the amounts for which the social costs of additional pollution equals the social benefits of additional pollution prevention. Given such conservation, the economically efficient charge for water is the marginal cost of a unit of water, if the fixed costs generate an equal or greater amount of site rentals. If the water infrastructure does not generate site rentals, this indicates that the amounts of the capital goods are excessive.

Water is a negative natural resource when its supply is excessive, causing crop failure and flooding. The economics of natural resources has to account for a resource as a good and as a bad. In a free market, bad waters reduce land values, and the incentives of communities are to protect lives and property with controls to prevent flooding. When these functions are shifted to national governments, the protection becomes vulnerable to national politics, and the moral hazard is created of residents not bearing the cost until there is a catastrophe, such with Hurricane Katrina in New Orleans.

Gaseous substances such as air and natural gas (land 2c) have pool characteristics similar to liquid natural resources. As a nonrenewable resource, natural gas's extraction is similar to that of oil. If natural gas is liquified, it becomes a capital good. Indeed, as soon as oil and gas are extracted, they become capital goods.

If one asks students for an example of a free good, they will often answer "air." But one does not breathe air in the abstract. One uses air in particular places. If the location in which one is breathing is scarce, then so is the air there. So if the site has a rent, the air there is not a free good; one must pay to be located there, or the host must pay.

Clean air is also not free if resources are used to prevent pollution. Advancing technology enables communities to directly charge for air pollution from automobiles with remote sensing devices. The benefit of clean air is subjective as, aside from the costs of illness, there is an aesthetic benefit to clean air, as well as the benefit to wildlife. Demand revelation could be used to measure the subjective benefits, as each person reveals his or her value and has to pay the social cost of changing the outcome. But atmospheric pollution can have an immense global social cost if it causes the warming of the earth, so the avoidance of pollution is not just a matter of local choice.

It has been claimed that the cost of avoiding air pollution is so high that it would stifle economic development and even crush current industry. This argument ignores the current excess burden of taxation. If current taxes were shifted away from income and sales and value added and toward site values and pollution charges, the added cost from pollution charges would be offset by reducing the cost imposed by current taxes on production. This shift would involve transition costs but no long-term costs on industry.

Air is a negative resource when it has a high velocity, as with hurricanes. As with flooding, when governments provide aid after a disaster, the short-run benefit has the long-run cost of a moral hazard, as the cost of residing in the location is borne by others, inducing excessive locating in the areas prone to disaster.

V

Living Beings

THE SUPPLY OF WILDLIFE (land 3a) can grow or be depleted depending on the habitat and the weather. As a renewable resource highly valued by many people today and surely in the future, the morally justified use is only of the annual yield, leaving the principal intact. The rent from fishing is similar to that of oil extraction, except that governance can limit the extraction to the yield based on the current wildlife population, with consideration to the age and sex of the animals.

Some animals, such as elephants, live in government-owned lands such as national parks where hunting is prohibited, while others of the same species are private property or the communal property of villages. Poaching is common for state-held elephants, while private and local communal owners have an incentive to protect their elephants and thus profit from foreign tourism and hunting. The population of elephants and other wildlife in private or communal reserves has been growing while that in government parks has been reduced.

Harmful wildlife—viruses, bacteria, protozoa, rats, mice, and mosquitos—has always been a plague on humanity and can now spread diseases quickly as human beings travel the globe. Currently, the bird flu virus is being spread by migrating birds such as ducks to poultry and then to other animals and human beings around the world. Prevention is less costly than curing the plagues after they hit, but the problem is the uncertainty of the future damage and the psychological tendency to weigh visible problems more than hypothetical ones.

The genetic base of life (land 3b) consists of the variety of organisms and their current and potential contributions to human welfare. Many drugs originated in wild plants, and there are potential scientific and health benefits in preserving the genetic heritage. This is complementary to conserving the wildlife (3a).

Ecology means the relationships among the interdependent species of wildlife, creating an overall structure of complementarities. Prey needs predators to keep their numbers in check and to prevent overgrazing the vegetation. The entire habitat requires preservation in order to maintain the viability of each particular species (3a).

VI

Human Action in the Production of Wealth

THE LABOR FACTOR can be regarded as human exertion in the production of wealth, although "exertion" implies applying oneself energetically, whereas a wage could conceivably be earned just sitting there looking pretty. (Sitting still is a choice, and thus also a human action.) The more general definition of labor is "human action that produces economic wealth."

The reason that labor is a separate factor is the moral and biological autonomy of human beings. If there is no slavery, the worker has not just consumer sovereignty but also production sovereignty. It is his or her choice to engage either in labor or in leisure, each being the opportunity cost of the other. A worker buys leisure by foregoing extra wages and the consumer goods it buys. In contrast, a capital good is directed by its owner. If slavery were accepted, then a human slave would function economically like a horse, a capital good. With no slavery, unlike capital goods and land, labor has no purchase price but only a rental if employed by others, and an implicit rental as well as economic profit if self-employed.

Just as a natural resource can be positive or negative for human welfare, an employee can reduce economic value by being wasteful and destructive, but in that case, this was not what he or she was hired to do. Thus labor implies an ex ante anticipation of a positive marginal product. When a worker destroys property, this is not *economic* labor. Labor is therefore always ex ante productive. A thief, for example, exerts effort in stealing, but this is not economic labor. The employee who steals is a thief, not a laborer. There are negative natural resources and negative capital goods, but not negative labor. The deliberately destructive worker ceases to engage in labor.

Some production functions split labor into the quantity of labor and the amount of human capital. Investment is then on human capital, rather than the quantity of labor. But in actuality, even "unskilled" work involves human capital, so in application, labor cannot be separated from human capital.

The wage of an ordinary worker is generally that of the worker's marginal product, but with variances due to personal relationships, arbitrary discrimination, union rules, and government law. Unions with the power to strike tend to raise union wages but depress nonunion wages, as the higher union wages reduces labor in the union sector, shifting the supply to the nonunion sector.

The responsiveness of labor supply to a change in the after-tax wage has to account for many margins: working extra hours, taking a second job, whether a spouse is employed, when to retire, and how much to invest in human capital. The elasticity of labor should be applied to the most elastic margin, just as the marginal utility of water is for its least important use.

Entrepreneurs also engage in labor, but they do not earn a marginal product. Their wage is rather a residual of revenue minus all costs, the entrepreneurial profit. This is a second residual, after the locational residual of land rent.

VII

Capital Goods

AS NOTED ABOVE, neoclassical theory homogenizes capital goods into the factor K. The deficiency is not just the melding of natural resources into capital goods, but also that even the homogenous treatment of capital goods alone leaves out important aspects of that factor.

Austrian School economists have recognized that capital goods have a structure based on two elements: stages and time. Production proceeds in stages. To make bread, production has to first raise the grain, then make the flour, and then bake the bread and sell it. The time involved in the stages for a product is the "period of production."

The time element also consists of the rate of growth of the value of the capital goods. This can be illustrated with planted trees. Some planted trees take many years to mature, while others, such as small Christmas trees, turn over quickly. Inventory in a retail store has a very small period of production; it is also referred to as circulating capital, as opposed to the long-lasting fixed capital.

A low rate of interest induces investment in slower-growing capital goods with a long period of production. A high rate of interest induces less higher-order investments and more in circulating capital, which is not as affected by the interest rate (Hayek 1931).

If the funds for investment come from savings, then the interest rate serves to allocate goods between investment and consumption, since savings equals investment. But when money is artificially injected into the banking system by a monetary authority, it acts like greater savings, reducing interest rates and inducing investment in slower-growing long-lived capital goods. But intended consumption has not been reduced, resulting in price inflation. Prices then rise, including land prices, which accelerate with speculation. Interest rises again as the monetary authority reduces the growth of the money supply. Rising costs reduce investments, which reduce demands for other goods, resulting in an economic downturn.

The land factor is thus linked to the capital goods factor and the financial markets in creating the boom-bust cycle (Foldvary 1997). Central banking has an inherent knowledge problem, as the future is too uncertain to be able to provide the optimal growth of the money supply. An alternative is free market banking, or "free banking," where the monetary base is a commodity (which today could be a frozen supply of government money), with future expansion with private bank notes convertible into the base money (Selgin 1988).

Various capital goods can be substitutes and complements. Complementary capital goods, like software and hardware, enhance their mutual productivity. Network externalities can increase the productivity of the individual items as more are used, as with fax machines. Thus, contrary to models of growth such as that of Robert Solow, where an increase in K with constant L has a diminishing marginal productivity of K, complementarities can make an increase in capital goods exhibit increasing returns.

Advancing technology is embedded in both human capital and in capital goods. The power to make the same inputs more productive comes from a better harnessing of the power of nature, using a better knowledge of the laws of physics, chemistry, and biology to produce more, so at bottom, nature is the foundation of technological advancement.

VIII

Conclusion

THE FACTORS OF PRODUCTION are much more complex than the simple two-factor neoclassical models suggest, or even the classical three factors. Land, for example, comes in multiple forms, each with its own characteristics. There is also a complex ecology of factor interaction. The inclusion of all the factors of production in their full complexity provides for a more complete understanding of economic reality, with important policy implications.

References

Foldvary, Fred. (1997). "The Business Cycle: A Georgist-Austrian Synthesis." *American Journal of Economics and Sociology* 56(4): 521–541.

——. (2005). "Geo-Rent: A Plea to Public Economists." *Econ Journal Watch* 2(1): 106–132.

George, Henry. ([1879] 1975). *Progress and Poverty*. New York: Robert Schalkenbach Foundation.

Hayek, F. A. (1931). *Prices and Production*. London: Routledge & Sons. Second revised edition, London: Routledge & Kegan Paul, 1935.

Hutt, William H. (1974). *A Rehabilitation of Say's Law*. Athens, OH: Ohio University Press.

Mises, Ludwig von. ([1949] 1966). *Human Action*. New Haven: Yale University Press, and Henry Regnery Company.

Selgin, George A. (1988). *The Theory of Free Banking*. Totowa, NJ: Rowman & Littlefield.

Heterogeneity and Time

From Austrian Capital Theory to Ecological Economics

By MALTE FABER and RALPH WINKLER*

ABSTRACT. Although heterogeneity and time are central aspects of economic activity, it was predominantly the Austrian School of economics that emphasized these two aspects. In this paper we argue that the explicit consideration of heterogeneity and time is of increasing importance due to the increasing environmental and resource problems faced by humankind today. It is shown that neo-Austrian capital theory, which revived Austrian ideas employing a formal approach in the 1970s, is not only well suited to address issues of structural change and of accompanying unemployment induced by technical progress but also can be employed for an encompassing ecological-economic analysis demanded by ecological economics. However, complexity, uncertainty, and real ignorance limit the applicability of formal economic analysis. Therefore, we conclude that economic analysis has to be supplemented by considerations of political philosophy.

*Professor Malte Faber is Professor Emeritus in economic theory at the Alfred-Weber-Institute of Economics, University of Heidelberg. He studied economics, statistics, and mathematics at the Free University Berlin before he took an MA in mathematical economics at the University of Minnesota. His Ph.D. was on stochastic programming from the Technical University of Berlin. He was director of the Interdisciplinary Institute of Environmental Economics at the University of Heidelberg. He has published widely in capital theory, public choice, the concept of entropy in environmental economics, and the philosophical foundations of ecological economics. Dr. Ralph Winkler is Marie Curie Research Fellow at the School of Politics, International Relations and the Environment, Keele University, UK. He studied physics and economics at the Technical University of Munich and the University of Heidelberg. His Ph.D., from the University of Heidelberg, was on the time-lagged accumulation of capital and pollution stocks. He has published in capital theory, the valuation of ecosystem goods and services, structural change, decisions under uncertainty, and intertemporal decision making and discounting. The authors are grateful to Kate Farrell and Thomas Peterson for valuable comments on an earlier draft.

American Journal of Economics and Sociology, Vol. 65, No. 3 (July, 2006).

I

Heterogeneity and Time

HETEROGENEITY AND TIME are central aspects of economic activity. By *heterogeneity* we mean the variety of different entities that are involved in the economic process. In particular, this encompasses all primary resources, all intermediate and capital goods, and all final outputs (consumption goods). This also includes natural resources, stocks of pollutants, and species populations, at least insofar as these influence or are influenced by economic activity. With the term *time* we want to emphasize that all economic activity is dynamic in a twofold manner. First, the different entities involved in the economic process do not stay constant, but change over time. Second, economic activity does not turn resources instantly into final outputs; the production process and consumption itself takes time.

In our understanding, during the last few centuries, both heterogeneity and time have become increasingly important for the analysis of economic activity as humankind has extracted more and more resources, produced an increasing number and amount of intermediate (i.e., capital goods) and final goods, and seemed insatiable in its consumption wants, leading to the depletion of natural resources and an increasing variety and amount of pollutants emitted into the environment. This is, on the one hand, because the human population has been (and still is) growing drastically and, on the other hand, because technological progress steadily increases humankind's means of extracting, producing, and consuming. As a consequence, economic activity depends more and more on the natural environment either directly, such as by land use and resource extraction, or indirectly, such as by environmental pollution and use of natural degradation capacity.

Thus, heterogeneity increases with the number of resources, pollutants, goods, and production techniques; at the same time, the number of natural systems influenced by human activity also increases. Time becomes more and more important as many natural resources are finite and many wastes and pollutants are durable over long time horizons and accumulate pollutant stocks, such as radioactive waste and greenhouse gases. In addition, for many durable pollutants, the so-called stock pollutants, it is not the flow of emissions that is relevant for environmental damage but the accumulated stock (Baumgärtner et al. 2002). As a consequence, we experience significant time lags between pollutant emissions and their consequences. Prime examples are climate change due to the anthropogenic emissions of greenhouse gases and the depletion of the ozone layer, induced by the emissions of a class of artificial chemicals, the chlorofluorocarbons (CFCs).

Due to dissatisfaction with the narrow view of environmental and resource economics on the "environment," the new field of ecological economics emerged in the 1980s. While environmental and resource economics employs the standard paradigm of neoclassical economics and views the environment as a subsystem of the economic system, ecological economics sees the opposite. It emphasizes an encompassing interdisciplinary view on both the diverse interactions between society, economy, and nature and the long timeframe of many environmental problems. As a consequence, heterogeneity and time play a crucial role in ecological economics.

Traditionally in economics, it is the field of capital theory that deals with the dynamics of durable stocks (cf. Bliss et al. 2005). In fact, heterogeneity and time, in the two-fold manner defined above, were first introduced into capital theory by the Austrian School of economics. Thus, in Section II, we first describe how the Austrian School of economics emerged from early marginalism in the 1870s. Although originally influential, Austrian economics fell into oblivion during the 1930s. The consideration of the vertical time structure of production, one of the distinctive characteristics of Austrian economics, experienced a rebirth in the 1970s in two different schools of thought, the Austrian subjectivist school and neo-Austrian capital theory, as outlined in Section III. In Section IV, we show how neo-Austrian capital theory has gradually been applied to the study of problems posed by ecological economics. Finally, in Section V, we address problems that are beyond capital theory.

II

The Emergence of the Austrian School of Economics and Its Classical Roots

THE AUSTRIAN SCHOOL OF ECONOMICS was founded by Carl Menger (1840–1921). At the same time, Menger co-founded marginalist theory with Jevons (1871) and Walras ([1874/1877] 1954). As outlined in his magnum opus *Grundzüge der Volkswirtschaftslehre* [*Principles of Economics*] (1871), the two distinctive characteristics of the Austrian School are:

1. A completely subjectivist approach to value theory. According to Menger, the value of goods is determined by the individual's expectations of their potential to satisfy present and, in particular, future needs and wants.
2. The explicit consideration of the time structure of production. Therefore, Menger ordered all goods in accordance to the time at which they entered the production process. Consumption goods, which satisfy human needs and wants directly, are goods of the "first order." Goods of higher order do not directly satisfy human needs and wants but are intermediate goods, used in production. Eventually, one reaches the nonproducible primary factors of production, which are goods of highest order.

Thus, heterogeneity and time were central aspects of the Austrian School of economics. In fact, the Austrians' concern about the vertical time structure of production finds its roots in the classical concept of circulating capital. Although the classical economists followed Adam Smith's ([1776] 1976) distinction between fixed capital (machines, buildings) and circulating capital (money, stocks of output), they emphasized circulating capital, which could be used to finance the labor needed in the production process over the period of production. Thus, both the classical economists and the Austrians emphasized circulating capital, while neoclassical economics concentrated on fixed capital.

While "[t]he Austrian method is indeed a Classical method" (Hicks 1985: 156), the Austrian value theory, however, was completely opposite to the classical theory of value. While the classical economists mainly advocated a labor theory of value in which the value of goods is determined by the costs of production in terms of labor, Menger took the opposite point of view, deriving the values of goods exclusively from consumers' expectations about their future potential to satisfy needs and wants.[1]

Austrian capital theory was established by Böhm-Bawerk ([1889]

1921), a follower of Menger. After having established himself as a leading economist, he took up the prestigious position of the Austrian Minister of Finance. In his first speech in the parliament as the minister of finance, he said: "We public servants are interested in the just and the good" (cited in Hennings 1997: 20). His profound interest in justice is why he "turned the *Soziale Frage* [social question] into a theoretical problem and attempted to discuss it within the framework of the Austrian theory of value, capital and interest, he developed" (Hennings 1997: 22). His main scientific achievement was, according to Hennings (1997: 155–167), the development of a theory of interest within an intertemporal model of general equilibrium, which enabled him to analyze two topics:

1. The existence and the determinants of the rate of interest, allowing him to study problems of income distribution, which had come into prominence since the appearance of Marx's *Capital* ([1867–1894] 1965–1967).
2. The repercussions of technical change and accompanying accumulations of capital leading to structural changes in the economy.

Schumpeter (1954: 847) assessed his achievement: "It is Böhm-Bawerk's model, or schema of the economic process adumbrated above which makes him one of the great architects of economic science, and this schema was quite outside Menger's as well as Jevon's range of vision. A few of the best minds in our field, Wicksell and Taussig in particular, have in fact considered him as such."

Böhm-Bawerk gave three famous and much debated reasons for a positive rate of interest: (1) consumers expect better provision of goods in the future; (2) consumers systematically underestimate future needs and wants; and (3) producers value present goods over future goods as capital to fund more roundabout and, thus, superior production techniques. While the first two were in the Austrian tradition, the third was strongly criticized by the Austrians because they considered it as a fallback to the classical cost of production approach to value theory.

To overcome the classical capital aggregation problem, he introduced the concept of the average period of production, which is supposed to be the timespan between the inputs and the corresponding final output. However, his mathematical presentation had fundamental flaws, for the concept of the average period of production can only be applied under very restrictive assumptions (e.g., Dorfman 1959).

Böhm-Bawerk's theory of capital and interest was both highly influential and controversial. In 1913 (Böhm-Bawerk 1913a, 1913b), he had an exchange with Schumpeter (1913) concerning whether the rate of interest in a stationary economy is positive or zero (cf. Faber 1979: ch. 7). In the 1930s, Knight (1934, 1935, 1936a, 1936b) debated with the Austrians Machlup (1935) and Hayek (1934, 1936) about the applicability of the average period of production.

After the capital controversies in the 1930s, the Austrian view on capital fell into oblivion for a variety of reasons. Among them were the formal deficiencies of the Austrian approach, the heterogeneity of different views within the Austrian School itself, and the rise of fascism in Europe on the eve of World War II, which led to the emigration of many Austrian economists.

Schumpeter (1954: 909) summarized the state of Austrian capital theory: "But so the reader might well ask, if we recognize all this and if we introduce all those corrections, what is left of Böhm-Bawerk's Capital Theory and in particular of his period of production? Well nothing is left of them except the essential idea. And this keeps on proving its vitality by every piece of criticism and every piece of constructive work it evokes." In the following, we show that this statement still stands today.

III

The Rebirth of the Austrian School of Economics

In the 1950s and 1960s, the Cambridge-Cambridge controversies on capital theory showed once more the impossibility of a suitable capital aggregate: many parables derived via the highly aggregated and mostly timeless neoclassical growth theory do not hold in general

(Samuelson 1966; Harcourt 1972; Bliss 1975). This finding demonstrated the importance of the explicit consideration of heterogeneous capital goods and the time structure of production.

As a consequence, the Austrian view on capital was revived in the 1970s. This renaissance occurred in two different strands:[2]

1. The Austrian subjectivist school, which dominates in the United States (and therefore is also known as the American Austrian School). It relies on the subjective value theory of Menger, Friedrich von Wieser (1851–1926), and Ludwig von Mises (1881–1973). According to the methodological underpinnings of this tradition, "economic theorizing must be rooted in the subjective evaluations of market participants, each employing means in pursuit of his or her separate ends" (Garrison 2000: 85). Their representatives believe in markets as the most appropriate coordination mechanism. They reject formal approaches and emphasize phenomena of uncertainty and real ignorance. Prominent representatives of this school include Lachmann (1973, 1976, 1977, 1986), Kirzner (1963, 1976, 1982, 1986, 1993, 1995, 1996), Moss (1976), Garrison (1979, 1985, 1991, 2005), Rothbart (1962, 1976a, 1976b, 1977, 1990, 1993, 1995), and O'Driscoll and Rizzo (1985).
2. Neo-Austrian capital theory, which is prevalent in Europe, harks back to Böhm-Bawerk's theory of capital and interest. It splits into three subgroups. The first is von Weizsäcker (1971) and Orosel (1979), who revived the concept of the period of production and formulated it consistently within a steady state economy with many sectors. Second is Hicks (1970, 1973), who introduced a vertical flow input flow output model in which capital is not explicitly considered. Hicks studied the repercussions of new innovations, which lead to a traverse between two steady states. And third is Bernholz (1971) and Bernholz and Faber (1973, 1976), who also emphasized the vertical time structure of production. In addition, and in contrast to Hicks's approach, they considered capital goods explicitly, thus combining the Austrian emphasis on circulating capital with the neoclassical consideration of fixed capital.

In the following, Bernholz and Faber's approach to neo-Austrian capital theory (abbreviated as BF) is discussed in detail. Neo-Austrian capital theory in the tradition of Bernholz (1971), Bernholz and Faber (1973, 1976), Bernholz et al. (1978), and Faber (1979) has its roots in von Stackelberg (1941, 1941/1943), whose 1941 article can be viewed as a culmination of traditional Austrian capital theory. Originally, BF's aim was to rigorously determine the conditions under which the rate of interest is positive and to analyze structural change in an economy due to technical progress. Thus, from an epistemological point of view, neo-Austrian capital theory is clearly in the tradition of Böhm-Bawerk's magnum opus.

From a methodological viewpoint, there also are some similarities. Like Böhm-Bawerk, BF explains the positive rate of interest by a combination of psychological and technological reasons. Consumers are assumed to be impatient to consume (i.e., they value present goods over future goods). Böhm-Bawerk's concept of the superiority of roundabout production methods is disentangled into superiority and roundaboutness. (Both are relative concepts; for example, one production technology is superior to another production technology if it allows the production of more output over a given time horizon, and one production technique is more roundabout than another production technique if it allows the production of less output in the short run but more output in the long run.)

However, there are differences as well. First, the dubious concept of the period of production was abandoned. Second, BF combined the emphasis on the vertical time structure of the Austrian School with the emphasis on disaggregated capital goods of the neoclassical general equilibrium theory. Moreover, it incorporated Schumpeter's ([1911] 1941) notion of innovation and creative destruction. This explicit consideration of the technological side of the model economy (which is in contrast to the original Austrian School and also to Hicks's approach to neo-Austrian capital theory) enabled BF to analyze structural change due to the changing composition of the portfolio of heterogeneous capital goods.

In addition, by utilizing an activity analysis approach, which roots back to Koopmans (1951), in which the technology is described by individual processes instead of a neoclassical production function, the model framework proved to be highly flexible and easily amended. Another important aspect, connected to the use of Koopman's activity analysis approach, is the realization of the irreversibility of the production process (1951: 48–50).

In the 1980s, BF was generalized in several directions (which are summarized in Faber 1986):

1. Reiß (1979, 1981) and Reiß and Faber (1986) generalized the two-period model of Bernholz et al. (1978) to n periods.
2. To circumvent specific problems of models with a finite time horizon, Malinvaud (1953) assumed an infinite time horizon. However, to guarantee the existence of an intertemporal price system within the neoclassical framework, additional assumptions concerning the technology like nontightness or reachability concerning the technology have to be made (cf. Stephan 1995: 163). Stephan (1983, 1985) shows that the neo-Austrian concept of roundaboutness is sufficient for a price system to exist and, in addition, that it is a weaker assumption than nontightness. Analogously, he shows that the superiority and roundaboutness of production methods are sufficient to establish a present-value price system but are, at the same time, a weaker condition than reachability.
3. Böge et al. (1986), Böge (1986), and von Thadden (1986) give up the assumption of either a central planning agency or the existence of perfectly competitive markets, which have predominated BF analysis. In a strategic game setting, they answer the question of how the market structure influences the rate of interest, the price, the quantity, and the intertemporal accumulation of capital goods.

Recently, Winkler (2003) and Winkler et al. (2005) transferred BF into a continuous time optimal control framework with infinite horizon, which further strengthens the vertical time structure of production. This is because the continuous approach allows a much richer time structure to be considered, since it disentangles the investment decision and the exogenously given timespan between the inputs and the final outputs. Within this framework, one can analyze how the length of this timespan influences price, quantity, and the intertemporal accumulation of capital. This influence indicates how important is the explicit consideration of the vertical time structure of production. It is shown that (even in a model with only one capital good) there is a qualitative difference between models with and without vertical time structure of production: While optimal accumulation paths are monotone in the case of a vanishing timespan, they are oscillatory for any positive timespan. However, it is shown that quantitatively there is a continuous transition between models with and without vertical time structure of production, in the sense that the oscillatory behavior of the optimal paths increases in accordance to the length of the timespan.

IV

From Neo-Austrian Capital Theory to Ecological Economics

NEO-AUSTRIAN CAPITAL THEORY in the versions of Hicks and BF both focus their analysis on the question of how economies change in the course of time when technical progress occurs. In the terminology of Hicks (1973: pt. II), it is the "traverse" from one steady state to another that is of primary concern because many social questions, such as distributive justice both within and among generations, can only be properly addressed by considering the dynamic change of an economy. To illustrate, the innovation of a new production technique brings about new capital goods and new consumption goods being introduced into the economy; in addition, capital stocks of old techniques are replaced by those associated with the new technique. This process is accompanied by decreasing employment in the sectors of the old techniques and leads to increasing employment in the sector of the new technique. Depending on the capital intensity of the old and the new techniques, permanent structural unemployment can occur, which, in turn, raises severe social problems.

The potential for social problems strongly increases if we embed our model economy into a natural environment and allow for interdependencies between the two. In this case, we have to consider two additional driving forces of structural change:

1. Natural resources amend the two input factors of labor and capital goods. As a consequence, the innovation of a new pro-

duction technique may employ a formerly unused natural resource as an input for production. Moreover, if the availability of this input is limited, then, given that this factor is essential and of limited substitutability, the new technique and its capital goods can be employed only for a certain period of time. This in turn triggers the innovation of new techniques, which use different natural resources.

2. All production processes manufacture not only the intended consumption goods but also unintended joint outputs that may be unwanted or even harmful. If the joint output is both unintended and unwanted, then we speak of an environmental pollutant. These pollutants often accumulate in the environment, for example, wastewater, CO_2, CFCs, heavy metals, and DDT. If the amount of an accumulated pollutant in the natural environment determines the direct or indirect harmfulness to humankind, one speaks of a stock pollutant. This introduces an additional type of stocks that exhibit their own dynamics, which are similar to those of capital goods.

These two additional momentums have a vertical time structure similar to the production system that is the focus of Austrian capital theory.

As noted in Section I, the increasing awareness among scientists of environmental and resource problems and the dissatisfaction with the state of environmental and resource economics based on the neoclassical paradigm has led to the emergence of the new field of ecological economics (for an overview, see, for example, Costanza 1991 and Faber et al. 1996). Ecological economics demands an encompassing view on the compound system of nature and economy. This is in line with an insight that Hicks in 1985 passed on to one of his students, Stefano Zamagni, now Professor of Economics at the University of Bologna, Italy, "that the concept of income [a main concept of capital theory] would be more useful in connection with exhaustible resources and environmental economics than it had been in the mainstream economics" (Hammond 1994: 193).

However, this change in perspective implies certain criteria for the theoretical and empirical analysis of environmental and resource problems:

1. The need for interdisciplinary cooperation.
2. The consideration of the heterogeneity of production techniques, capital goods, natural resources, species, consumption goods, and pollutants.
3. The explicit recognition of physical constraints of economic activity, in particular the laws of thermodynamics.
4. The consideration of time in two dimensions, the internal time structure of individual processes and the external time structure of the related biodynamic systems, which together determine the time structure of the ecological-economic system.
5. Capacity to deal with uncertainty and real ignorance because of the complexity of the compound system and the long time horizons involved.
6. Questions of intra- and intergenerational equity and responsibility, posed by the strong interconnectedness within and over time.

Obviously, neo-Austrian capital theory in general, with its emphasis on heterogeneity and time, is well suited to analyze resource and environmental problems in line with the demands of ecological economics. Moreover, the approach of BF has two additional favorable prerequisites: (1) the combination of fixed and circulating capital is similar to the "flow-fund" approach applied by Georgescu-Roegen (1971), who introduced thermodynamic considerations into economic analysis; and (2) the activity analysis approach (Koopmans 1951) is, on the one hand, highly flexible and easily amendable and, on the other hand, already incorporates irreversibility.

How have the representatives of neo-Austrian capital theory dealt with these six criteria of ecological economics? It all started with the application of neo-Austrian capital theory to a study on the innovation of wastewater treatment in southern Germany (Faber et al. 1983, 1987; Stephan 1989). It soon became clear that an adequate analysis needed a sound physical underpinning. This underpinning was found in the laws of thermodynamics, specifically, the concept of entropy.[3] The First Law of Thermodynamics states that energy and matter can neither be created nor destroyed. The Second Law of Thermodynamics claims that the entropy in an isolated system cannot decrease.

In particular, this implies that production and consumption are irreversible processes. Entropy, interpreted as the concentration of the resource, built the conceptual foundation for integrating the energy requirement for the resource extraction, the resource use in economic activity, and the emission of pollutants created in the production and consumption processes into an intertemporal, unified framework. This approach made it possible to take into account the physical and intertemporal constraints of the ecological-economic system.

Furthermore, it can be shown that from a thermodynamic point of view all production must result in joint production. That is, each production process produces more than one output (Faber et al. 1998; Baumgärtner et al. 2001; Baumgärtner et al. 2006). Often, these outputs include unintended and harmful substances. This implies for the formal analysis that joint production is the rule rather than the exception. As neo-Austrian capital theory in the tradition of BF uses an activity analysis approach, joint production can easily be incorporated into the model framework (Baumgärtner et al. 2002; Winkler 2005).

Again, it was the activity analysis approach that made it relatively easy to introduce resource extraction, pollution emissions, and abatement into the original BF framework by introducing additional processes. Although these topics are exhaustively analyzed in environmental and resource economics, the neo-Austrian emphases on the vertical time structure and the transition path between steady states led to new insights. Maier (1984) studied the transition of techniques due to the exhaustion of natural resources that are essential and nonsubstitutable for a specific production technique. Speck (1997) took a more encompassing point of view by integrating resource extraction, capital accumulation, recycling, pollution emissions, and abatement into a neo-Austrian model framework.

Moreover, the activity analysis approach naturally suggests another application of neo-Austrian capital theory. Faber and Proops (1991) and Faber et al. (1999: ch. 7) show how to represent the neo-Austrian approach in an input-output framework that is also the basis for national accounts. Incorporating the dynamics of resource extraction and depletion, capital accumulation, and pollution emissions, a national accounting system can be established that accounts for environmental issues. In particular, Samuelson's (1961) and Weitzman's (1976) arguments for considering the consumption stream, rather than GDP, occurs as a natural conclusion.

Jöst (1996), Faber et al. (1999: ch. 9), and Proops (2004) address questions of intra-generational distributive justice in a neo-Austrian framework. They show how neo-Austrian capital theory may be extended to integrate international trade of resources and capital goods. This allows the study of trade and environment relationships between rich (developed) and poor (developing) countries, in particular, the distributional effects of climate change taxes (Jöst 1996) and the trade implications of falling resource rents (Proops 2004).

The long time horizons involved in many environmental and resource studies call into question the usefulness of an intertemporal optimization framework. The assumptions of static expectations and perfect foresight are hardly met in reality because of novelty and complexity. For this reason, so-called rolling myopic plans are used for the decision process. According to these, plans are made for a few periods ahead and are steadily reevaluated as time goes by. This allows flexible adaptation to new circumstances (Faber et al. 1999: 96–98).

However, rolling myopic plans do not solve the problems associated with uncertainty and ignorance. A study of the anatomy of surprise and ignorance (Faber et al. 1992) questions if it is feasible at all to develop "a general tool for the operationalization of ignorance" (Funtowicz and Ravetz 1991: 7).

In our understanding, there is no doubt that formal models can help us in analyzing the complex interconnectedness of ecology-economy interaction (e.g., Winkler 2006a). However, we acknowledge that formal modeling is ultimately limited by complexity; that is, the world is generally too complex to be perceived by humankind with indefinite accuracy. This is especially true because of the occurrence of novelty, that is, the potentialities of a system change in an a priori, unknown, and unforeseeable way (e.g. Winkler 2006b). Nevertheless, we follow Hayek (1972: 33, our translation) "[that] it is high time that we took our ignorance more seriously." In this respect, we sympathize with the Austrian subjectivist school, which emphasizes the importance of uncertainty and real ignorance and, therefore, rejects formal analysis. However, we take a different approach to deal with this issue, as is outlined in our concluding section.

V

Beyond Capital Theory

BÖHM-BAWERK'S ATTITUDE toward economics may be expressed in a nutshell: "In the social science the heart precedes the head. . . . In our days it is the fate of the working class, the emerging fourth estate, which like a magnet attracts the social scientists" (cited in Hennings 1997: 22). During the last decades of the 20th century, we have experienced environmental problems that tend to endanger the natural living conditions of humankind. The environmental crisis is one of the great challenges of the 21st century. Therefore, we may extend Böhm-Bawerk's insight to the problems posed by the environmental crisis.

In our paper, we have shown that neo-Austrian capital theory is a general suitable method to study changes in time. We have demonstrated its potential and its limits, which apply to formal analysis in general. This leads to the following conclusions and insights.

Planning and decision should not be a one-period activity but a continuing process that values and takes care of flexibility. Therefore, intertemporal optimization should be supplemented by myopic optimization, which is open for revision when novelty occurs.

However, uncertainty and ignorance pose severe problems for any analytical framework. Aspects of ignorance have to be given much more priority than in conventional economic analysis, as they are important aspects of contemporary environmental problems; for example, the depletion of the ozone layer by CFCs took humankind completely by surprise. Here, we can learn from the Austrian subjectivists, who have related problems of uncertainty and ignorance to the market system to a greater extent than other economic school of thought. They have stressed again and again that the price mechanism leads individual economic actors to learn and thus induces inventions. For instance, a very effective means of overcoming the scarcity of an exhaustible natural resource is to increase its price. This first leads to a more efficient use of it, finding new exploration sites, and, in the course of time, developing substitutes for it. However, it is not unimportant to note that there exists an asymmetry between natural resources and environmental pollution. This is the case because, in general, a market for scarce natural pollution degradation capacities of the environment does not exist. There is therefore no market price for the latter and, as many examples in the past have shown, it often takes a long time until the harmful effects of an economic activity are recognized by the public; often, it takes even longer for politicians to react. Political action is not restricted to command and control approaches but may also internalize the harmful effect within the market system. To this end, the price system has to be supplemented by levying charges or taxes or by introducing licenses.

Moreover, ignorance in combination with the ubiquitous phenomenon of joint production leads to problems of complexity. This raises the question of responsibility: How can we act responsibly in the face of radical ignorance? Take as an example the production and consumption of CFCs, which have resulted in the depletion of the ozone layer, endangering the basis for our survival. Here we have to differentiate between individual responsibility, which, according to the tradition of political philosophy, always has its limits, and collective responsibility, which is in principle unlimited in respect to all anthropogenic effects on the environment (Baumgärtner et al. 2006: pt. III). If these recognitions are taken seriously, this would imply that politics has to look at nature in a new way. This, in turn, implies for economic theory that the standard economic concept of humankind, the *homo economicus,* has to be supplemented by an additional concept of man, which we call *homo politicus* (Faber et. al. 2002). Ecological economics has been characterized as the "science and management of sustainability" (Costanza 1991). This is a normative aim that cannot be derived from individual or collective welfare, as is the case in environmental economics, which is based on the neoclassical paradigm, as mentioned above. In fact, this normative aim of sustainability has to be derived from the task of securing the basic foundations of living for humankind in the long run. *Homo economicus* and *homo politicus* likewise have their sources in political philosophy. Whereas *homo economicus* refers to the anthropology of Thomas Hobbes, *homo politicus* is founded in a tradition of political philosophy that focuses on human interest in justice and the well-being of the community (Faber et al. 1997, 2002). Thus, at the end of our considerations, we hark back to justice, the central concept of Böhm-Bawerk's motivation and thinking.

Notes

1. In his *Principles of Economics,* Marshall ([1890] 1964) took a combined view, incorporating both marginal utility and marginal productivity, to determine prices and production costs simultaneously. This concept still determines the neoclassical paradigm today.
2. The relationship between the two strands was investigated by Pellengahr (1986a, 1986b). See also Garrison (1980, 2000), Moss (1980), Moldofsky (1982), and Kirzner (1987).
3. In fact, entropy was introduced in economic analysis by Georgescu-Roegen (1971).

References

Baumgärtner, S., H. Dyckhoff, M. Faber, J. Proops, and J. Schiller. (2001). "The Concept of Joint Production and Ecological Economics." *Ecological Economics* 36: 365–372.

Baumgärtner, S., M. Faber, and J. Proops. (2002). "How Environmental Concern Influences the Investment Decision: An Application of Capital Theory." *Ecological Eonomics* 40: 1–12.

Baumgärtner, S., M. Faber, and J. Schiller. (2006). *Joint Production and Responsibility in Ecological Economics: Foundations of Environmental Policy*. Cheltenham, UK: Edward Elgar.

Bernholz, P. (1971). "Superiority of Roundabout Processes and Positive Rate of Interest." *Kyklos* 24: 687–721.

Bernholz, P., and M. Faber. (1973). "Technical Productivity of Roundabout Processes and Positive Rate of Interest: A Capital Model with Depreciation and n-Period Horizon." *Zeitschrift für die gesamte Staatswissenschaft* 129: 46–61.

——. (1976). "Time-Consuming Innovation and Positive Rate of Interest." *Zeitschrift für Nationalökonomie* 36: 347–367.

Bernholz, P., M. Faber, and W. Reiß. (1978). "A Neo-Austrian Two Period Multisector Model of Capital." *Journal of Economic Theory* 17: 38–50.

Bliss, C. J. (1975). *Capital Theory and the Distribution of Income.* Oxford and Amsterdam: North-Holland/American Elsevier.

Bliss, C. J., A. J. Cohen, and G. C. Harcourt (eds.). (2005). *Capital Theory*, vols. I–III. Cheltenham, UK: Edward Elgar.

Böge, W. (1986). "Remarks on a Dynamic Game with Macroeconomic Investment." In *Studies in Austrian Capital Theory: Investment and Time*. Ed. M. Faber. Berlin: Springer.

Böge, W., M. Faber, and W. Güth. (1986). "A Dynamic Game with Macroeconomic Investment Decisions under Alternative Market Structures." In *Studies in Austrian Capital Theory: Investment and Time*. Ed. M. Faber. Berlin: Springer.

Böhm-Bawerk, E. V. ([1889]1921). *The Positive Theory of Capital*. London: Macmillan.

——. (1913a). "Eine dynamische Theorie des Kapitalzinses." *Zeitschrift für Volkswirtschaft, Sozialpolitik und Verwaltung* 22: 1–62.

——. (1913b). "Eine dynamische Theorie des Kapitalzinses. Schlußbemerkung." *Zeitschrift für Volkswirtschaft, Sozialpolitik und Verwaltung* 22: 640–656.

Burmeister, E. (1980). *Capital Theory and Dynamics*. Cambridge: Cambridge University Press.

Costanza, R. (ed.). (1991). *Ecological Economics: The Science and Management of Sustainability*. New York: Columbia University Press.

Dorfman, R. (1959). "Waiting and the Period of Production." *Quarterly Journal of Economics* 73: 351–372.

Faber, M. (1979). *Introduction into Modern Austrian Capital Theory*. Berlin: Springer.

——. (ed.). (1986). *Studies in Austrian Capital Theory: Investment and Time*. Berlin: Springer.

Faber, M., R. Manstetten, and T. Petersen. (1997). "Homo Oeconomicus and Homo Politicus: Political Economy, Constitutional Interest and Ecological Interest." *Kyklos* 50: 457–483.

Faber, M., R. Manstetten, and J. R. Proops. (1992). "Humankind and the Environment: An Anatomy of Surprise and Ignorance." *Environmental Values* 1: 217–241.

——. (1996a). "On the Conceptual Foundations of Ecological Economics: A Teleological Approach." *Ecological Economics* 12: 41–54.

——. (1996b). *Ecological Economics: Concepts and Methods*. Cheltenham, UK: Edward Elgar.

Faber, M., H. Niemes, and G. Stephan. (1983). *Umweltschutz und Input-Output Analyse. Mit zwei Fallstudien aus der Wassergütewirtschaf*. Tübingen: Mohr (Paul Siebeck).

——. (1987). "Entropy, Environment and Resources." In *Physico-Economics*. Heidelberg: Springer.

Faber, M., T. Petersen, and J. Schiller. (2002). "Homo Oeconomicus and Homo Politicus in Ecological Economics." *Ecological Economics* 40: 323–333.

Faber, M., and J. R. Proops. (1991). "National Accounting, Time and the Environment." In *Ecological Economics: The Science and Management of Sustainability*. Ed. R. Costanza. New York: Columbia University Press.
——. (1998). *Evolution, Time, Production and the Environment*, 3rd ed. Heidelberg: Springer.
Faber, M., J. R. Proops, and S. Baumgärtner. (1998). "All Production Is Joint Production: A Thermodynamic Analysis." In *Sustainability and Firms: Technological Change and the Changing Regulatory Environment*. Eds. S. Faucheux, J. Gowdy, and I. Nicolaï. Cheltenham, UK: Edward Elgar.
Faber, M., J. R. Proops, and S. Speck. (1999). *Capital and Time in Ecological Economics*. Cheltenham, UK: Edward Elgar.
Funtowicz, S. O., and J. R. Ravetz. (1991). "Global Environmental Issues and the Emergence of Second Order Science." In *Ecological Economics: The Science and Management of Sustainability*. Ed. R. Costanza. New York: Columbia University Press.
Garrison, R. W. (1979). "In Defense of the Misesian Theory of Interest." *Journal of Libertarian Studies* 3: 141–150.
——. (1980). "Review of Faber (1979), Introduction to Modern Austrian Capital Theory." *Austrian Economics Newsletter* 2: 5, 11.
——. (1985). "A Subjectivist View of a Capital-Using Economy." In *The Economics of Time and Ignorance*. Eds. G. P. O'Driscoll Jr. and M. J. Rizzo with R. W. Garrison. Basil Blackwell, Oxford.
——. (1991). "Austrian Capital Theory and the Future of Macroeconomics." In *Austrian Economics: Perspectives on the Past and Prospects for the Future*. Ed. R. M. Ebeling. Hillsdale: Hillsdale College Press.
——. (2000). "Review of Faber, Proops and Speck (1999), Capital and Time in Ecological Economics: Neo-Austrian Modelling." *Quarterly Journal of Austrian Economics* 3: 85–88.
——. (2005). "The Austrian School." In *Modern Macroeconomics: Its Origins, Development and Current State*. Eds. B. Snowdon and H. R. Vane. Aldershot, UK: Edward Elgar.
Georgescu-Roegen, N. (1971). *The Entropy Law and the Economic Process*. Cambridge: Harvard University Press.
Hammond, P. (1994). "Is There Anything New in the Concept of Sustainable Development?" In *The Environment After Rio: International Law and Economics*. Eds. L. Campiglio, L. Pineschi, D. Siniscalco, and T. Treves. London: Graham & Trotman.
Harcourt, G. C. (1972). *Some Cambridge Controversies in the Theory of Capital*. Cambridge: Cambridge University Press.
Hayek, F. A. (1934). "On the Relationship Between Investment and Output." *Economic Journal* 44: 207–231.
——. (1936). "The Mythology of Capital." *Quarterly Journal of Economics* 50: 199–228.

——. (1972). *Die Theorie komplexer Phänomene*. Tübingen: Mohr (Paul Siebeck).

Hennings, K. H. (1997). *The Austrian Theory of Value and Capital.* Cheltenham, UK: Edward Elgar.

Hicks, J. R. *Capital and Time: A Neo-Austrian Approach*. Oxford: Clarendon Press.

——. (1985). *Methods of Dynamics.* Oxford: Oxford University Press.

Jevons, W. S. (1871). *The Theory of Political Economy*. London: Macmillan.

Jöst, F. (1996). "Climate Change and Economic Development: A Neo-Austrian Approach." In *Models of Sustainable Development*. Eds. S. Faucheux, D. Pearce, and J. R. Proops. Cheltenham, UK: Edward Elgar.

Kirzner, I. M. (1963). *Market Theory and the Price System*. New York: Van Nostrand.

——. (1976). "Ludwig von Mises and the Theory of Capital and Interest." In *The Economics of Ludwig von Mises: Towards a Critical Reappraisal*. Ed. L. Moss. Kansas City: Sheed Andrews and McMeel

——. (ed.). (1982). *Method, Process and Austrian Economics: Essays in Honour of Ludwig von Mises*. Lexington: Lexington Books.

——. (1986). *Subjectivism, Intelligibility and Economic Understanding: Essays in Honor of Ludwig M. Lachmann on his Eightieth Birthday.* New York: New York University Press.

——. (1987). "Austrian School of Economics." In *The New Palgrave: A Dictionary of Economics.* Eds. J. Eatwell, M. Mulgate, and P. Newman. London: Macmillan, London.

——. (1993). "The Pure Time Preference Theory of Interest: An Attempt at Clarification." In T*he Meaning of Ludwig von Mises: Contributions in Economics, Sociology, Epistemology, and Political Philosophy*. Ed. J. M. Herbener. Boston: Kluwer Academic Publishers.

——. (1995). "The Subjectivism of Austrian Economics." In *New Perspectives on Austrian Economics*. Ed. G. Meijer. London: Routledge.

——. (1996). *Essays on Capital and Interest*. Cheltenham, UK: Edward Elgar.

Knight, F. A. (1934). "Capital, Time and the Interest Rate." *Economics* 1: 257–286.

——. (1935). "Professor Hayek and the Theory of Investment." *Economic Journal* 45: 75–94.

——. (1936a). "The Quantity of Capital and the Rate of Interest: Part I." *Journal of Political Economy* 44: 433–463.

——. (1936b). "The Quantity of Capital and the Rate of Interest: Part II." *Journal of Political Economy* 44: 612–642.

Koopmans, T. C. (1951). "Analysis of Production as an Efficient Combination of Activities." In *Activity Analysis of Production and Allocation*. Ed. T. C. Koopmans. New York: Wiley.

Lachmann, L. M. (1973). *Macro-Economic Thinking and the Market Economy.* Menlo Park, CA: Institute for Humane Studies.

——. (1977). *Capital, Expectations and the Market Process.* Kansas City: Sheed Andrews and McMeel.

——. (1986). *The Market as an Economic Process.* Oxford: Basil Blackwell.

Machlup, F. (1955). "Professor Knight and the Period of Production." *Journal of Political Economy* 43: 577–624.

Maier, G. (1984). *Rohstoffe und Innovation: Eine dynamische Untersuchung.* Königstein: Athenäum.

Malinvaud, F. (1953). "Capital Accumulation and Efficient Allocation of Resources." *Economoetrica* 21: 233–286.

Marx, K. ([1867–1894] 1965–1967). *Capital.* Moscow: Progress Publishers.

Menger, C. (1871). *Grundzüge der Volkswirtschaftslehre.* Wien: Hölder-Pichler-Tempsky.

Moldofsky, N. (1982). "Review of Faber (1979), Introduction to Modern Austrian Capital Theory." *Economic Record* 58: 295–297.

Moss, L. S. (ed.). (1976). *The Economics of Ludwig von Mises: Towards a Critical Reappraisal.* Kansas City: Sheed Andrews and McMeel.

——. (1980). "Review of Faber (1979), Introduction to Modern Austrian Capital Theory." *Journal of Economic Literature* 18: 1095–1098.

O'Driscoll, G. P., and M. J. Rizzo. (1985). *The Economics of Time and Ignorance.* Oxford: Basil Blackwell.

Pellengahr, I. (1986a). "Austrians Versus Austrians I: A Subjectivist View of Interest." In *Studies in Austrian Capital Theory: Investment and Time.* Ed. M. Faber. Berlin: Springer.

——. (1986b). "Austrians Versus Austrians II: Functionalist Versus Essentialist Theories of Interest." In *Studies in Austrian Capital Theory: Investment and Time.* Ed. M. Faber. Berlin: Springer.

Proops, J. R. (2004). "The Growth and Distributional Consequences of International Trade in Natural Resources and Capital Goods: A Neo-Austrian Analysis." *Ecological Economics* 48: 83–91.

Reiß, W. (1979). "Substitution in a Neo-Austrian Model of Capital." *Zeitschrift für Nationalökonomie* 39: 33–52.

——. (1981). *Umwegproduktion und Positivität des Zinses: Eine neo-österreichische Analyse.* Berlin: Duncker & Humbolt.

Reiß, W., and M. Faber. (1986). "Own Rates of Interest in a General Multi-sector Model of Capital." In *Studies in Austrian Capital Theory: Investment and Time.* Ed. M. Faber. Berlin: Springer.

Rothbard, M. N. (1962). *Man, Economy and State.* Princeton: van Nostrand.

——. (1976a). "The Austrian Theory of Money." In *The Foundations of Modern Austrian Economics.* Ed. E. G. Dolan. Kansas City: Sheed Andrews and McMeel.

——. (1976b). "New Light on the Prehistory of the Austrian School." In *The Foundations of Modern Austrian Economics.* Ed. E. G. Dolan. Kansas City: Sheed Andrews and McMeel.

——. (1977). "Introduction." In *Frank A. Fetter: Capital, Interest and Rent. Essays in the Theory of Distribution*. Ed. M. N. Rothbard. Kansas City: Sheed Andrews and McMeel.

——. (1990). "Time Preference." In *Capital Theory*. Eds. J. Eatwell, M. Milgate, and P. Newman. New York: W. W. Norton & Company.

——. (1993). "Mises and the Role of the Economist in Public Policy." In *The Meaning of Ludwig von Mises*. Ed. J. M. Herbener. Norwell, MA: Kluwer Academic Publishers.

——. (1995). "The Present State of Austrian Economics." *Journal des Economistes et des Etudes Humaines* 6: 43–89.

Samuelson, P. A. (1961). "The Evaluation of 'Social Income': Capital Formation and Wealth." In *The Theory of Capital*. Eds. F. A. Lutz and D. C. Hague. London: Macmillan.

——. (1966). "A Summing Up." *Quarterly Journal of Economics* 80: 568–583.

Schumpeter, J. A. ([1911] 1941). *The Theory of Economic Development*. Cambridge: Harvard University Press.

——. (1913). "Eine dynamische Theorie des Kapitalzinses. Eine Entgegnung." *Zeitschrift für Volkswirtschaft, Sozialpolitik und Verwaltung* 22: 599–639.

——. (1954). *History of Economic Analysis*. London: George Allen & Unwin.

Smith, A. ([1776] 1976). *An Inquiry into the Nature and the Causes of the Wealth of Nations*. Oxford: Clarendon Press.

Speck, S. (1997). "A Neo-Austrian Five Process Model with Resource Extraction and Pollution Abatement." *Ecological Economics* 21: 91–103.

Stackelberg, H. V. (1941). "Elemente einer dynamischen Theorie des Kapitals (Ein Versuch)" ["Elements of a Dynamic Theory of Capital (A Trial)"]. *Archiv für mathematische Wirtschafts- und sozialforschung* 7: 8–29, 70–73.

——. (1941/43). "Kapital und Zins in der stationären Verkehrswirtschaft" ["Capital and Interest in the Stationary Market Economy"]. *Zeitschrift für Nationalökonomie* 10: 25–61.

Stephan, G. (1983). "Roundaboutness, Nontightness and Malinvaud Prices in Multisector Models with Infinite Horizon." *Zeitschrift für die gesamte Staatswissenschaft* 131: 660–677.

——. (1985). "Competitive Finite Value Prices: A Complete Characterization." *Zeitschrift für Nationalökonomie* 45: 35–45.

——. (1989). *Emission Standards, Substitution of Techniques and Imperfect Markets: A Computable Equilibrium Approach*. Heidelberg: Springer.

——. (1995). *Introduction into Capital Theory: A Neo-Austrian Perspective*. Berlin: Springer.

Thadden, E-L. V. (1986). "A Dynamic Macroeconomic Investment Game with Non-Linear Saving Behaviour." In *Studies in Austrian Capital Theory: Investment and Time*. Ed. M. Faber. Berlin: Springer.

Walras, L. ([1874/1877] 1954). *Elements of Pure Economics.* London: Allen & Unwin.

Weitzman, M. (1976). "Welfare Significance of National Product in a Dynamic Economy." *Quarterly Journal of Economics* 90: 156–162.

Weizsäcker, C. C. V. (1971). *Steady State Capital Theory.* Berlin: Springer Verlag.

Winkler, R. (2003). *Zeitverzögerte Dynamik von Kapital- und Schadstoffbeständen: Eine österreichische Perspektive.* Marburg: Metropolis.

——. (2005). "Structural Change with Joint Production of Consumption and Environmental Pollution: A Neo-Austrian Approach." *Structural Change and Economic Dynamics* 16: 111–135.

——. (2006a). "Valuation of Ecosystem Goods and Services. Part 1: An Integrated Dynamic Approach." Forthcoming in *Ecological Economics.*

——. (2006b). "Valuation of Ecosystem Goods and Services. Part 2: An Integrated Dynamic Approach." Forthcoming in *Ecological Economics.*

Winkler, R., U. Brandt-Pollmann, U. Moslener, and J. Schlöder. (2005). "On the Transition from Instantaneous to Time-Lagged Capital Accumulation. The Case of Leontief-Type Production Functions." Discussion Paper No. 05-30, Centre for European Economic Research (ZEW), Mannheim.

Reconciling Gray and Hotelling

Lessons from Early Exhaustible Resource Economics

By Richard J. Brazee and L. Martin Cloutier*

Abstract. Early exhaustible resource economics provides an important foundation for recent suggestions that firm-level economic modeling plays a larger role in the analysis of resource scarcity. The lack of empirical support for Hotelling's *r*-percent rule, introduced in 1931, and recent suggestions that industry behavior may not be reducible to firm behaviors are the primary motivating factors for examining the relative value of Gray's contribution to the field of exhaustible resource economics relative to Hotelling's contribution. Specifically, Gray's papers that appeared in the 1910s provide insight into the heterogeneity of deposits and their spatial dimensions, and offer the possibility that firms will be subject to fixed costs carried over between periods. In this paper, the arguments presented by Gray are formalized in a dynamic model, which allows the differences between Gray's and Hotelling's assumptions to be more fully explored. The results of the paper illustrate that by considering spatially identifiable heterogeneous deposits, fixed costs, and entry costs, in general Hotelling's *r*-percent rule is not a sufficient condition for firm-level decision making and that firms' extraction behavior cannot be linearly aggregated to describe industry behavior.

*Richard J. Brazee is an Associate Professor in the Department of Natural Resources and Environmental Sciences at the University of Illinois at Urbana-Champaign, Urbana, IL; e-mail: brazee@uiuc.edu. L. Martin Cloutier is an Associate Professor in the Department of Management and Technology at the University of Quebec at Montreal, Montreal, QC; e-mail cloutier.martin@uqam.ca. We are grateful to the workshop participants of the program in Environmental and Resource Economics (pERE) at the University of Illinois at Urbana-Champaign for helpful comments.

American Journal of Economics and Sociology, Vol. 65, No. 3 (July, 2006).

I

Introduction

CONSIDERABLE TIME AND ENERGY have been devoted into the economic study of exhaustible resources (ER) since the early 20th century. The results of these efforts provide a basis for a modern field in ER economics. Like progress in many scholarly fields, ER economics research is characterized by waves in which efforts increase, peak, decrease, bottom out, and then rise again. One such peak of effort came shortly after the 1973 oil embargo of the United States. During this peak, the paper by Harold Hotelling (1931) was identified as the seminal work from earlier periods that had stood the test of time and provided a solid foundation for ER economics: "Hotelling's elegant and comprehensive analysis, his wide ranging conjectures and asides, make the 1931 analysis very nearly the sole source of work in a vigorously growing branch of economics" (Devarajan and Fisher 1981: 71).

The contribution of Lewis Cecil Gray (1913, 1914) to the field of ER economics has only recently begun to be more fully recognized by economists (see, for example, Crabbé 1983 and Smith 1986). The acknowledgment of Gray's work, however, comes at a cost: the literature often bundles and characterizes his contributions as similar to Hotelling's contributions. Nevertheless, the contributions of Gray and Hotelling, other than focusing on ER economics, are significantly different and cannot, except perhaps in special cases, be thought of as a unified economic argument. As a result, the potential for applying ideas elucidated by these respective frameworks that support empirical applications of the economics of ER has not been fully developed in previous research. In this paper, we submit that Gray's and Hotelling's frameworks have the potential to provide primarily distinct contributions, and that this awareness could influence empirical research or the search for evidence in support of Hotelling's r-percent rule. Hotelling's r-percent rule states that during a period of continuous extraction, the price of a resource over time would follow an equilibrium path equivalent to the value society places on the remaining resource stock. Thus, the price gap between the price and the marginal cost of extraction over time would rise with the discount rate. Cairns (1994) notes that Gray's work, by contrast, is quantity, or output, focused, and incorporates a possibility of a firm's adjusting its rate of production, so the difference between the anticipated price and the marginal cost, assuming a U-shaped cost curve, would increase with the rate of interest.

The aim of this paper is twofold: first, to explore a reevaluation of Gray's relative value to Hotelling's and, second, to use a model as an illustration that helps establish key distinctions between Gray and Hotelling. The remainder of the paper is structured as follows. In Section II, conceptual issues in the economics of ER associated with Gray and Hotelling are surveyed, and general differences between the works by Gray and Hotelling are highlighted. A model, formally examining the implications of Gray's assumptions in contrast to Hotelling's asssumptions, which have not been explored before, is introduced. This model, presented in Section III, is in fact Gray's example of spatially identifiable, heterogeneous deposits with fixed costs. Results and policy implications are presented in Section IV.

II

Concepts in Economics of Exhaustible Resources: Gray and Hotelling

GRAY'S WORK has been associated more with classical economics, although his contribution is mostly neoclassical due to the marginal analysis presented and the emphasis of economic substitution in input use (Crabbé 1983). The literature refers to one or both authors, and only a few contributions have contrasted differences between the analyses presented by Gray and Hotelling (Crabbé 1983, 1986; Smith 1986; Cairns 1994). Hotelling's theoretical framework had a major impact in the early 1970s, about 40 years after its publication, at a time when economists were focused on global resource issues and the management of natural resources was on the verge of becoming a full-fledged research field (for example, see Meadows et al. 1972). At the time, economists were interested in industry-level analyses and engaged in the econometric estimation of aggregate production functions (Cloutier and Rowley 2003). The profession's increased reliance on abstract mathematical models also contributed to the appeal and diffusion of Hotelling's readily available calculus-of-variations framework among economists who were learning optimal control theory and dynamic programming. The dynamic modeling work and the *r*-percent rule of Hotelling have paved the way for research strands on various dimensions of ER economics (see, for example, Gordon 1966, 1967). In recent years, contributions in the field have offered only limited help to the management of ER, and some have even concluded that this stream of research has reached an "impasse" (Cairns 1994).

By contrast to Hotelling's mathematical formulation, Gray's concepts were presented in words, and his economic analyses and assumptions regarding the conservation of the resource were illustrated with simple, yet intuitive, numerical examples. The recent attention to Gray's work is motivated in part by the lack of, or the existence of at best only mixed, empirical support for Hotelling's *r*-percent rule (Barnett and Morse 1963; Slade 1982; Farrow 1985; Miller and Upton 1985; Withagen 1988; Halvorsen and Smith 1991; Berck and Roberts 1996; Chermak and Patrick 2000, 2002a, 2002b). Although Gray's impact on the literature to date is modest compared to Hotelling's, in recent years, Gray's value from an ER economics standpoint may have increased relative to Hotelling's. The divergence between social and individual decision making were clearly emphasized by Gray (1913), although he did not conduct a formal analysis of the heterogeneity of deposits or of information asymmetry in support of his idea of conservation. The need to better understand firm-level decision making, the spatial dimension of deposits in the hands of the mine owner, warrants further formal exploration of Gray's ideas that, in contrast to Hotelling, insist on the firm as a distinct unit of analysis. Crabbé (1983, 1986) and Cairns (1994) have suggested refocusing the economic analysis to the firm level, which could encompass the heterogeneity of the resource, its exploration and development, and the asymmetry of information in contracts, but no formal analysis or indication of how this could be accomplished has been presented.

Hotelling's analysis suggests that firms and industries behave identically. In the basic Hotelling model, the linear aggregation of firms into an industry is immediate, through the assumption of identical firms and arbitrage arguments. This makes for a very "tight" link between the mine owner and the industry, reducing one to the other. To achieve identical behavior of both the firm and the industry levels requires the homogeneity of individual decision-making units, homogeneity of deposits, and perfect arbitrage. If, as discussed in Gray (1914: 477), deposits are of heterogeneous quality and the spatial component of extraction of the ER is considered, the firm-industry aggregation is not so straightforward. With heterogeneous starting times or heterogeneous ordering of deposit quality or "mines" with different distributions of deposit quality, Hotelling's rule cannot be expected to hold because the scarcity rent differs between mines.

These observations lead us to a more explicit examination of Gray's assumptions and to the individual firm's decision-making problem. In particular, we identify three assumptions in Gray's work that differ from Hotelling's assumptions:

1. Deposits of exhaustible resources may be of different quality;
2. There is a spatial component to the extraction of exhaustible resources, in the sense that deposits are removed in a spatial order rather than in quality order; and
3. The cost function in any period is likely to be U-shaped.

First, in the basic Hotelling model with all deposits of the same quality, with the possible exception of increasing or decreasing marginal costs, the spatial sequence of deposit extraction has no impact on the firm or industry extraction path. Hartwick addresses in detail the quality or heterogeneity of the deposit raised in Assumption 1. When deposits vary in quality or grade and all deposits are accessible, deposits are extracted in order of quality, grade, or least cost (Hartwick 1978). The conclusion that Hartwick demonstrates through analyzing arbitrage and comparative statics is that, at any point in time, the best quality (that is, the least-cost) deposits will be removed first.

Second, unlike Assumption 2, Hartwick's analysis assumes that all deposits are immediately accessible; that is, there is no spatial component to the extraction of the ER. Gray (1914: 476–477) explicitly addresses the issue and comments on "the possibility that the removal of coal in the first year may change entirely the conditions of removal in the second year. . . . This may be true because of greater depth or special difficulties encountered, such as water, gas or the thinness of the vein of coal." However, as Gray conjectured decades earlier, with a spatial component to extraction, Hartwick's conclusion will not in general hold. Indeed, when deposit quality varies and all deposits are not immediately accessible, in general, the individual firm's extraction path must change. Gray also conjectures that the change in the extraction path arises from firms' responses; that is, firms will respond by basing current extraction decisions not only on the quality of the currently extracted deposit but also the quality of deposits that may be extracted in the future, provided the current deposit and more accessible future deposits are extracted first. Deposits of different grades and currently inaccessible deposits should be incorporated directly into the value function of the firm that values future extraction.

Finally, Assumption 3 implies that a firm's average costs in any period are U-shaped rather than just a function of variable costs, as in a standard Hotelling model. The assumption of the U-shaped curve implies that firms make different decisions regarding the production level in any one period. Average costs may be made U-shaped by both adding fixed costs and increasing marginal costs in a period. The inclusion of fixed costs may influence the firm's decision to shut down in a period. If a firm is extracting at time t with fixed costs, at $t + 1$ it will produce whenever total revenues exceed total variable costs. In general, the inclusion of increasing marginal costs in a period impacts a firm's decisions regarding quantity extracted in any period.[1]

In the next section, we propose a simple model of exhaustible resources that focuses on the introduction of spatially identifiable heterogeneity in deposits, fixed costs, and startup costs incurred by firms. This analysis integrates dimensions introduced by Gray that have been underemphasized in previous analyses that followed from Hotelling's work, including clarification of the firm and industry extraction paths.

III

The Model

In this section, we present a model that highlights Gray's assumptions discussed in the previous section. To more completely compare the impacts of Gray's assumptions with Hotelling's assumptions, we include startup costs; that is, if a firm does not extract in period t and decides to extract in period $t + 1$, the firm incurs an additional cost. Although not necessary to produce U-shaped average costs, startup costs preserve U-shaped average costs and provide a measure of symmetry between decisions to enter and exit.

Let stock size be $S(i)$, where $S(i)$ is the stock size at level i. i runs from 0 to I, where $S(I)$ is the smallest allowable stock size. Since Gray assumes that deposits are extracted in order, I places each deposit or fraction of a deposit in the order of extraction. That is, if $i1 < i2$, then the $i1$ deposit must be extracted before the $i2$ deposit. S_0 is the initial stock size, and $S(i) \in [S_I, S_0]$.

In addition to being defined by stock level, stock size may also be defined by time. Let X_t be the stock size at time t. Note that for each t, $X_t = S(i)$ for some i. For clarity and notational ease, $S(i)$ is assumed to be a continuous function, while X_t is assumed to be a discrete function.

The equation of motion may be defined in terms of stock or time:

$$X_{t+1} = X_t - Z_t, \text{ or } S(i2) = S(i1) - Z_t, \tag{1}$$

where Z_t is the quantity of the stock extracted at time t, $i1 < i2$, $X_t = S(i1)$, and $X_{t+1} = S(i2)$.

Each level and interval of the stock is of a specific quality grade. Let $g(S(i))$ be the grade at stock level i, and $G(S(i2),S(i1))$ be the grade on interval $i1$ to $i2$. We assume that $g(S(i))$ and $G(S(i2),S(i1))$ exist for every stock level and every interval between 0 and I. For simplicity, we also assume $g((S(i))$ and $G(S(i2),S(i1))$ range on $[G_{\min}, G_{\max}]$, where $G_{\min}$ is the minimum possible grade and $G_{\max}$ is the maximum possible grade. For many resources, grades on an interval $[i1, i2]$ will be the average grade between $S(i1)$ and $S(i2)$; that is:

$$G(S(i2), S(i1)) = \frac{1}{S(i1) - S(i2)} \int_{S(i2)}^{S(i1)} g(S(i))di.$$

Q_t is defined as a combined measure of quantity and quality extracted during period t as a function of deposit quality, initial stock level at time t, and quantity of ore extracted at time t, that is, $Q_t = Q(G(X_{t-1}, X_t), Z_t)$. Q_t is a measure of how much ore or "effective" resource material quantity is extracted during period t. In many cases, Q_t will equal grade quality as percentage of ore or resource material in the deposit extracted multiplied by amount of ore extracted; that is, $Q_t = G(X_t - Z_t, X_t) \cdot Z_t$, where from Equation (1), $X_t - Z_t$ replaces X_{t+1}. This implies, $Q_t = \int_{S(i2)}^{S(i1)} g(S(i))dS(i)$. For simplicity, we make this assumption.

The firm's goal is to maximize the net present value of its deposits. The firm faces two decisions at every time t. First, the firm must decide whether to extract during period t. Second, if the firm decides to extract at time t, it must decide how much to extract. The firm's problem is solved using dynamic programming to first determine what level should be extracted during a period of extraction, and then to determine whether it should extract during the period. The firm's general objective to determine the optimal level of extraction is defined as:

$$V(X_t, t) = \underset{w.r.t.Z_t}{Maximium}\, P_t Q_t - C(Z_t) + \beta V(X_{t+1}, t+1), \qquad (2)$$

where β is the discount factor, that is, $\beta = 1/1 + r$, with r being the firm's discount rate; $C(Z_t)$ is total costs at time t; P_t is the price of ore at time t; and $V(X_t, t)$ is a value function based upon stock size and time period. The firm finds the optimal level of Z_t by maximizing the sum of total revenue at time t minus the costs incurred at time t plus the discounted value of the stock at time $t + 1$.

The firm's total costs are:[2]

$$C(Z_t) = FC + VC(Z_t) + EC, \qquad (3)$$

where FC are fixed costs; $VC(Z_t)$ are the variable costs of extracting Z_t at time t; and EC are startup or entry costs. By assumption, variable costs are increasing, that is, $dVC(Z_t)/dZ_t > 0$ and $dVC(Z_t)^2/dZ_t^2 > 0$, which implies that total costs are increasing, $dC(Z_t)/dZ_t > 0$ and $dC(Z_t)^2/dZ_t^2 > 0$. We add startup costs to Gray's assumptions for symmetry with fixed costs and to more fully illustrate entry and exit decisions. For simplicity, we assume that fixed costs last one period, that is, if $Z_{t-1} > 0$, then $FC > 0$ at time t, regardless of whether Z_t equals

0 or is positive.[3] Similarly, if $Z_{t-1} = 0$ and $Z_t > 0$, then $EC > 0$ at time t. If $Z_{t-1} > 0$, then EC always equals 0 at time t. That is, fixed costs are incurred during a period in which extraction is positive and during the period after extraction stops; startup costs are incurred during periods in which extraction is positive and in which there was no extraction in the previous period.

With respect to fixed costs and entry costs, four combinations for period $t - 1$ and t fully describe costs in period t. Let N represent a nonmining period in which extraction does not occur, and let M denote a mining period in which extraction is positive; these four combinations are presented in Table 1.

If extraction did not occur in periods $t - 1$ and t, then the firm incurs no costs at time t, while if extraction is positive in period t, then the firm incurs both entry costs and fixed costs. If extraction was positive in period $t - 1$, the firm always incurs fixed costs in period t, but never incurs startup costs in period t.

The existence of either fixed costs or entry costs introduces two complications to the solution process for the objective function defined in Equation (2). First, the objective function in Equation (2),

Table 1

Characterization of Fixed and Entry Costs in Two Consecutive Periods

ID	$t - 1$	t	Z_t	Characterization	$C(Z_t)$
I	N	N	0	Two consecutive nonmining periods	0
II	M	N	0	A mining period followed by a nonmining period	FC
III	N	M	+	A nonmining period followed by a mining period	$FC + VC(Z_t) + EC$
IV	M	M	+	Two consecutive mining periods	$FC + VC(Z_t)$

$V(X_t, t)$, becomes path dependent based on whether extraction was positive or zero in the previous period. That is, fixed costs and startup costs link total costs, and hence, net revenue in period t to the firm's extraction decision in period $t-1$. To solve for the optimal level of extraction when positive extraction is optimal, the objective function in Equation (2) is separated into two objective functions:

$$V(X_t, t|N) = \underset{w.r.t. Z_t}{Maximium}\ P_tQ_t - EC - FC - VC(Z_t) + \beta V(X_{t+1}, t+1|N), \tag{4A}$$

$$V(X_t, t|M) = \underset{w.r.t. Z_t}{Maximium}\ P_tQ_t - FC - VC(Z_t) + \beta V(X_{t+1}, t+1|M), \tag{4B}$$

where $V(X_t, t|N)$ is the value function in period t based upon stock size and zero extraction in period $t-1$, and $V(X_t, t|M)$ is the value function in period t based upon stock size and positive extraction in the previous period.

Second, the existence of fixed and entry costs introduces a discontinuity in the first derivative of costs, and hence the value function, at $Z_t = 0$, that is:

$$\lim_{Z_t \to 0} \frac{\partial V(X_t, t|N)}{\partial Z_t} \neq \left. \frac{\partial V(X_t, t|N)}{\partial Z_t} \right|_{Z_t = 0} \tag{5A}$$

$$\lim_{Z_t \to 0} \frac{\partial V(X_t, t|M)}{\partial Z_t} \neq \left. \frac{\partial V(X_t, t|M)}{\partial Z_t} \right|_{Z_t = 0}. \tag{5B}$$

To determine a maximum, given the discontinuity in the first derivative of the value function at $Z_t = 0$, requires a direct comparison of the optimal level of positive extraction found by standard derivative methods with the value of zero extraction.

Constraints on the choice variables are important in the analysis of mining eras. We assume that the quantity of stock extracted is nonnegative and bounded by the maximum allowable extraction in period t:

$$\forall t, \quad 0 \leq Z_t \leq Z_t^{\max} \leq X_t, \tag{6}$$

where $Z_t^{\max}$ is maximum allowable extraction in period t.

The first step of the optimization is to maximize the value functions in Equations (4A) and (4B) subject to the stock size at time t,

Equations (1), (3), and (6). By substituting for X_{t+1} from Equation (1), and for Q_t from its definition into Equations (4A) and (4B), the firm's objective functions become:

$$V(X_t, t|N) = \underset{w.r.t.Z_t}{Maximium}\ P_t \int_{X_t - Z_t}^{X_t} g(S(i))dS(i) - EC - FC - VC(Z_t) + \beta V(X_t - Z_t, t+1|N) \tag{7A}$$

$$V(X_t, t|M) = \underset{w.r.t.Z_t}{Maximium}\ P_t \int_{X_t - Z_t}^{X_t} g(S(i))dS(i) - FC - VC(Z_t) + \beta V(X_t - Z_t, t+1|M). \tag{7B}$$

Including the constraints for Z_t from Equation (6) in the firm's objective functions and forming Langrangean functions yields:

$$L(Z_t, \delta_t^N, \gamma_t^N|N) = P_t \int_{X_t - Z_t}^{X_t} g(S(i))dS(i) - EC - FC - VC(Z_t) + \beta V(X_t - Z_t, t+1|N) + \delta_t^N Z_t + \gamma_t^N (Z_t^{\max} - Z_t) \tag{8A}$$

$$L(Z_t, \delta_t^M, \gamma_t^M|M) = P_t \int_{X_t - Z_t}^{X_t} g(S(i))dS(i) - FC - VC(Z_t) + \beta V(X_t - Z_t, t+1|M) + \delta_t^M Z_t + \gamma_t^M (Z_t^{\max} - Z_t), \tag{8B}$$

where δ_t^M, δ_t^N, γ_t^M, and γ_t^N are the Langrangean multipliers on the nonnegativity and maximum constraints for Z_t, respectively.

The firm's optimal extraction level in period t, if extraction is positive, is found by totally differentiating Equations (8A) and (8B) with respect to Z_t, setting the result equal to 0, and including the conditions for the constraints:[4]

$$\frac{\partial L(Z_t, \delta_t^N, \gamma_t^N|N)}{\partial Z_t} = P_t g(X_t - Z_t) - \frac{dVC(Z_t)}{\partial Z_t} - \frac{\beta \partial V(X_t - Z_t, t+1|N)}{\partial X_{t+1}} + \delta_t^N - \gamma_t^N = 0$$

$$Z_t \geq 0; \quad \delta_t^N \geq 0; \quad Z_t \cdot \delta_t^N = 0; \quad Z_t^{\max} \geq Z_t \quad \gamma_t^N \geq 0; \quad Z_t \cdot \gamma_t^N = 0; \tag{9A}$$

$$\frac{\partial L(Z_t, \delta_t^M, \gamma_t^M|M)}{\partial Z_t} = P_t g(X_t - Z_t) - \frac{dVC(Z_t)}{\partial Z_t}$$
$$- \frac{\beta \partial V(X_t - Z_t, t+1|M)}{\partial X_{t+1}} + \delta_t^M - \gamma_t^M = 0$$
$$Z_t \geq 0; \quad \delta_t^M \geq 0; \quad Z_t \cdot \delta_t^M = 0; \quad Z_t^{\max} \geq Z_t \quad \gamma_t^M \geq 0; \quad Z_t \cdot \gamma_t^M = 0. \tag{9B}$$

Note that the right-hand sides of Equations (9A) and (9B) are identical. The identical right-hand sides follow directly from the fact that fixed costs and constant entry costs do not impact interior marginal decisions. If the constraints are nonbinding, then $\delta_t^M = \delta_t^N = \gamma_t^M = \gamma_t^N = 0$. This implies that extraction will be between 0 and $Z_t^{\max}$, that is, $Z_t^* \in (0, Z_t^{\max})$, where Z_t^* is the positive level of extraction that solves Equations (9A) and (9B); the marginal revenue product of extraction equals the sum of marginal costs in period t and marginal user costs from extraction in future periods. If the capacity constraints are binding, then $Z_t^* = Z_t^{\max}$ and $\gamma_t^M = \gamma_t^N > 0$; the marginal revenue product of extraction exceeds the sum of marginal costs in period t and marginal user costs from extraction in future periods. If the non-negativity constraints are binding on extraction, then $Z_t^* = 0$, $\delta_t^M = \delta_t^N > 0$; the marginal revenue product of extraction is less than the sum of marginal costs in period t and marginal user costs from extraction in future periods.

Since the firm faces fixed costs and exit costs after the optimal level of extraction is determined for periods in which extraction is positive, the firm needs to decide whether to extract or not during period t. $V(X_t, t|N)$ or $V(X_t, t|M)$ need to be compared to $BV(X_t, t+1|N)$, the value function at time $t+1$, if extraction is zero in period t. Substituting the optimal level of extraction when extraction is positive, Z_t^*, from Equations (9A) and (9B) into the objective functions of Equations (7A) and (7B) produces:

$$V(X_t, t) = \arg\max[V(X_t, t|N), B(VX_{t+1}, t+1|N) \tag{10A}$$

$$V(X_t, t) = \arg\max[V(X_t, t|M), B(VX_{t+1}, t+1|N), \tag{10B}$$

where $V(X_t, t|N)$

$$= P_t \int_{X_t - Z_t}^{X_t} g(S(i))dS(i) - EC - FC - VC(Z_t^*) + \beta V(X_t - Z_t^*, t + 1|M)$$

and $V(X_t, t|M) = P_t \int_{X_t - Z_t}^{X_t} g(S(i))dS(i) - FC - VC(Z_t^*)$

$+ \beta V(X_t - Z_t^*, t + 1|M)$.

The value of the remaining stock at time t is the maximum of the firm's decision to mine or not in period t. If $Z_t^* = 0$, then the solution to Equations (10A) and (10B) are trivial, that is, $Z_t^{**} = Z_t^* = 0$, where Z_t^{**} is the level of extraction that maximizes $V(X_t, t)$.

Conditions (10A) and (10B) each offer three possibilities:

$$\begin{aligned} &i)\ V(X_t, t|N) < B(VX_{t+1}, t+1|N) \quad Z_t^{**} = 0 \\ &ii)\ V(X_t, t|N) = B(VX_{t+1}, t+1|N) \quad 0 < Z_t^{**} < Z_t^{\max} \\ &iii)\ V(X_t, t|N) > B(VX_{t+1}, t+1|N) \quad Z_t^{**} = Z_t^{\max} \end{aligned} \quad (11A)$$

$$\begin{aligned} &i)\ V(X_t, t|M) < B(VX_{t+1}, t+1|N) \quad Z_t^{**} = 0 \\ &ii)\ V(X_t, t|M) = B(VX_{t+1}, t+1|N) \quad 0 < Z_t^{**} < Z_t^{\max} \\ &iii)\ V(X_t, t|M) > B(VX_{t+1}, t+1|N) \quad Z_t^{**} = Z_t^{\max} \end{aligned} \quad (11B)$$

When condition (*i*) of Equation (11A) or (11B) holds, the value of extraction in period t for all positive values of Z_t^* is less than the discounted value of the current stock in period $t + 1$, and no extraction is optimal; that is, $Z_t^{**} = 0$. When condition (*iii*) of Equation (11A) and (11B) holds, the value of extraction in period t for all positive values of Z_t^* is greater than the discounted value of the current stock in period $t + 1$, and the stock is extracted as fast as possible; that is, $Z_t^{**} = Z_t^{\max}$. When condition of (*ii*) of Equation (11A) or (11B) holds, the net revenue from the extraction of the marginal unit of stock extracted in period t plus the discounted value of the remaining stock next period equals the discounted value of the current stock period $t + 1$, and the firm is indifferent between no extraction and some positive level of extraction on the interval; that is, $Z_t^{**} = (0, Z_t^{\max})$.

IV

Results

In this section, we present comparative static results for the firm model developed from Gray's assumptions in the previous section. The model is used to develop conditions for optimal extraction paths. We show that Hotelling's *r*-percent rule is a subcase of Gray's conditions for optimal extraction paths and qualitatively compare Gray's conditions to Hotelling's *r*-percent rule for the firm. We also aggregate the extraction path from the firm-level model to the extraction path for the industry-level model and contrast the industry extraction path developed from Gray's and Hotelling's assumptions.

Since comparative statics results for Z_t^{**} depend on the comparative statics for Z_t^*, $V(X_t, t|N)$ and $V(X_t, t|M)$, we present results for Z_t^*, $V(X_t, t|N)$, and $V(X_t, t|M)$. Comparative statics results for the optimal level of extraction, Z_t^*, when extraction is optimal, are derived from Equations (9A) and (9B) using standard comparative statics methods. That is, the impact of a parameter shift is evaluated by invoking the implicit function theorem to implicitly define Z_t^* as function of the parameters, setting the total derivative of the new identity with respect to the parameter of interest equal to zero, and solving for the partial derivative of Z_t^* with respective to the parameter of interest (Silberberg 1990).

Extraction is unconstrained in condition (*ii*) of both Equations (11A) and (11B), and these conditions provide the nontrivial comparative results for Z_t^*. Since comparative statics results are defined in a neighborhood Z_t^*, when extraction is constrained as in conditions (*i*) and (*iii*) of Equations (11A) and (11B), most parameters shifts have no impact on Z_t^*. The exception is $Z_t^{\max}$ in condition (*iii*) of both Equations (11A) and (11B), since $Z_t^* = Z_t^{\max}$; when $Z_t^{\max}$ increases, Z_t^* increases. The results of conditions (*ii*) from Equations (11A) and (11B) are presented in Table 2.

Most of the comparative statics results for Z_t^* are unsurprising. If deposit quality in period *t*, price, or the discount rate increase, then extraction in period *t* increases. If marginal costs or the discount rate increase, then extraction in period *t* decreases. Additional stock at time $t + 1$ will increase the value of the stock at time *t* if the stock

Table 2

Comparative Statics Results for the Optimal Extraction at Time $t^{\dagger}$

ID	$t-1$	t	Choice Variable	β	$g(X_t - Z_t)^5$	P_t	r	$\frac{\partial V(X_{t+1}, t+1)}{\partial X_{t+1}}$	$\frac{dVC(Z_t)}{dZ_t}$
III	N	M	$0 < Z_t < Z_{max}$	–	+	+	+	–/0	–
IV	M	M	$0 < Z_t < Z_{max}$	–	+	+	+	–/0	–

†The interpretation of the symbols used to define the relationships with the choice variable are: "+" for a direct relationship with the choice variable; "–" for an inverse relationship; and "0" for no relationship.

constraint is eventually binding. That is, if there exists $j > 0$ such that $X_{t+j} = S(I)$, then the additional stock has value. However, if the stock constraint is never binding, there does not exist j and the additional stock has no value. Since exit costs or fixed costs do change with the level of extraction, neither has any impact on the optimal extraction level.

Fixed costs and entry costs complicate the comparative statics results for $V(X_t, t)$. Each of the four cases described in Table 1 must be evaluated. Comparative statics results are derived by substituting Z_t^{**} into all four value functions of Equations (11A) and (11B) and then differentiating with respect to the parameter of interest. Results are presented in Table 3.

Similar to Z_t^*, the comparative statics results for $V(X_t, t|N)$ or $V(X_t, t|M)$ are straightforward. If current grade or current price increases, then $V(X_t, t|N)$ or $V(X_t, t|M)$ increases. If the discount rate or variable costs increases, then $V(X_t, \mathrm{t})$ decreases. Similar to Z_t^*, additional stock at time $t + 1$ will increase the value of the stock at time t if the stock constraint is eventually binding.

Fixed costs are incurred if mining occurs at time $t - 1$ or t, that is, in Cases II, III, and IV. If fixed costs increase, then $V(X_t, t|N)$, or $V(X_t, t|M)$ in Cases II, III, and IV, decrease. Entry costs are incurred if a nonextraction period is followed by an extraction period, which occurs in Case III. If entry costs increase, then $V(X_t, t|\mathrm{N})$ in Case III decreases.

Table 3

Comparative Statics Results for NPV of Extraction at Time $t^{\dagger}$

ID	$t-1$	t	Value in Period t	Z_t^{**}	β	EC	FC	$g(X_t - Z_t)$	P_t	r	$\frac{\partial V(X_{t+1}, t+1)}{\partial X_{t+1}}$	$\frac{dVC(Z_t)}{dZ_t}$
I	N	N	$V(X_t, t\mid N)$	0	+	0	0	+	+	–	+/0	–
II	M	N	$V(X_t, t\mid M)$	0	+	0	–	+	+	–	+/0	–
III	N	M	$V(X_t, t\mid N)$	+	+	–	–	+	+	–	+/0	–
IV	M	M	$V(X_t, t\mid M)$	+	+	0	–	+	+	–	+/0	–

†The interpretation of the symbols used to define the relationships with the choice variable are: "+" for a direct relationship with the choice variable; "–" for an inverse relationship; and "0" for no relationship.

A. Firm Extraction Path

The firm's optimal level of extraction in any period t is described by Equations (9)–(11). Here we describe the firm's extraction path over time. Two trivial extraction paths from Equation (11) are:

$$\begin{array}{lll} \forall t & V(X_t, t|N) < \max[B(VX_{t+1}, t+1|N), 0] & Z_t^{**} = 0 \\ \forall t & V(X_t, t|M) > B(VX_{t+1}, t+1|N) & Z_t^{**} = Z_t^{\max}. \end{array} \quad (12)$$

These extraction paths are trivial because either the resource is never extracted or the resource is extracted as fast as possible until exhaustion. Extraction paths in which extraction is positive in some periods and zero in other periods are nontrivial.

We begin the analysis of firms by examining the firm's extraction decisions in two consecutive periods. To fully characterize two consecutive periods with fixed costs and entry costs, three consecutive periods must be analyzed. Conditions (9)–(11) give rise to 2^3 (= 8) subcases. The characterizations from Table 1 are expanded in Table 4 to include period $t + 1$.

Similar to the results in Table 1, each of the eight subcases may be

Table 4

Description of Fixed and Entry Costs in Three Consecutive Periods[†]

ID	$t-1$	t	$t+1$	Z_t^*	Z_{t+1}	$C(Z_t)$	$C(Z_{t+1})$
Ia	N	N	N	0	0	0	0
Ib	N	N	M	0	+	0	$FC + VC(Z_{t+1}) + EC$
IIa	M	N	N	0	0	FC	0
IIb	M	N	M	0	+	FC	$FC + VC(Z_{t+1}) + \text{EC}$
IIIa	N	M	N	+	0	$FC + VC(Z_t) + EC$	FC
IIIb	N	M	M	+	+	$FC + VC(Z_t) + EC$	$FC + VC(Z_{t+1})$
IVa	M	M	N	+	0	$FC + VC(Z_t)$	FC
IVb	M	M	M	+	+	$FC + VC(Z_t)$	$FC + VC(Z_{t+1})$

[†]The interpretation of the symbols used to define the relationships with the choice variable are: "+" for a direct relationship with the choice variable; "–" for an inverse relationship; and "0" for no relationship.

characterized. A simple characterization is that Subcase Ia is three consecutive nonmining periods, Subcase IVb is three consecutive mining periods, and Subcases Ib–IVa are combinations of mining and nonmining periods.

B. Gray's Rule

Hotelling's *r*-percent rule is the best-known result in ER economics. The rule states that during a mining era, the marginal scarcity rent of an exhaustible resource should follow an equilibrium path in which the gap between the price and the marginal cost of extraction over time rises at the discount rate. To evaluate Gray's and Hotelling's contributions, we present "Gray's rule," show that Hotelling's *r*-percent rule is a special case of Gray's rule, and then use Hotelling's *r*-percent rule as benchmark by which to contrast the results from Gray's rule.

Analogously to Hotelling's rule, a key aspect of Gray's rule is determining when the firm will be indifferent between extracting in consecutive periods. Under Gray's assumptions, the extraction decision is characterized by conditions (9)–(11). The *r*-percent aspect of Gray's rule arises directly from Equation (9). Equation (9A) is a necessary but not sufficient condition for the firm to start a mining era. Similarly, Equation (9B) is a necessary but not sufficient condition to continue a mining era. The firm will only be indifferent between extracting in period t and period $t + 1$ if the nonnegativity and capacity constraints are nonbinding; that is, $Z_t^* \in (0, Z_t^{\max})$ and $\delta_t^N = \gamma_t^N = 0$ in Equation (9A), and $Z_t^* \in (0, Z_t^{\max})$ and $\delta_t^M = \gamma_t^M = 0$ in Equation (9B). Since fixed costs and entry costs are not present in Equations (9A) and (9B), the right-hand sides of the equations are identical and may be analyzed as one condition.[6] If the nonnegativity constraint is binding, the marginal scarcity rent minus the discounted marginal value of the stock is less than zero, $Z_t^* = 0$, and the firm would prefer to extract in future periods. If the capacity constraint is binding, the marginal scarcity rent minus the discounted marginal value of the stock is less than zero, $Z_t^* = Z_t^{\max}$, and the firm would prefer to extract in the current period. With nonbinding constraints, Equation (9) becomes:

$$P_t g(X_t - Z_t) - \frac{\partial VC(Z_t)}{\partial Z_t} = \frac{\beta \partial V(X_t - Z_t, t+1|M)}{\partial X_{t+1}}. \qquad (13)$$

The marginal scarcity rent equals the discounted marginal value of the stock. Similarly, t may be replaced by $t + 1$:

$$P_{t+1}g(X_{t+1} - Z_{t+1}) - \frac{\partial VC(Z_{t+1})}{\partial Z_{t+1}} = \frac{\beta \partial V(X_{t+1} - Z_{t+1}, t+2|M)}{\partial X_{t+2}}. \quad (14)$$

Hotelling's rule follows from noting that:

$$\frac{\beta \partial V(X_t - Z_t, t+1|M)}{\partial X_{t+1}} = \frac{\beta^2 \partial V(X_{t+1} - Z_{t+1}, t+2|M)}{\partial X_{t+2}}. \quad (15)$$

This equality is established by contradiction. The left-hand side of Equation (15) is the value of another unit of stock at period $t + 1$ discounted to period t. The right-hand side of Equation (15) is the value of another unit of stock at period $t + 2$ discounted to period t. For a mining era to exist, the net present value of the marginal unit of stock between consecutive periods must be constant. If the discounted value of the stock is higher in $t + 1$, then optimality requires that more stock be extracted in period $t + 1$. Similarly, if the discounted value of the stock is higher in $t + 2$, then optimality requires that more stock be extracted in period $t + 2$.

Multiplying Equation (14) by β and substituting from Equation (15) provides:

$$P_t g(X_t - Z_t) - \frac{\partial VC(Z_t)}{\partial Z_t} = \beta \left[P_{t+1}g(X_{t+1} - Z_{t+1}) - \frac{\partial VC(Z_{t+1})}{\partial Z_{t+1}} \right]. \quad (16)$$

Recalling that $\beta = 1/1 + r$ and simplifying provides:

$$P_{t+1}g(X_{t+1} - Z_{t+1}) - \frac{\partial VC(Z_{t+1})}{\partial Z_{t+1}} = (1+r)\left[P_t g(X_t - Z_t) - \frac{\partial VC(Z_t)}{\partial Z_t} \right]$$

$$\frac{P_{t+1}g(X_{t+1} - Z_{t+1}) - \frac{\partial VC(Z_{t+1})}{\partial Z_{t+1}} - \left[P_t g(X_t - Z_t) - \frac{\partial VC(Z_t)}{\partial Z_t} \right]}{P_t g(X_t - Z_t) - \frac{\partial VC(Z_t)}{\partial Z_t}} = r. \quad (17)$$

In the first part of Equation (17), the scarcity rent in period $t + 1$ equals the scarcity rent in period t multiplied by $1 + r$. In the second part of Equation (17), the growth rate in scarcity rent equals the discount rate. These equations are the r-percent aspect of Gray's rule and are analogous to Hotelling's rule. Although the r-percent aspect of Gray's rule in Equation (17) is analogous to Hotelling's rule, due to the existence of fixed costs, entry costs, and the spatial aspect of extraction under Gray's assumptions, Equation (17) is not equivalent to Hotelling's rule.

C. Hotelling's Rule

To complete the derivation of Hotelling's rule, simplifying assumptions and associated revisions of Equations (4)–(11) and (17) are necessary. The first assumption is that both fixed costs and entry costs are zero. As discussed in the model section, if both fixed costs and entry costs are zero, then the derivative of the value function with respect to the extraction level is continuous throughout its range, including at 0, and the value function is not path dependent. With a continuous derivative, comparison of the value of the optimal level of extraction when extraction is positive no longer needs to be compared with zero extraction. This implies that conditions (10) and (11) are no longer necessary, and the optimal level of extraction may be found from Equation (9) only. That is, Z_t^{**} always equals Z_t^*. With a lack of path dependence, the two conditions in Equations (4), (5), (7), and (8) collapse to one condition each, and $V(X_t, t) = V(X_t, t|N) = V(X_t, t|M)$.

The second difference between Equation (17) and Hotelling's rule is that deposits are mined in a specific spatial order, rather than from high quality to low quality. To remedy this difference in extraction order, we adopt Hotelling's original assumption that stock quality is homogenous, and Equation (17) becomes:[7]

$$P_{t+1}g - \frac{\partial VC(Z_{t=1})}{\partial Z_t} = (1+r)\left[P_t g - \frac{\partial VC(Z_t)}{\partial Z_t}\right] \quad and$$

$$\frac{[P_{t+1} - P_t]g - \left[\frac{\partial VC(Z_{t+1})}{\partial Z_t} - \frac{\partial VC(Z_t)}{\partial Z_t}\right]}{P_t g - \frac{\partial VC(Z_t)}{\partial Z_t}} = r, \tag{18}$$

where g is the homogenous stock grade. Condition (18) is a version of Hotelling's rule with marginal scarcity rent rising at the discount rate.

We adopt Hotelling's original assumption that marginal costs of extraction are constant. Letting k equal the constant marginal costs, Equation (18) becomes:

$$P_{t+1}g - k = (1+r)[P_t g - k] \quad and \quad \frac{[P_{t+1} - P_t]g}{P_t g - k} = r. \qquad (19)$$

The modified objective function that directly produces condition (19) is linear in the level of extraction at time t and not path dependent; that is, Equation (4) collapses to:

$$V(X_t, t) = \underset{w.r.t. Z_t}{Maximium}\ P_t Q_t - k + \beta V(X_{t+1}, t+1|M). \qquad (20)$$

Condition (20) implies that the extraction level in each period is a bang-bang control (Clark 1990). That is, under Hotelling's rule expressed in Equation (19), the extraction level is either 0, $Z_t^{\max}$, or a point on the singular path. Only along the singular path does a mining era exist in which the firm is indifferent to extracting in period t or other periods in the mining era.

D. Comparison of Gray's Rule and Hotelling's Rule

Since Hotelling's rule is a special case of Gray's rule, Hotelling's rule is a natural benchmark to illustrate the impact of Gray's assumptions on the optimal extraction path. As discussed in the model section, the firm's decisions to determine its extraction path are when to extract and how much to extract when extraction is positive. Here we contrast the qualitative impacts of increasing marginal costs, fixed costs, entry costs, and heterogeneous stock grade on the extraction path.

As described in Equations (4)–(12) and (17)–(19), each of Gray's assumptions reintroduce nonlinearity and/or path dependence into the objective function. This implies that there is no singular path under Gray's assumptions. Nontrivial mining eras under Gray's assumptions arise when the firm is indifferent between extracting the marginal unit of stock in two or more consecutive periods. This implies that, even with the assumption of homogenous stock, Equations (17) and (19) are qualitatively different due to the increasing marginal extraction costs each period in Equation (17). First, since Equation (17) arises from Equation (4), a nonlinear objective function, the control arising from Equation (17) is not a bang-bang control; that is, in contrast to Equation (19), Z_t^* may assume any value on $[0, Z_t^{\max}]$ and not merely 0, $Z_t^{\max}$, or the value of Z_t on the singular path. Second, increasing marginal extraction costs increase the likelihood of more periods in which extraction is positive. With increasing marginal extraction costs beyond some level of extraction, average costs increase with the extraction level. With average costs increasing with the level of extraction, lower levels of extraction are relatively less expensive than higher levels of extraction, and periods in which extraction is low but positive are more appealing. This implies that there are incentives for the firm to spread extraction out over more periods; that is, the optimal extraction level is less likely to be 0 or $Z_t^{\max}$.

As demonstrated in Equations (9) and (13)–(18), for a mining era to exist, the resource scarcity rent rises at the discount rate between consecutive periods. In the absence of fixed and entry costs, scarcity rent rising at the discount rate is a sufficient condition for a mining era to exist. With the presence of fixed or entry costs, scarcity rent rising at the discount rate is a necessary but not sufficient condition for a mining era to exist.

To more completely evaluate the impact of fixed and entry costs on the decision to extract or not in every period, Equations (10) and (11) need to be evaluated for the four cases of mining and nonmining cases described in Table 1. In addition to knowing the costs at time t, it is also necessary to know the impact of fixed costs and entry costs on the stock value. We present these costs and benefits from Equation (11) in Table 5.

We now substitute arguments from Table 5 into conditions (*i*) and (*ii*) of Equations (11A) and (11B):

$$
\begin{array}{lll}
\textit{Case I} & V(X_t, t|N) = 0 + BV(X_t, t+1|N) & Z_t^{**} = 0 \\
\textit{Case II} & V(X_t, t|M) = -FC + BV(X_t, t+1|N) & Z_t^{**} = 0 \\
\textit{Case III} & V(X_t, t|N) = P_t Q_t - EC - FC - VC(Z_t) & 0 < Z_t^{**} < Z_t^{\max} \\
 & \quad + BV\left(X_t - Z_t^{**}, t+1|M\right) & \\
\textit{Case IV} & V(X_t, t|M) = P_t Q_t - FC - VC(Z_t) & 0 < Z_t^{**} < Z_t^{\max} \\
 & \quad + BV\left(X_t - Z_t^{**}, t+1|M\right). &
\end{array}
\tag{21}
$$

Table 5

Fixed Costs, Entry Costs, Revenues, and Stock Values in Two Consecutive Periods[†]

ID	$t-1$	t	$V(.,t,.)$	Z_t^{**}	TR	$C(Z_t)$	$V(.,t+1.)$
I	N	N	$V(X_t,t,N)$	0	0	0	$V(X_t,t+1/N)$
II	M	N	$V(X_t,t,M)$	0	0	FC	$V(X_t,t+1/N)$
III	N	M	$V(X_t,t,N)$	+	P_tQ_t	$FC+VC(Z_t)+EC$	$V(X_{t+1},t+1/M)$
IV	M	M	$V(X_t,t,M)$	+	P_tQ_t	$FC+VC(Z_t)$	$V(X_{t+1},t+1/M)$

[†]The interpretation of the symbols used to define the relationships with the choice variable are: "+" for a direct relationship with the choice variable; "–" for an inverse relationship; and "0"for no relationship.

In Cases I–IV, net present value in period t equals net revenue in period t plus the net present value in period $t+1$ discounted to period t.

Two observations from Equation (21) prove the result. With positive fixed costs, if there is no extraction in period t, then $V(X_t, t|N) > V(X_t, t|M)$ from Cases I and II, respectively, since $V(X_t, t|N) = BV(X_t, t+1|N) > -FC + BV(X_t, t+1|N) = V(X_t, t|M)$. This implies that zero extraction is more likely to be optimal in period t if extraction was zero in period $t-1$. Similarly with positive entry costs, if there is positive extraction in period t, then $V(X_t, t|M) > V(X_t, t|N)$ from Cases IV and III, respectively, since $V(X_t, t|M) = P_tQ_t - FC - VC(Z_t) + BV(X_t - Z_t^{**}, t+1|M) > P_tQ_t - EC - FC - VC(Z_t) + BV(X_t - Z_t^{**}, t+1|M) = V(X_t, t|N)$. This implies that positive extraction is more likely to be optimal in period t if extraction was positive in period $t-1$.

These inequalities and the optimality of the chosen extraction level in period t imply that if zero extraction is optimal in period t when extraction was positive in period $t-1$ (Case II), then zero extraction is optimal in period t when extraction was zero in period $t-1$ (Case I). The converse is not true, that is, if zero extraction is optimal in period t when extraction was zero in period $t-1$ (Case I), then zero extraction may or may not be optimal in period t when extraction was positive in period $t-1$ (Case II).

If positive extraction is optimal in period t when extraction was zero in period $t-1$ (Case III), then positive extraction is optimal in period t when extraction was positive in period $t-1$ (Case IV). Once again, the converse is not true; that is, if positive extraction is optimal in period t when extraction was positive in period $t-1$ (Case IV), then positive extraction may or may not be optimal in period t when extraction was zero in period $t-1$ (Case III).

Collectively, these relationships imply that with fixed costs and entry costs, if the firm extracted in period $t-1$, it is more likely to extract in period t than if it did not extract in period $t-1$. This confirms Gray's assertion that mining eras are more likely to exist than was later suggested by Hotelling. These relationships also imply a similar result; that is, with entry costs, if the firm is not extracting in period $t-1$, it is more likely not to extract in period t than if it were extracting in period $t-1$.

E. Gray's Conjecture on Future Stock Grade

Based on the discussion of mining eras, we now evaluate a conjecture by Gray (1914) regarding future stock grade. Under the assumptions of heterogeneous deposits that must be extracted in a spatial order, Gray conjectures (1914) that if the stock grade increases in the current or a future period of the current mining era, then current extraction will increase. From the column labeled $g(X_t - Z_t)$ of Table 2, the comparative statics result shows that if the stock grade increases during the current period of extraction, then current extraction increases.

Since Gray's conjecture for shifts in future stock grades requires tracing impacts over more than one period, demonstrating comparative statics results from a shift in a future stock grades on current extraction is more difficult than demonstrating the result for a shift in the current period. An increase in grade at some future period of the current mining era implies that there is an interval $i2$ to $i1$ in which grade quality shifts, and some or all of the interval is during the current mining era.

To demonstrate the impacts of changes in stock grade in future periods of extraction, let $S(i2), S(i1)$ be the interval upon which grade increases with $S(i1) < X_{t+1}$. To show the result, it is necessary to expand $\frac{\partial \beta V(X_t - Z_t, t+1|N)}{\partial X_{t+1}}$ or $\frac{\partial \beta V(X_t - Z_t, t+1|M)}{\partial X_{t+1}}$ through the periods in which stock grade increases; that is, the appropriate derivative is expanded to evaluate future shifts in stock grade. If there exists $S(i1) \leq X_{t+j}, < S(i2)$, $j > 1$, then the marginal stock grade is higher in some period $t + j$. This increase in stock grade increases $P_t g(X_{t+j} - Z_{t+j})$, which increases the left-hand sides of Equations (9A) and (9B). Since during a mining era $\frac{\partial \beta V(X_t - Z_t, t+1|M)}{\partial X_{t+1}}$ is constant, other extraction levels during the mining era must be adjusted to keep these derivatives constant. To make this adjustment, Z_t^* and hence Z_t^{**} must increase. That is, an increase in stock grade at some future time increases the optimal level of current extraction, which confirms Gray's conjecture.

F. Market Extraction Path

Our analysis has focused on firm decision making and the firm's extraction path. To describe the aggregate extraction path for an ER market requires both market demand and market supply. Similar to many other analyses in the area, for simplicity we assume a fixed down-sloping market demand function, that is, $\Theta_t^d = D(P_t)$, where Θ_t^d is quantity demanded at price in period t, and $\mathrm{d}D(P_t)/\mathrm{d}P_t \leq 0$.

At market equilibrium, $\Theta_t^s = \Theta_t^d$; that is, quantity supplied equals quantity demanded in period t. Constructing a market supply function requires aggregating extraction over firms. Under Hotelling's assumptions, including constant marginal costs as in condition (19) describes the firm's decision. The firm's extracts either 0, $Z_t^{**} \in (0, Z_t^{\max})$, or $Z_t^{\max}$. Under Hotelling's assumptions, all firms behave identically. If no firm is extracting in period t, then aggregate extraction is zero, and $\Theta_t^s = 0$. If all firms are extracting $Z_t^{\max}$, then aggregate extraction is the sum of each firm's $Z_t^{\max}$, and some demand may not be met.

During a mining era, each firm is on a singular path and is indifferent to extracting in period t or any other period of the mining era. This implies that the quantity that each firm extracts in period t is unknown. Since supply is infinitely elastic, demand determines the equilibrium quantity; that is, $\Theta_t^d = D(P_t)$ and $\Theta_t^s = \Theta_t^d$.

Under Gray's assumptions, constructing supply and equilibrium quantity is more complex. With firms exploiting spatially identifiable heterogeneous deposits with fixed and entry costs, each firm will have a different price at which it is willing to produce in period *t*. With each firm having its own reservation price, supply is upward sloping.

To determine the upward sloping supply function in each period *t*, the firms need to be identified in ascending order of the price that they are willing to accept to extract in period *t*. Let $j = 1$ to J be this ordering. The supply function, $S(P_t)$, equals extraction in period *t*; that is, $S(p_t) = \sum_{j=1}^{E} Q_t^E$ where E is the marginal firm and Q_t^E is the quantity extracted by the marginal firm. At equilibrium, $D(p_t) = \Theta_t^d = \Theta_t^s = S(p_t) = \sum_{j=1}^{E} Q_t^E$.

With nonidentical firms and an upward sloping supply curve, the tight link of Hotelling's rule between firms and industry does not hold. Market behavior in period *t* must be determined through aggregation of quantity extracted by the each firm.

From previous analysis, we know that under Gray's assumptions, during a mining era firms are more likely to continue either extracting or not extracting than under Hotelling's assumptions. This implies that some firms will be accepting prices that grow more slowly in scarcity rent than the discount rate. In turn, this implies that the marginal scarcity rent is likely to increase at a rate lower than the discount rate.

V

Conclusions

THE GOAL OF THIS PAPER was to revisit the concepts that Lewis Cecil Gray (1913, 1914) introduced and analyzed with basic economic intuition and a numerical example with more formal methods used in ER economics analysis, introduced in part during the 1930s by Harold Hotelling. The motivation for undertaking this work was twofold. First, the considerable efforts to uncover evidence for Hotelling's r-percent rule has produced at best mixed results. Second, given the recognition by economists for a more complete firm-level approach to the analysis of ER deposit heterogeneity, the effort seems timely.

In this paper, we proposed a model based on Gray's assumptions of varying quality, spatially identifiable deposits, and fixed costs to derive and obtain Gray's rule. The formalization of Gray's assumptions and concepts has allowed us to use Hotelling's (1931) work as a benchmark to explore the distinctions between the concepts expounded by Gray and Hotelling.

An important result using this model is that Hotelling's r-percent rule is shown to be a special case of Gray's rule under Gray's assumptions of fixed costs and our assumption of entry costs. Indeed, under Gray assumptions, we show that mine owners are more likely than under Hotelling's assumptions to keep doing what they are doing. That is, under Gray's assumptions, if firms are mining, they are more likely to keep mining, and if they are not mining, they are more likely to keep not mining than under Hotelling's assumptions. An implication of this result is that under Gray's rule, industry behavior does not directly follow from firm behavior. In particular, the industry extraction path is significantly more complex under Gray's assumptions of fixed costs and spatially located heterogeneous deposits than under Hotelling's assumptions. The model also allows the confirmation of Gray's conjecture, in which he hypothesized that an increase in stock grade in the future will increase current extraction.

Given the current emphasis on firm-level analyses through many fields of economic scholarship, and calls to focus more on firm-level analyses within ER economics, our analysis suggests that the relative value of Gray to Hotelling has increased since the 1970s. Perhaps the existence of a Gray's rule that generalizes Hotelling r-percent rule may help explain the lack of empirical evidence for Hotelling's rule.

It is interesting, nonetheless, to notice that Hotelling had the benefit of building on Gray's work but chose not to follow him in several respects, such as the heterogenous quality of deposits, the spatial dimension of the deposits, and the U-shaped cost curve. In any case, it seems that the fields of ER economics and natural resource economics would be a lot different had Hotelling followed on Gray's assumptions!

This comparison of Gray and Hotelling is far from complete; much work remains to be done. A next step would be to analyze available data to test whether Gray's rule is more consistent with existing data than Hotelling's *r*-percent rule. However, given the required data needed to test Hotelling's *r*-percent rule, let alone the more specific requirements to test Gray's rule, this may not be immediately feasible. Given the limited available data, the use of dynamic simulations may be the most promising way to learn more about Gray's assumptions and to further explore how Gray and Hotelling compare. Several possible risks may best be explored using methods of Dixit and Pindyck (1994). An irony of history here could be that Gray's focus on numerical examples and nuance may have made his analysis less appealing to both Hotelling and later mathematically inclined natural resource economics because of its lack of elegance and detail. Yet, nearly one century later, a renewed attention to numerical examples and nuance and the use of the descendants of Hotelling's calculus-of-variations analysis may best demonstrate the deep and broad applicability of Gray's ideas.

Notes

1. Increasing marginal costs within a single period have been included in several extensions of Hotelling's model. Here we link increasing marginal costs with fixed costs. Since our primary goal is to contrast Gray's and Hotelling's work, we focus on the assumptions of Hotelling (1931).

2. It is easy to add shutdown costs. However, we already have fixed costs, so shutdown costs do not have a significant qualitative impact.

3. We could have fixed costs decrease by a percentage in each period, or some other scheme. However, these other schemes do not significantly impact the qualitative results, and greatly complicate derivations.

4. We assume that the sufficient conditions for a maximum are met. Second-order conditions will always be met unless the grade at the end of the period is declining at a fast rate.

5. To formally derive the comparative statics for the grade function, $g(\cdot)$ and marginal costs $dVC(t)/dZ_t$, we defined each as a specific increasing non-linear function and then took the first derivative with respect to the parameter on the lowest power of Z_t greater than 1.

6. Note that condition (9) is the only one in which extraction in period $t - 1$ does not change the optimality condition in period t.

7. Condition (17) is equivalent to Hotelling's r-percent rule if stock grade is homogenous and costs only vary by the current extraction level, and not spatial location. Constant marginal costs is a sufficient but not necessary condition to avoid differences in spatial location.

References

Barnett, C., and C. Morse. (1963). *Scarcity and Growth: The Economics of Natural Resource Availability.* Baltimore, MD: Johns Hopkins University Press.

Berck, P., and M. Roberts. (1996). "Natural Resource Prices: Will They Ever Turn Up?" *Journal of Environmental Economics and Management* 30: 65–78.

Cairns, R. D. (1994). "On Gray's Rule and the Stylized Facts of Non-Renewable Resources." *Journal of Economic Issues* 28: 777–798.

Chermak, J. M., and R. H. Patrick. (2000a). "A Microeconomic Test of the Theory of Exhaustible Resources." *Journal of Environmental Economics and Management* 43: 47–70.

——. (2002b). "Comparing Tests of the Theory of Exhaustible Resources." *Resource and Energy Economics* 24: 301–325.

——. (2002c). "Reconciling Tests of the Theory of Exhaustible Resources." Unpublished manuscript.

Clark, C. W. (1990). *Mathematical Bioeconomics: The Optimal Management of Renewable Resources,* 2nd ed. New York: John Wiley & Sons.

Cloutier, L. M., and R. Rowley. (2003). "Simulation, Quantitative Economics and Econometrics: Electronic Infrastructure and Challenges to Methodological Standards." *European Journal of Economic and Social Systems* 15: 221–239.

Crabbé, P. J. (1983). "The Contribution of L. C. Gray to the Economic Theory of Exhaustible Natural Resources and Its Roots in the History of Economic Thought." *Journal of Environmental Economics and Management* 10: 195–220.

——. (1986). "Gray and Hotelling: A Reply." *Journal of Environmental Economics and Management* 13: 295–300.

Devarajan, S., and A. C. Fisher. (1981). "Hotelling's 'Economics of Exhaustible Resources': Fifty Years Later." *Journal of Economic Literature* 14: 65–73.

Dixit, A. K., and R. S. Pindyck. (1994). *Investment under Uncertainty.* Princeton, NJ: Princeton University Press.

Farrow, S. (1985). "Testing the Efficiency of Extraction from a Stock Resource." *Journal of Political Economy* 93: 452–487.

Gordon, R. L. (1966). "Conservation and the Theory of Exhaustible Resources." *Canadian Journal of Economics and Political Sciences* 32: 319–326.

——. (1967). "A Reinterpretation of the Pure Theory of Exhaustion." *Journal of Political Economy* 75: 274–286.

Gray, L. C. (1913). "The Economic Possibilities of Conservation." *Quarterly Journal of Economics* 27: 497–519.

——. (1914). "Rent under the Assumption of Exhaustibility." *Quarterly Journal of Economics* 28: 466–489.

Halvorsen, R., and T. R. Smith. (1991). "A Test of the Theory of Exhaustible Resources." *Quarterly Journal of Economics* 123–140.

Hartwick, J. M. (1978). "Exploitation of Many Deposits of an Exhaustible Resource." *Econometrica* 46: 201–218.

Hotelling, H. (1931). "The Economics of Exhaustible Resources." *Journal of Political Economy* 39: 137–175.

Meadows, D. H., D. L. Meadows, J. Randers, and W. Behren. (1972). *The Limits to Growth.* New York: Universe Press.

Miller, M. H., and C. W. Upton. (1985). "A Test of the Hotelling Valuation Principle." *Journal of Political Economy* 93: 1–25.

Silberberg, E. (1990). *Structure of Economics: A Mathematical Analysis.* New York: McGraw-Hill.

Slade, M. E. (1982). "Trends in Natural Resource Commodity Prices: An Analysis of the Time Domain." *Journal of Environmental Economics and Management* 9: 122–137.

Smith, G. A. (1986). "Gray and Hotelling: A Comment." *Journal of Environmental Economics and Management* 13: 292–294.

Withagen, C. (1988). "Untested Hypotheses in Non-Renewable Resource Economics." *Environmental and Resource Economics* 11: 623–634.

Index

UNUSUAL PERSPECTIVES ON A CLASSIC TOPIC